Birds of the Dominican Republic and Haiti

The authors, artists, and publishers express their sincere gratitude to the following organizations for providing sponsorship and support in the production of this book:
Vermont Institute of Natural Science
PRBO Conservation Science
U.S.G.S. Biological Resources Division
Sociedad Ornitológica de la Hispaniola
U.S. Fish and Wildlife Service

With additional support from:
Wendling Foundation
Smithsonian Institution's National Museum of Natural History

Birds of the Dominican Republic & Haiti

Steven Latta
Christopher Rimmer
Allan Keith
James Wiley
Herbert Raffaele
Kent McFarland
and Eladio Fernandez

Principal Illustrators: Barry Kent MacKay,
Tracy Pedersen, and Kristin Williams
Supporting Illustrators: Cynthie Fisher, Bart Rulon

PRINCETON UNIVERSITY PRESS
PRINCETON AND OXFORD

Published by Princeton University Press, 41 William Street, Princeton, New Jersey 08540

Library of Congress Cataloging-in-Publication Data

Birds of the Dominican Republic and Haiti / Steven Latta . . . [et al.] ; principal illustrators, Barry Kent MacKay, Tracy Pedersen, and Kristin Williams ; supporting illustrators, Cynthie Fisher . . . [et al.].
p. cm. — (Princeton field guides)
Includes bibliographical references.
ISBN-13: 978-0-691-11890-1 (alk. paper)—ISBN-13: 978-0-691-11891-8 (pbk. : alk. paper)
ISBN-10: 0-691-11890-6 (alk. paper)—ISBN-10: 0-691-11891-4 (pbk. : alk. paper)
1. Birds—Dominican Republic—Identification. 2. Birds—Haiti—Identification. I. Latta, Steven C. II. Series.
QL688.D6B57 2006
598'.097293—dc22 2005048817

British Library Cataloging-in-Publication Data is available

This book has been composed in Optima

Printed on acid-free paper. ∞

pup.nathist.edu

Printed in Italy by Eurografica

Composition by Bytheway Publishing Services, Norwich NY

10 9 8 7 6 5 4 3 2 1

CONTENTS

ACKNOWLEDGMENTS

This guide is the result of many years of work by the coauthors, with the help and support of a large number of others. We thank in particular the many members of the Sociedad Ornitológica de la Hispaniola who contributed recent reports and observations of birds, including Stephen Brauning, Sandra Brauning, Nicolás Corona, Elvis Cuevas, Esteban Garrido, Miguel A. Landestoy, Danilo Mejía, Vinicio Mejía, Marisabel Paulino, Francisco Rivas, Pedro Genaro Rodríguez, and Kate Wallace. Other important contributors of records included Jesús Almonte, J. R. Crouse, André Dhondt, Peter Nash, Rina Nichols, Russell Thorstrom, Filip t'Jollyn, Andrea Townsend, Jason Townsend, and Lance Woolaver. Kate Wallace, Dennis G. Crouse, Jr., Florence E. Sergile, Stephen Brauning, and Sandra Brauning also contributed significant portions of Appendix A; Birdwatching on Hispaniola.

This guide would not haven been possible without the early support of the Wendling Foundation, whose enthusiasm for this project we gratefully acknowledge. We are also thankful for the many individuals who put their faith in us and in the talents of Barry Kent MacKay and purchased plates before they were even painted; without their support we could not have proceeded. We recognize the generosity of Herb Raffaele, his coauthors, and the artists of *A Guide to the Birds of the West Indies*, as well as Princeton University Press, for allowing us the use of many of the fine plates from that guide in this work. In addition, we recognize Allan Keith, his coauthors of *The Birds of Hispaniola: Haiti and the Dominican Republic*, as well as the British Ornithologists' Union, for permission to publish data and descriptions that originally appeared in their annotated checklist. The elevation map was developed by Kent McFarland using DTED® Level 0, a product of the National Imagery and Mapping Agency.

We thank the museums and curators who loaned bird specimens from their collection, including Paul Sweet at the American Museum of Natural History; David Willard at the Field Museum of Natural History in Chicago; Nathan Rice at the Academy of Natural Sciences in Philadelphia; Steven W. Cardiff and J. Van Remsen at Louisiana State University Museum of Natural Science; James Dean and Gary Graves at the Smithsonian Institution's National Museum of Natural History; and especially Mark Peck and Glenn Murphy at the Royal Ontario Museum in Toronto who helped coordinate the loans.

Several people carefully read portions or all of the text and provided valuable comments. These included Jason Townsend, Kate Wallace, and Joseph Wunderle, Jr. We also appreciate the help of our editors at Princeton University Press, Robert Kirk and Ellen Foos, and especially of our copyeditor, Elizabeth Pierson.

Steven Latta's work in the Dominican Republic has been supported by the USDA Forest Service–International Institute of Tropical Forestry, University of Missouri Research Board, National Fish and Wildlife Foundation, Wildlife Conservation Society, National Geographic Society, Association of Avian Veterinarians, The Nature Conservancy, and USDA Forest Service North-Central Forest Experiment Station. Latta was also supported by a STAR Graduate Fellowship from the U.S. Environmental Protection Agency. The work of Chris Rimmer and Kent McFarland on Hispaniola has been supported by the American Bird Conservancy, Blake Fund of the Nuttall Ornithological Club, Carolyn Foundation, Conservation and Research Foundation, National Geographic Society, National Fish and Wildlife Foundation, Stewart Foundation, The Nature Conservancy, Thomas Marshall Foundation, Wildlife Conservation Society, U.S. Fish and Wildlife Service, USDA Forest Service International Program, Wendling Foundation, and friends of the Vermont Institute of Natural Science. Jim Wiley's work has been supported by the USDA Forest Service–International Institute of Tropical Forestry, U.S. Fish and Wildlife Service, Puerto Rico Department of Natural Resources, USGS Biological Resources Division, National Science Foundation, World Parrot Trust, and Wildlife Preservation Trust International. Permission for Latta, Rimmer, McFarland, and Wiley to work in the Dominican Republic was provided by the Directorate of National Parks and the Department of Wildlife; permission to work in Haiti was provided by the Haitian Ministry of the Environment.

INTRODUCTION

Our goal in writing *Birds of the Dominican Republic and Haiti* is to fill a large void in the birdwatching, conservation, and environmental education needs of Hispaniola. There has never been a comprehensive field guide devoted to the birds of Hispaniola, and the only existing guide, by Annabelle Dod, is almost 30 years old, covers only 226 species, and is illustrated with black-and-white line drawings. Here we describe and illustrate all 306 species known to have occurred on the island. But our intention is to provide more than just a means of identifying bird species; our guide also provides information on the biology and ecology of the birds, with the hope that we can help inspire a new generation of birdwatchers, ornithologists, and conservationists. With this guide in hand, we hope that more Dominicans and Haitians will become as fascinated as we are by the diversity of the island's avifauna.

Our guide is based on *A Guide to the Birds of the West Indies* by Herb Raffaele and others, and it incorporates detailed information on the status and range of species from the annotated checklist *The Birds of Hispaniola: Haiti and the Dominican Republic* by Allan Keith and coauthors. Our guide features expanded species accounts, and it provides new information from our personal research on the biology and ecology of Hispaniolan avifauna. Thanks to the generosity of the publisher and artists of the West Indies guide, we have been able to use many of the fine plates from that guide in this work. We also include more than 105 new images of Hispaniolan species painted by Canadian artist Barry Kent MacKay, as well as new, detailed range maps of unsurpassed accuracy and precision prepared by Kent McFarland.

We are confident that by dramatically expanding possibilities for the appreciation of birds in the Dominican Republic and Haiti, this guide will promote conservation of migratory and resident birds, and build support for environmental measures to conserve and protect their habitats. The guide is certain to be used in the many educational, outreach, and training activities by environmental organizations such as the Sociedad Ornitológica de la Hispaniola and the Société Audubon Haiti. We sincerely hope that it will increase public awareness throughout Hispaniola and internationally for the unique birds of the island, and underscore the need to protect these special species and their habitats for the enjoyment of future generations.

MAP OF HISPANIOLA

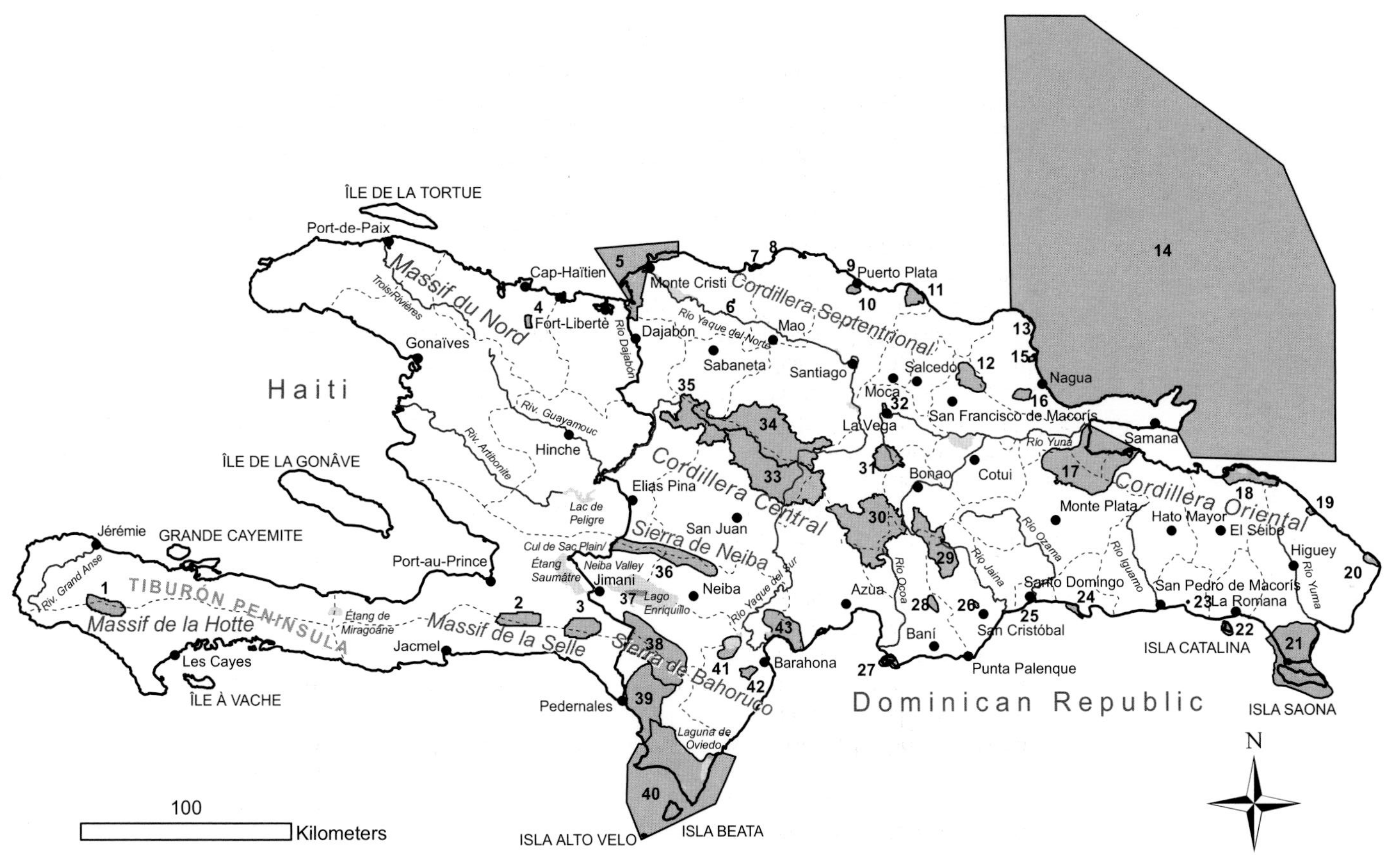

Figure 1. Map of the Dominican Republic and Haiti showing principal cities, physical features, and major protected areas.

1 Macaya Biosphere Reserve
2 La Visite National Park
3 Forêt des Pins
4 National Historic Park of the Citadelle, Sans Souci, and the Ramiers
5 Monte Cristi National Park and Cayos Siete Hermanos
6 Villa Elisa Scientific Reserve
7 Estero Hondo
8 La Isabela National Park
9 Litoral Norte de Puerto Plata National Park
10 Isabel de Torres Natural Monument
11 El Choco
12 Loma Quita Espuela Scientific Reserve
13 Cabo Francés Viejo National Park
14 Banco de la Plata (Santuario de Mamíferos Marinos)
15 La Gran Laguna Wildlife Refuge
16 Miguel Canela Lázaro (Loma Guaconejo) Scientific Reserve
17 Los Haitises National Park
18 Lagunas Redonda y Limón Natural Reserve
19 Albufera de Maimón Natural Monument
20 Laguna Bávaro Wildlife Refuge
21 Del Este National Park
22 Isla Catalina Natural Monument
23 Cueva de las Maravillas Anthropological Reserve
24 Submarino La Caleta National Park
25 Litoral Sur de Santo Domingo National Park
26 Cuevas de Borbón (del Pomier) Anthropological Reserve
27 Bahía de las Calderas Natural Monument
28 Erik Leonard Eckman (Loma La Barbacoa) Scientific Reserve
29 Eugénio Jesús Marcano (Loma La Humeadora) National Park
30 Juan B. Pérez Rancier (Valle Nuevo) National Park
31 Ebano Verde Scientific Reserve
32 La Vega Vieja National Park
33 José del Carmen Ramírez National Park
34 Armando Bermúdez National Park
35 Nalga de Maco National Park
36 Sierra de Neiba National Park
37 Isla Cabritos National Park
38 Sierra de Bahoruco National Park
39 Aceitillar-Cabo Rojo Panoramic Way
40 Jaragua National Park
41 Laguna de Rincón (Cabral) Wildlife Refuge
42 Padre Miguel D. Fuertes (Bahoruco Oriental) Biological Reserve
43 Sierra de Martín García National Park

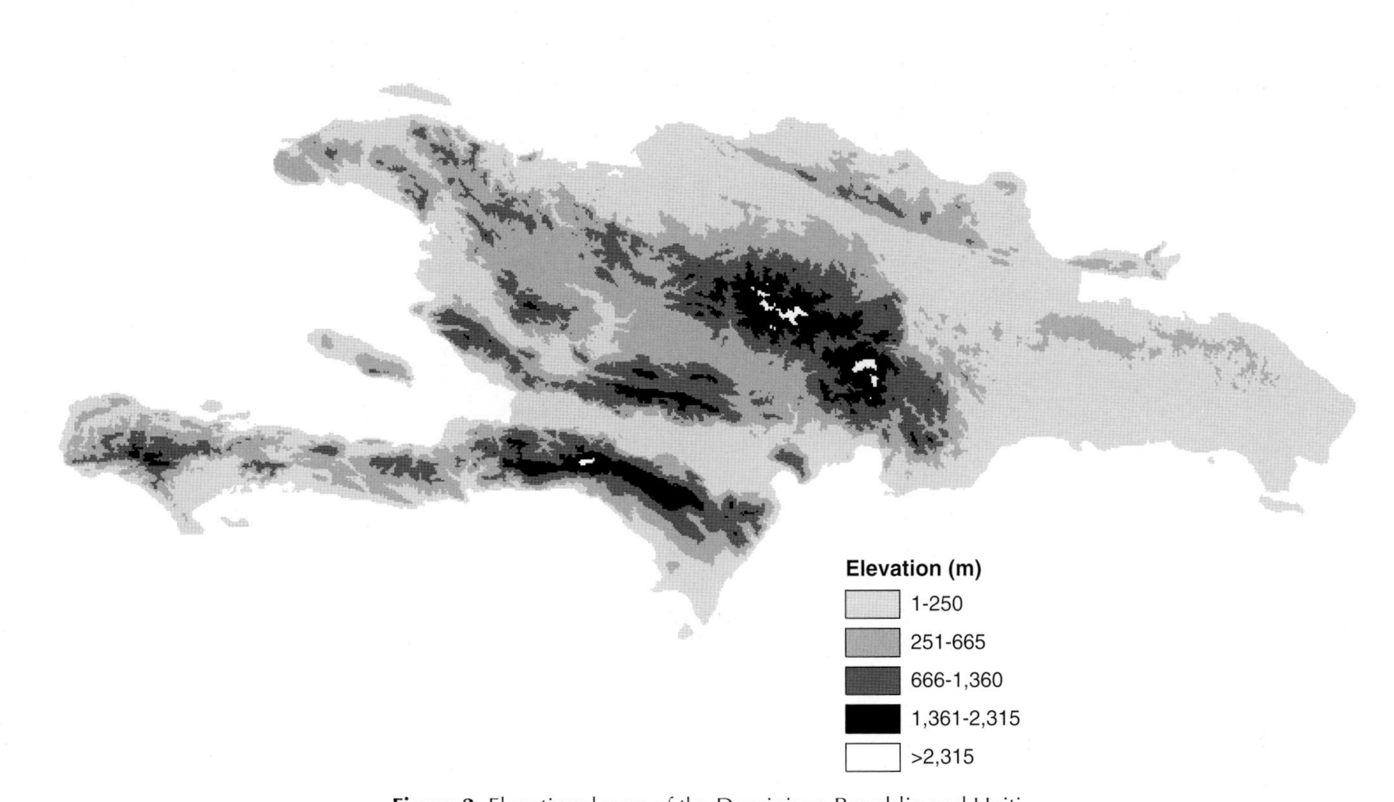

Figure 2. Elevational map of the Dominican Republic and Haiti.

PLAN OF THE GUIDE

Names. In the species accounts, scientific and English names, and the sequence of species, are those of the *Check-list of North American Birds,* seventh edition (American Ornithologists' Union [AOU] 1998) and its supplements (AOU 2000, 2002, 2003, 2004). Most subspecies names are those given in Dickinson 2003 or Keith et al. 2003. We have introduced two changes to common names; changed the sequence of species in two cases; and recognize three previously proposed splits in species, resulting in three more endemic species for Hispaniola. We recognize the changes proposed by Lovette and Bermingham 2001 and Klein et al. 2004 suggesting that the genera *Microligea* and *Xenoligea* are not wood-warblers but are closely aligned with the tanager genus *Phaenicophilus*. As such we have renamed the Green-tailed Warbler (also known as the Green-tailed Ground-Warbler) as the Green-tailed Ground-Tanager, and we have renamed the White-winged Warbler as the Hispaniolan Highland-Tanager. Both we now place in the Thraupidae. We also recognize proposed splits of three species. We split the Hispaniolan Nightjar from the Cuban Nightjar (both were formerly united as the Greater Antillean Nightjar), based on distinct vocalizations and other characteristics as noted by Hardy et al. (1988), Garrido and Reynard (1993), and the AOU (1998). We follow Garrido et al. (1997) and Raffaele et al. (1998) in recognizing the Hispaniolan Palm Crow as an endemic species, distinct from the Cuban Palm Crow, and we follow Garrido et al. (in press) in distinguishing the Hispaniolan Oriole from others in the Greater Antillean Oriole assemblage. Immediately following the name at the head of each species account, we note the status of each species (see "Status" below) and highlight endemic species and those considered threatened or endangered.

Description. We provide size measurements for all species, including length (from bill tip to tail tip) and mass. Where size varies between sexes, or for example with the presence of tail plumes, more than one measurement is provided. The mass presented here for each species is an average and is taken from Dunning 1993 or the authors' own data. Descriptions of all commonly encountered plumages focus on key characteristics allowing field identification. In general, the most commonly encountered plumages are described first. For example, non-breeding visitors are described in their non-breeding plumage first; breeding residents are described in their breeding plumage first. Other plumages, including juvenal and immature, are subsequently described.

Age terminology of avian plumages can be confusing, as birders use several different systems. In this book we distinguish primarily between immatures and adults when plumages of the two differ markedly. We further discriminate between juveniles and immatures for those species that have a distinct juvenal plumage (the first true, nondowny plumage) that is likely to be seen by birders on Hispaniola. Many species retain their juvenal plumage for only a short period after leaving the nest and are seldom encountered by birders in this plumage; we do not describe these short-lived plumages. Other species (e.g., grebes, shorebirds, gulls, terns, and some passerines) retain their juvenal plumage for several months before molting into a subsequent plumage, which may or may not be distinguishable from the definitive adult plumage. We recognize those prolonged juvenal plumages in the species accounts. For those species that retain a juvenal plumage during their entire first year (e.g., some herons and hawks), we simply use the term "immature." We also use "immature" to describe the distinct plumages of many first-year birds (e.g., many passerines) between their juvenal and adult plumages. Thus, for simplicity, we recognize three typical age plumages in this book: juvenal, immature, and adult.

Similar species. Here we highlight differences among the species being described and any others occurring on Hispaniola with which it might be confused.

Voice. The calls, songs, and notes as known on Hispaniola are described. In the case of winter visitors that rarely sing or otherwise vocalize while on the island, their songs and calls are also described but are noted as rare.

Hispaniola. Here we describe where on Hispaniola the species is likely to be encountered. This includes major habitats occupied by the species, range of elevations where it has been found, and specific locales. For species with 10 or fewer reports, all sightings are listed. For more commonly occurring species, distributions are generalized based on habitat and elevational range. For species that visit Hispaniola seasonally, we give general dates of arrival and departure. We also list all of the larger outlying islands where the species has been found. Distributions of all but the rarest or most locally distributed species are illustrated in range maps that depict where on Hispaniola the species might be expected in appropriate habitat.

Status. We distinguish among breeding residents, breeding visitors, non-breeding visitors, vagrants, and passage migrants as follows. We also note if species are endemic or introduced to the island.

Breeding resident: A species known to breed on Hispaniola and that remains on the island year-round.
Breeding visitor: A species known to breed on Hispaniola but that generally migrates off-island during the non-breeding period.
Non-breeding visitor: A species that breeds elsewhere but resides on Hispaniola during the non-breeding season, generally from September to April.
Vagrant: A species known to have occurred on Hispaniola fewer than five times or likely to occur less frequently than once every five years.
Passage migrant: A species that migrates through Hispaniola on a seasonal basis but does not generally reside on the island for extended periods of time. Sometimes referred to as "transient"; also includes wanderers that may move throughout the West Indies or beyond at irregular intervals.
Endemic: A species confined to Hispaniola and associated islands and found nowhere else in the world.
Introduced: A species that is not native to Hispaniola, but that occurs as a population of escaped or intentionally released birds.

In some cases a species may be represented by more than a single distinct population. For example, a breeding resident population may be joined in the non-breeding season by a migratory population from the north. In such cases, both populations are described, with the more common situation listed first.

For each species we characterize population status as abundant, common, uncommon, or rare on Hispaniola. All abundance categories refer to a birdwatcher's chance of observing the species in its preferred habitat:

Abundant: Species is invariably encountered without much effort in large numbers.
Common: Species is invariably encountered singly or in small numbers.
Uncommon: Species is occasionally encountered but not to be expected each trip.
Rare: Species has 10 or fewer records and is not likely to occur more than once or twice a year.

We also describe species' population trends where possible, drawing particular attention to species thought to be declining in abundance, and noting likely factors associated with that decline. Species that are threatened with extinction are listed as threatened, endangered, or critically endangered. Determination of such status is based on a variety of published accounts, including BirdLife International 2000, Keith et al. 2003, Latta and Lorenzo 2000, and the authors' personal experience. Finally, we address taxonomic questions when appropriate, such as alternative treatment of species and the presence of endemic subspecies on the island.

Comments. In this section we may comment on the biology and ecology of the species. This is intended to provide the reader with a better appreciation for the species, and may help in identification. Comments may include, for example, information on foraging, social behavior, or courtship. Because little is known about the ecology of many Hispaniolan species, many of these comments incorporate the authors' own data.

Nesting. This is a brief description of the nest, nest site, number of eggs laid, egg color, and breeding season for those species that breed on Hispaniola. Nesting data are not provided if the species is not known to have bred on the island. Nesting biology of many Hispaniolan birds is not well known, although recent studies by Latta and Rimmer have begun to contribute the first quantifiable data for a variety of species. Some of these data are summarized for the first time in these species accounts.

Range. We summarize the worldwide range of each species, including, for migratory species, the breeding and wintering grounds. We also draw particular attention to a species' occurrence in other portions of the West Indies. Abbreviations used here include: n. (northern), s. (southern), e. (eastern), w. (western), ne. (northeastern), nw. (northwestern), se. (southeastern), sw. (southwestern), c. (central), nc. (north-central), sc. (south-central), ec. (east-central), and wc. (west-central).

Local names. We provide local names in the Dominican Republic and Haiti for each species when possible. In many cases a variety of local names are used, and we list these in approximate order of popularity of use; in some cases, no local names are known.

TOPOGRAPHY AND HABITATS OF HISPANIOLA

Topographic Features

Hispaniola is a diverse island with many habitats and a rich assemblage of bird species, in part a result of its complex geologic history. Although its geologic history is not well understood, Hispaniola is thought to have formed by the merging of at least three land blocks, with two of these formerly attached to what are now Cuba and Puerto Rico. These three blocks probably came together about nine million years ago, but change continued to take place even then. Global cycles of glacial and interglacial periods caused rising and lowering of sea levels, and the alternation of dry and moist environments, resulting in drastic environmental changes and repeated isolation of higher elevation sites by the rising seas. Cyclic climatic changes contributed to the repeated separation of Hispaniola into two "paleo-islands" by a marine canal along the current Neiba Valley and Cul de Sac Plain during much of the Pliocene and portions of the Pleistocene. These two paleo-islands are generally referred to as the North Island and the South Island of Hispaniola. In addition, the South Island was likely divided in pre-Pleistocene times by an intermittent sea passage across the peninsula at the Jacmel-Fauché depression. This would have effectively separated the Massif de la Hotte to the west from the Massif de la Selle and Sierra de Bahoruco to the east.

Cyclic climatic changes in the Pleistocene are likely to have contributed significantly to speciation and extinction events. Unique flora and fauna are thought to have existed on the two paleo-islands, as evidenced by the several pairs of bird species that are today found on the north and south paleo-islands. For example, the Eastern Chat-Tanager is found in the Cordillera Central and the Sierra de Neiba, whereas the Western Chat-Tanager is found in the Sierra de Bahoruco and the southern peninsula of Haiti. Similar processes may have contributed to the speciation of the Gray-crowned and Black-crowned palm-tanagers, the two tody species, and two subspecies of La Selle Thrush.

Cyclic climatic changes also had great impacts on the island's vegetation. It is clear that vegetation types such as conifers, now confined to higher elevations, occurred much lower during the cooler, drier periods, when glaciation occurred on Hispaniola down to the level of 1,800 m. It was also during such periods that sea levels were significantly lower, allowing the appearance of a broad expanse of savanna and thorn scrub habitat in the Hispaniolan lowlands. During these periods of cold and aridity, the wet slopes of the Massif de la Hotte in particular are thought to have served as a refugium for plants and animals adapted to mesic environments. The mountain range's geography with respect to winds and weather fronts positioned it to receive naturally high levels of rainfall. Today the Massif de la Hotte displays extraordinary levels of endemism in orchids, other plants, and amphibians.

Geographically, Hispaniola is the second largest island in the Caribbean, covering 77,842 km^2. The island is longer (650 km) than it is wide (260 km), and it is split politically between the larger Dominican Republic (48,442 km^2) and the smaller Haiti (29,400 km^2). The island is dominated by a series of roughly parallel mountain ranges and valleys that are aligned east to west. These ranges change names between Haiti and the Dominican Republic but essentially bridge both countries. The southern paleo-island features, from west to east, the Massif de la Hotte-Massif de la Selle-Sierra de Bahoruco range. High points in this range include Pic Macaya (2,347 m) in the Massif de la Hotte, Pic la Selle (2,574 m), and Loma de Toro (2,367 m) in the Sierra de Bahoruco. North of the Neiba Valley and the Cul de Sac Plain, on the northern paleo-island, lies the second major east-west range of mountains. These are the Montagnes de Trou-d'Eau in Haiti and the Sierra de Neiba in the Dominican Republic. At its summit, Monte Neiba reaches 2,279 m. Somewhat isolated to the east of the Sierra de Neiba, and southwest of Azua, is the Sierra de Martín García. Farther north, the Plateau Central and the Valle de San Juan separate this range from the next east-west range, the Cordillera Central, which extends into Haiti as the Massif du Nord. This is the largest mountain range on the island, and it includes Pico Duarte, at 3,098 m the highest elevation in the Caribbean. North of the Cibao Valley lies the Cordillera Septentrional, which runs from Monte Cristi to

Samaná Bay and rises to 1,250 m. Two additional, minor ranges include the Cordillera Oriental, southeast of Samaná Bay, and the Montagnes du Nord-Ouest in the northwestern peninsula of Haiti.

Hispaniola has several lakes and lagoons, many of which lie along the current Neiba Valley and Cul de Sac Plain. These include the hypersaline Lago Enriquillo (which can vary from 180 to 265 km^2) in the western Dominican Republic and, to its east, the largest freshwater lake on the island, Laguna de Rincón (30 km^2) at Cabral; and in Haiti, the slightly brackish Étang Saumâtre (113 km^2) and marshy freshwater Trou Caïman (7 km^2). Other large water bodies include Laguna de Oviedo (25 km^2) in the southeast of the Barahona Peninsula, Laguna Redonda (7 km^2) and Laguna Limón (5.1 km^2) on the northeastern coast, and Étang de Miragoâne, consisting of two freshwater lakes (combined 8 km^2) and adjacent marshes on the northern coast of the Tiburón Peninsula.

There are several significant river systems on the island, including the Río Yaque del Norte, Río Yaque del Sur, Río Ozama, and Río Dajabón in the Dominican Republic, and in Haiti the Guayamouc, Les Trois Rivières, and Artibonite. At 400 km, the Artibonite is the longest river in the Caribbean.

Bisected by mountain ranges and rivers, and dotted with lakes and lagoons, Hispaniola contains a diversity of habitats. Most of the mountains are steep and rugged, and frequently cut by deep gorges or valleys. Mountain valleys tend to be cool and moist, supporting either pine or broadleaf forests, but lower elevations are dominated by dry forest and thorn scrub habitats. There are extensive areas of limestone karst in the southern paleo-island, including the Tiburón Peninsula, Barahona Peninsula, Sierra de Bahoruco, and Sierra de Neiba. In addition, much of the eastern Dominican Republic is limestone karst. Along the northern coast, limestone karst forms tower formations in Los Haitises National Park, on the Samaná Peninsula, and along the Cordillera Septentrional. Sand dunes are found in more than 20 coastal locations, and those near Baní on the southern coast are the largest in the Caribbean.

Ten offshore islands contribute to Hispaniola's avifauna. These islands tend to be relatively low, small, and dry but are often of high importance to birds. Many are crucial nesting sites for seabirds and other species, and some are home to endemic subspecies of land birds. Associated with the southern paleo-island are Isla Beata (47 km^2, 100 m elevation); Isla Alto Velo (1 km^2, 152 m elevation); Île Grande Cayemite and Île Petite Cayemite, with the larger being 45 km^2 and 152 m in elevation; and Île à Vache (52 km^2, 30 m elevation). Associated with the northern paleo-island are Isla Saona (111 km^2, 35 m elevation); Isla Catalina (18 km^2); the Cayos Siete Hermanos which are seven small, low, and sandy islands; Île de la Tortue (180 km^2, 325 m elevation); and Île de la Gonâve (658 km^2, 755 m elevation). Navassa Island (5 km^2, 77 m elevation), a U.S. possession 55 km due west of the westernmost point of Haiti, is included in this guide because of its zoogeographic association with Hispaniola.

Major Habitats

For the purposes of this guide, nine major habitats are identified based on Tolentino and Peña 1998 and Keith et al. 2003.

Mangroves. This habitat type is found at coastal sites around river mouths and lagoons where the soil is flooded most or all of the year, and also inland along the margins of both freshwater and saline lakes where the soil may only be flooded seasonally. In some places the mangrove forest reaches heights of 20 m and a density covering 70 to 85 percent of the ground surface. Dominant species are buttonwood mangrove (*Conocarpus erectus*), red mangrove (*Rhizophora mangle*), white mangrove (*Laguncularia racemosa*), and black mangrove (*Avicennia germinans*). In the Dominican Republic, mangroves cover less than 1% of the land area; in Haiti, mangroves cover about 0.5% of the land area.

Freshwater swamps. This is an uncommon lowland habitat type on Hispaniola, usually occurring below 20 m elevation. It is sometimes forested, primarily with swamp bloodwood

(*Pterocarpus officinalis*), or may occur in the form of marshlands characterized by dense growth of cattail (*Typha domingenis*). Some of the marshlands in this category may have significant moisture for only part of each year. On Hispaniola, freshwater swamps cover less than 0.5% of the land area.

Grasslands. This habitat type includes natural savannas at all elevations. They are mostly in the lowlands but are also found in several intermountain valleys. On Hispaniola, grasslands cover less than 1% of the land area.

Agricultural lands. Included here are all lands cleared for agriculture, whether for large-scale farming enterprises such as sugarcane plantations and truck gardens or for subsistence agriculture, even at relatively high elevations in the foothills and mountains in many parts of the island, especially Haiti. Land cleared for pasture is also included here. In the Dominican Republic, agricultural lands and pastures cover about 55% of the land area; in Haiti, about 42% of the land is under cultivation, and another 19% is considered pasture.

Shrublands. This habitat type is typically dry and results from the recent removal of forest cover or because environmental or geological substratum conditions limit plant growth. It is now a widespread habitat type in both countries from sea level to, at least locally, 500 m. Depending on the elevation and original forest type, typical shrub species may include mahogany (*Swietenia mahagoni*), botoncillo (*Ternstroemia peduncularis*), mastic (*Sideroxylon cubensis*), waltheria (*Waltheria indica*), escobón (*Eugenia maleolens*), logwood (*Haematoxylon campechianum*), cordia (*Cordia globosa*), and sensitive plant (*Mimosa pudica*). Especially typical of thorny shrublands are *Jacquinia berterii*, capertree (*Capparis ferruginea*), damiana (*Turnera diffusa*), and another sensitive plant species (*Mimosa azuensis*). In the Dominican Republic, shrublands cover about 6% of the land area; in Haiti, where the forest cover has been removed from more than 95% of the land area and 60% of the land is mountainous, shrublands and low dense vegetation cover about 35% of the land area.

Dry scrub. This forest type now consists primarily of secondary growth of semideciduous trees growing at 40 to 500 m elevation in areas receiving 50 to 100 cm of annual rainfall. The canopy is largely open at a typical height of 10 m. Most of these forests are disturbed because of cutting by humans. This vegetation type is widespread in the lowlands of both the Dominican Republic and Haiti. Indicator species are gumbo limbo (*Bursera simaruba*), acacia (*Acacia sckeroxyla*), boxwood (*Phyllostylon brasiliensis*), tamarindo (*Acacia macracantha*), and white leadtree (*Leucaena leucocephala*). In the Dominican Republic, dry scrub covers about 8% of the land area; in Haiti, dry scrub is reduced to shrubland.

Dry forest. Typically found at elevations of 400 to 900 m on the coastal plain and in the foothills of mountains, this habitat type is often bordered by dry scrub at its lower edge and broadleaf forest at its upper edge. It occurs in areas with a distinct annual arid period and rainfall in the range of 100 to 180 cm. It is a common natural forest type over much of lower elevation Dominican Republic and Haiti but has been widely cut, especially in Haiti. In its undisturbed form it has a canopy density of 60% or greater; the canopy typically ranges from 3 to 10 m in height, less often to 20 m in wetter situations. Indicator species in drier areas are leadwood (*Krugiodendron ferreum*), mahogany (*Swietenia mahagoni*), seagrape (*Coccoloba diversifolia*), gumbo limbo (*Bursera simaruba*), lignumvitae (*Guaiacum sanctum*), poisontree (*Metopium brownei*), and crabwood (*Ateramnus lucidus*). Moister habitats usually contain oxhorn bucida (*Bucida buceras*), pond-apple (*Annona glabra*), and mara (*Calophyllum calabra*). In the Dominican Republic, dry forest covers about 8% of the land area; in Haiti, most dry forest has been converted to shrubland.

Broadleaf evergreen forest. Humid evergreen forest or rainforest is typically found below 500 m but locally up to elevations of 1,500 m. It is found in all Dominican Republic mountain ranges and very locally in Haiti, though extensive stands are now quite scarce. Typical canopy height is up to 25 m, and canopy density is 60% or greater. This forest type receives annual precipitation of 200 cm or more. Many humid evergreen forests are also mixed with pine or with shade coffee. Indicator species include wild mamee (*Clusia rosea*), myrtle laurelcherry (*Prunus myrtifolia*), lancewood (*Oxandra laurifolia*), manac palm (*Calyptronoma plumeriana*), tree-fern (*Cyathea arborea*), butterbough (*Exothea paniculata*), miconia (*Miconia dodecandra*), and coi (*Mora abbottii*).

At higher elevations up to 2,300 m, this habitat type is known as montane broadleaf forest or cloud forest. These humid forests are found in parts of the Cordillera Central, Cordillera Septentrional, Sierra de Neiba, and Sierra de Bahoruco; remnant stands in Haiti are found primarily in the Massif de la Hotte and Massif de la Selle. Canopy density is 80% or greater, and indicator canopy species include wind tree (*Didymopanax tremulus*), parrot-tree (*Brunellia comocladifolia*), bitter tree (*Garrya fadyenii*), tachvela (*Podocarpus aristulatus*), palms (*Coccothrinax* spp.), green ebony (*Magnolia pallescens* and *M. hamori*), rose-apple (*Clusia clusioides*), sierra palm (*Prestoea montana*), bone-tree (*Haenianthus salicifolius*), trumpet-tree (*Cecropia schreberiana*), swamp cyrilla (*Cyrilla racemiflora*), florida trema (*Trema micrantha*), tabebuia (*Tabebuia berterii*), and laurel (*Ocotea* sp). In the Dominican Republic, broadleaf evergreen forest covers about 13% of the land area; in Haiti broadleaf evergreen forests have probably been reduced to less than 1% of the land area.

Pine forest. Pine forest habitats include both pure pine stands and pine mixed with some broadleaf species. Pine forests can also be either closed pine forest, with a canopy density of 60% or greater, or open pine forest, with a canopy density between 40 and 60%. Virtually all closed pine habitat remaining in Hispaniola is in the Sierra de Bahoruco or above 2,000 m in the Cordillera Central of the Dominican Republic. Examples of open pine habitat are found in parts of the Cordillera Central, Sierra de Bahoruco, and Sierra de Neiba; small stands occur in the Macaya Biosphere Reserve and La Visite National Park, Haiti. Indicator species include Hispaniolan pine (*Pinus occidentalis*) in the canopy, and in the understory bitter tree (*Garrya fadyenii*), *Eupatorium illitium*, holly (*Ilex tuerckheimii*), and species of the genera *Fuchsia*, *Ambrosia*, and *Senecio*. In the Dominican Republic, pine forest covers about 6% of the land area; in Haiti the pine forests have been reduced to less than 1.5% of the land area.

ENDEMIC SPECIES AND SUBSPECIES

We recognize a total of 31 species endemic to Hispaniola and associated satellite islands and 50 endemic subspecies. Here we list those species and subspecies and provide general ranges of the subspecies. Subspecies are those identified by Dickinson 2003 and Keith et al. 2003. Although we recognize that the subspecies concept is sometimes controversial, it does provide an initial measure of geographic variation within a species, and serves as a preliminary reference for identifying genetic diversity and the uniqueness of populations which may be useful in conservation planning.

ENDEMIC SPECIES OF HISPANIOLA

Ridgway's Hawk (*Buteo ridgwayi*)
White-fronted Quail-Dove (*Geotrygon leucometopia*)
Hispaniolan Parakeet (*Aratinga chloroptera*)
Hispaniolan Parrot (*Amazona ventralis*)
Hispaniolan Lizard-Cuckoo (*Saurothera longirostris*)
Bay-breasted Cuckoo (*Hyetornis rufigularis*)
Ashy-faced Owl (*Tyto glaucops*)
Least Pauraque (*Siphonorhis brewsteri*)
Hispaniolan Nightjar (*Caprimulgus eckmani*)
Hispaniolan Emerald (*Chlorostilbon swainsonii*)
Hispaniolan Trogon (*Priotelus roseigaster*)
Broad-billed Tody (*Todus subulatus*)
Narrow-billed Tody (*Todus angustirostris*)
Antillean Piculet (*Nesoctites micromegas*)
Hispaniolan Woodpecker (*Melanerpes striatus*)
Hispaniolan Pewee (*Contopus hispaniolensis*)
Flat-billed Vireo (*Vireo nanus*)
Hispaniolan Palm Crow (*Corvus palmarum*)
White-necked Crow (*Corvus leucognaphalus*)
La Selle Thrush (*Turdus swalesi*)
Palmchat (*Dulus dominicus*)
Green-tailed Ground-Tanager (*Microligea palustris*)
Hispaniolan Highland-Tanager (*Xenoligea montana*)
Black-crowned Palm-Tanager (*Phaenicophilus palmarum*)
Gray-crowned Palm-Tanager (*Phaenicophilus poliocephalus*)
Western Chat-Tanager (*Calyptophilus tertius*)
Eastern Chat-Tanager (*Calyptophilus frugivorus*)
Hispaniolan Spindalis (*Spindalis dominicensis*)
Hispaniolan Oriole (*Icterus dominicensis*)
Hispaniolan Crossbill (*Loxia megaplaga*)
Antillean Siskin (*Carduelis dominicensis*)

ENDEMIC SUBSPECIES OF HISPANIOLA AND ASSOCIATED ISLANDS

Subspecies	Range
Sharp-shinned Hawk (*Accipiter striatus striatus*)	Hispaniola
American Kestrel (*Falco sparverius dominicensis*)	Hispaniola, associated islands
Double-striped Thick-knee (*Burhinus bistriatus dominicensis*)	Hispaniola
Common Ground-Dove (*Columbina passerina navassae*)	Navassa Island
Hispaniolan Lizard-Cuckoo (*Saurothera longirostris longirostris*)	Hispaniola, Isla Saona
Hispaniolan Lizard-Cuckoo (*Saurothera longirostris petersi*)	Île de la Gonâve
Burrowing Owl (*Athene cunicularia troglodytes*)	Hispaniola, Île de la Gonâve, Isla Beata
Stygian Owl (*Asio stygius noctipetens*)	Hispaniola, Île de la Gonâve

Short-eared Owl (*Asio flammeus domingensis*)	Hispaniola
Northern Potoo (*Nyctibius jamaicensis abbotti*)	Hispaniola, Île de la Gonâve
Antillean Mango (*Anthracothorax dominicus dominicus*)	Hispaniola, associated islands
Vervain Hummingbird (*Mellisuga minima vieilloti*)	Hispaniola, associated islands
Antillean Piculet (*Nesoctites micromegas micromegas*)	Hispaniola
Antillean Piculet (*Nesoctites micromegas abbotti*)	Île de la Gonâve
Greater Antillean Elaenia (*Elaenia fallax cherriei*)	Hispaniola
Hispaniolan Pewee (*Contopus hispaniolensis hispaniolensis*)	Hispaniola
Hispaniolan Pewee (*Contopus hispaniolensis tacitus*)	Île de la Gonâve
Stolid Flycatcher (*Myiarchus stolidus dominicensis*)	Hispaniola, associated islands
Loggerhead Kingbird (*Tyrannus caudifasciatus gabbii*)	Hispaniola
Thick-billed Vireo (*Vireo crassirostris tortugae*)	Île de la Tortue
Golden Swallow (*Tachycineta euchrysea sclateri*)	Hispaniola
Cave Swallow (*Petrochelidon fulva fulva*)	Hispaniola, associated islands
Rufous-throated Solitaire (*Myadestes genibarbis montanus*)	Hispaniola
La Selle Thrush (*Turdus swalesi swalesi*)	Southern Hispaniola
La Selle Thrush (*Turdus swalesi dodae*)	Northern Hispaniola
Palmchat (*Dulus dominicus dominicus*)	Hispaniola, Isla Saona
Palmchat (*Dulus dominicus oviedo*)	Île de la Gonâve
Yellow Warbler (*Dendroica petechia chlora*)	Cayos Siete Hermanos
Yellow Warbler (*Dendroica petechia solaris*)	Île de la Gonâve, Petite Gonâve
Yellow Warbler (*Dendroica petechia albicollis*)	Hispaniola, associated islands
Pine Warbler (*Dendroica pinus chrysoleuca*)	Hispaniola
Bananaquit (*Coereba flaveola bananivora*)	Hispaniola, associated islands
Bananaquit (*Coereba flaveola nectarea*)	Île de la Tortue
Green-tailed Ground-Tanager (*Microligea palustris palustris*)	Hispaniola
Green-tailed Ground-Tanager (*Microligea palustris vasta*)	Southwestern Hispaniola, Isla Beata
Gray-crowned Palm-Tanager (*Phaenicophilus poliocephalus poliocephalus*)	Hispaniola, Grand Cayemite
Gray-crowned Palm-Tanager (*Phaenicophilus poliocephalus coryi*)	Île de la Gonâve
Gray-crowned Palm-Tanager (*Phaenicophilus poliocephalus tetraopes*)	Île à Vache
Western Chat-Tanager (*Calyptophilus tertius tertius*)	Massif de la Hotte
Western Chat-Tanager (*Calyptophilus tertius selleanus*)	Massif de la Selle, Sierra de Bahoruco
Eastern Chat-Tanager (*Calyptophilus frugivorus frugivorus*)	Cordillera Central
Eastern Chat-Tanager (*Calyptophilus frugivorus neibei*)	Sierra de Neiba
Eastern Chat-Tanager (*Calyptophilus frugivorus abbotti*)	Île de la Gonâve
Greater Antillean Bullfinch (*Loxigilla violacea affinis*)	Hispaniola, associated islands
Greater Antillean Bullfinch (*Loxigilla violacea maurella*)	Île de la Tortue
Grasshopper Sparrow (*Ammodramus savannarum intricatus*)	Hispaniola
Rufous-collared Sparrow (*Zonotrichia capensis antillarum*)	Hispaniola
Tawny-shouldered Blackbird (*Agelaius humeralis humeralis*)	Hispaniola
Greater Antillean Grackle (*Quiscalus niger niger*)	Hispaniola, associated islands
Antillean Euphonia (*Euphonia musica musica*)	Hispaniola, Île de la Gonâve

AVIAN CONSERVATION ON HISPANIOLA

Conservation Issues

Hispaniola's contribution to global biodiversity has earned the island the highest ranking of biological importance in a worldwide assessment of bird-protection priorities (Stattersfield et al. 1998). Whereas habitats of Hispaniola are vital to the survival of many endemic and migrant bird species, a variety of commonly confronted environmental issues threaten the sustainability of bird populations in both the Dominican Republic and Haiti. In many ways, the conservation issues faced in the Dominican Republic and Haiti are identical, except that Haiti faces a much more extreme situation, owing to its severe economic, social, and political problems, many of which have been compounded by poor conservation efforts in the past. Bird conservation in Haiti must still be considered embryonic (Paryski et al. 1989, Keith et al. 2003), although recent momentum provides grounds for optimism (e.g., Rimmer et al. 2004).

By all accounts, the loss and degradation of habitats are the principal problems facing birds on Hispaniola. In the Dominican Republic recent estimates place forest loss at greater than 90% in the last 20 years (Food and Agricultural Organization 1991, Ottenwalder 2000), and most currently forested areas are fragmented and under continuing heavy pressure. A series of more than 60 laws protects forests and watersheds, including a 1967 order to close all existing sawmills and a ban on the cutting of all trees. The primary issue has been lack of enforcement. The principal government agencies responsible for forest administration and management, the General Directorate of Forests and the Directorate of National Parks, are underfunded and understaffed, and transportation of employees out of their offices and into the field where abuses occur has always been difficult. Reforestation programs have been proposed and executed from time to time by both government agencies and private sector organizations, but program implementation and enforcement of regulations have been hampered by small budgets and insufficiently trained personnel.

In Haiti, the landscape is already almost entirely deforested (Paryski et al. 1989, Ottenwalder 2000, Rimmer et al. 2005). Some see little chance under present conditions for recovery of environmental damage already done (Grupo Jaragua 1994). The Division of Natural Resources within the Ministry of Agriculture is responsible for protection and regulation of all forests and for reforestation efforts. However, the country's high population density, poverty, and political instability, compounded by small budgets, absence of trained staff, lack of clear policies, and shifting government priorities, have prevented any sustained conservation efforts. National parks in Haiti are few and essentially unprotected. There are agents responsible for the parks, and basic offices exist in the parks, but personnel seem to be present only intermittently and access is entirely uncontrolled.

Many Hispaniolan habitats are severely affected by deforestation and other human pressures. The 1998 National Planning Workshop for Avian Conservation in the Dominican Republic concluded that cloud forest and moist broadleaf forests were the most threatened habitats in the country (Latta and Lorenzo 2000), but every major native habitat has been adversely affected by human influences. Because population growth has been highest on the coasts and in the lowlands, these areas are the most heavily affected, with lowland forests, beaches, coastal swamps and lagoons, and mangroves all suffering from multiple threats. A corollary to outright destruction of habitat by people is the problem of human-introduced exotic predators, including dogs, cats, pigs, and mongoose, all of which have had enormous impacts on a variety of bird species, especially those nesting low to the ground.

The conditions of rivers and watersheds across Hispaniola are also poor, and declining as the result of heavy silting from erosion and, in some areas, severe water pollution. Erosion results from deforestation in the mountains, seriously affecting the lower portions of watersheds, as well as estuaries, coastal regions, and coral reefs. Pollution results primarily from the lack of adequate sanitation systems and from contamination by agricultural chemicals and industrial wastes.

A second major problem, alluded to above, is the lack of enforcement of environmental laws. Associated with this is the lack of funding, personnel, and vehicles and the inability of enforcement officers, managers, and researchers from government agencies to get into the field on a regular basis. Adequate law enforcement could not only better control illegal cutting and slash-and-burn agriculture but could restrict hunting and the cage-bird trade. Hunting of most species is illegal in the Dominican Republic, and rifles and shotguns are relatively scarce, but the use of slingshots, snares, and baiting is rampant. Collection of birds for the cage-bird trade has long been the most serious threat to parrots and parakeets, but many other species have also been found in captivity, including Greater Flamingos, Little Blue Heron, Hispaniolan Palm and White-necked crows, Hispaniolan Lizard-Cuckoo, Greater Antillean Bullfinch, and Village Weaver.

A third major area of avian conservation concerns involves the general lack of an established environmental education program, especially in schools, and the lack of a national environmental ethic. These two issues were identified as priority action items at the National Planning Workshop for Avian Conservation in the Dominican Republic (Latta and Lorenzo 2000). But it is also here that the conservation community may have exerted its greatest influence in recent years. Community-based non-governmental organizations (NGOs) that focus on avian issues have a strong presence in the island's conservation community, especially in the Dominican Republic. Groups such as the Sociedad Ornitológico de la Hispaniola, Fundación Moscoso Puello, Grupo Jaragua, Observadores de Aves Annabelle Dod, Grupo Ecologista Tinglar, and Société Audubon Haiti have been active in a wide variety of environmental education efforts, especially in communities that border protected areas. These groups have worked diligently to foster an entirely new perspective on the importance of natural resource protection as part of the island's national patrimony.

National Protected Areas in the Dominican Republic

In response to the environmental crisis, the Dominican government has created 70 protected areas covering more than 13,000 km^2. Approximately 8,000 km^2, or more than 16% of the country, is designated as protected terrestrial ecosystems. The Directorate of National Parks (DNP) recognizes 11 categories of management: national parks (22), panoramic ways (10), natural monuments (9), wildlife refuges (7), ecological corridors (6), scientific reserves (5), biological reserves (4), recreation areas (3), anthropological reserves (2), natural reserves (1), and special ecological reserves (1). Management plans have been written for 10 national parks, of which only 6 have found some level of implementation; 24 protected areas have designated personnel (Ottenwalder 2000). In addition, a number of personnel from NGOs are assigned to national parks under comanagement agreements between those organizations and the DNP. Examples include the management of Valle Nuevo by Fundación Moscoso Puello and of certain activities within Jaragua National Park by Grupo Jaragua. Attempts in 2004–2005 by the President and the legislature to eviscerate the national park system through the sell-off of protected lands for tourism and development activities underscore the fragility of the parks in the Dominican Republic, and the on-going need for building an environmental ethic that sanctifies parks as the national treasures that they are. Major protected areas in the Dominican Republic include the following.

Monte Cristi National Park. On the extreme northwestern coast of the country, Monte Cristi National Park covers 561 km^2, including the small offshore islands of the Cayos Siete Hermanos. Offering protection primarily for its diverse and abundant coral reefs, the park also protects important estuarine habitats, lagoons, and mangroves, as well as dunes, beaches, and coastal scrub forest.

Armando Bermúdez and José del Carmen Ramírez National Parks. These twin parks of the Cordillera Central comprise 766 km^2 and 764 km^2, respectively. Armando Bermúdez includes Pico Duarte (3,098 m), the highest point in the Caribbean, and both parks contain extensive tracts of pine forest, savanna, and montane humid broadleaf forest. Twelve of the country's most important rivers flow through or have their origin in these parks.

Juan B. Pérez Rancier National Park. Formerly the Valle Nuevo Scientific Reserve, and still commonly referred to as Valle Nuevo, Juan B. Pérez Rancier National Park is a 657-km^2 protected area in the heart of the Hispaniolan pine forest region. This region contains some of the best representations of pine, montane broadleaf, and cloud forests on Hispaniola.

Los Haitises National Park. Situated on Samaná Bay, and at 1,375 km^2 one of the Dominican Republic's largest parks, Los Haitises includes densely vegetated, moist broadleaf forest on limestone karst, secondary forest, and extensive mangroves.

Del Este National Park. Located in the extreme southeastern corner of the island, Del Este National Park covers 430 km^2 of coastal habitats and extensive woodlands and includes Isla Saona (110 km^2). Principal habitats in the park include lagoons and mangroves as well as coastal scrub forest.

Sierra de Neiba National Park. This 407-km^2 park is characterized by montane broadleaf forest, but much of the landscape has been heavily disturbed, such that secondary forest in varying stages of regrowth, and open pastures, are common. Pine forest is extremely reduced in the Sierra de Neiba, and nearly all forest below 1,600 m has been cut for agriculture and timber.

Laguna de Cabral Wildlife Refuge. Located at the eastern end of the Neiba Valley, the refuge (240 km^2) includes the largest (30 km^2) pool of freshwater on the island. Also referred to as Laguna de Rincón at Cabral, the lagoon still serves local communities as an important source for commercial fishing, and it provides critical habitat for many resident breeding, as well as migratory, bird species.

Sierra de Bahoruco National Park. This 800-km^2 park, and the adjoining 427-km^2 **Aceitillar–Cabo Rojo Panoramic Way**, are in extreme southwestern Dominican Republic and protect an important center of Hispaniolan endemism. Studies conducted in four major habitat types across an elevational gradient (Latta et al. 2003) have also documented the Bahoruco's importance to North American migrants.

Jaragua National Park. This is one of the Dominican Republic's largest protected areas, covering 1,374 km^2. The park includes scrub forest and dry forest, as well as the large lagoon at Oviedo and surrounding mangroves. Jaragua National Park also includes Isla Beata (47 km^2) and Isla Alto Velo (1.5 km^2), which lie 6 and 27 km, respectively, southwest of the Barahona Peninsula. Both islands are dominated by scrub forest and dry forest, which are interrupted in places by bare rock, beaches, lagoons, and mangroves.

National Protected Areas in Haiti

In Haiti there are only three national parks.

National Historic Park of the Citadelle, Sans Souci, and the Ramiers. Situated 18 km south of Cap Haïtien in the western Massif du Nord, this small, 2.5-km^2 park lies at 500 to 875 m elevation. The area immediately adjacent to the Citadelle is degraded by coffee, cocoa, and other agricultural crops, but the ridge of Bonnet-a-l'Evêque and the Ramiers area includes limestone pinnacles and ridges and contains xerophytic broadleaf forests, as well as moist broadleaf forests at upper elevations. Steep and rocky terrain has discouraged disturbance of the habitat.

Macaya Biosphere Reserve. In the Massif de la Hotte, 195 km southwest of Port-au-Prince, this 55-km^2 reserve rises from 950 to 2,347 m in elevation and includes the forested ridges and deep ravines of Morne Macaya and Morne Formon, and the moderately high Plain of Formon, which includes extensive areas of exposed karst. Five major rivers originate in the park:

the Grande Ravine de Sud, Port-a-Piment, L'Acul, Roseaux, and Glace. Principle habitats in the park include pine forest, savanna, montane broadleaf forest, karst forest, and disturbed habitats. The forest at Rak Bwa (1,100 m elevation) is of particular interest to conservation because of its diverse bird and orchid populations, but ease of access has resulted in agricultural encroachment. This area in particular should be prioritized for protection.

La Visite National Park. In the Massif de la Selle, 22 km south of Port-au-Prince, La Visite includes about 30 km^2 of pine forest, savanna, and montane broadleaf forest, all above 1,600 m elevation, and includes much more than just Morne La Visite. The Massif de la Selle is the westward extension in Haiti of the Sierra de Bahoruco in the Dominican Republic. The proximity of this park to Haiti's main population center has resulted in significant habitat loss and disturbance at all elevations. Very little montane broadleaf forest remains, and the ecological future of this park is very much in question.

Threatened and Endangered Species

On Hispaniola, 38 taxa are considered threatened or endangered or appear to have been extirpated from the island. Threatened or endangered species are identified in this guide as any species named as threatened or endangered by the Dominican Department of Wildlife (Secretaría de Estado de Agricultura 1990), the National Planning Workshop for Avian Conservation (Latta and Lorenzo 2000), *Threatened Birds of the World* (BirdLife International 2000), *The Birds of Hispaniola: Haiti and the Dominican Republic* (Keith et al. 2003), or the opinion of the authors based on the most recent data. Alarmingly, nearly half (15 of 31) of the endemic species (names italicized below) are considered to be threatened with extinction, and three of these endemics are critically endangered.

CRITICALLY ENDANGERED
- Black-capped Petrel
- *Ridgway's Hawk*
- Stygian Owl
- *Western Chat-Tanager*
- *Eastern Chat-Tanager*

ENDANGERED
- *White-fronted Quail-Dove*
- *Bay-breasted Cuckoo*
- Loggerhead Kingbird
- Golden Swallow
- *La Selle Thrush*
- *Hispaniolan Highland-Tanager*
- *Hispaniolan Crossbill*

POSSIBLY EXTIRPATED
- Wood Stork

THREATENED
- West Indian Whistling-Duck
- White-cheeked Pintail
- Sharp-shinned Hawk
- Black Rail
- Caribbean Coot
- Limpkin
- Double-striped Thick-knee
- Snowy Plover
- Piping Plover
- Roseate Tern
- Scaly-naped Pigeon
- White-crowned Pigeon
- Plain Pigeon
- Key West Quail-Dove
- Ruddy Quail-Dove
- *Hispaniolan Parakeet*
- *Hispaniolan Parrot*
- *Least Pauraque*
- Northern Potoo
- *Hispaniolan Trogon*
- *Hispaniolan Palm Crow*
- *White-necked Crow*
- Bicknell's Thrush
- *Gray-crowned Palm-Tanager*
- Tawny-shouldered Blackbird

ORNITHOLOGICAL HISTORY OF HISPANIOLA

The earliest written records of Hispaniola's avifauna date from Christopher Columbus's expeditions to the island between 1492 and 1504. These and subsequent explorations of Hispaniola during the following two centuries, however, provided little more than general narrative accounts and incidental natural history observations. France's occupation of Haiti in the early 1700s spawned advances in bird study and led to a number of published works, many of them based on specimen collections which were subsequently lost or destroyed. The first ornithological explorations of eastern Hispaniola were conducted by a French entomologist, Auguste Sallé, who published in 1857 a thorough account of his collections, which included 61 species.

Systematic documentation of Hispaniola's avifauna began with collecting expeditions by the U.S. ornithologist Charles Cory, who visited both Haiti and the Dominican Republic from 1881 to 1883 and published an important reference work, *The Birds of Haiti and San Domingo*, in 1885. Several follow-up collecting trips through 1896, under Cory's direction, amassed several thousand bird specimens, most of which are housed in U.S. museums. From 1916 to1923, William Abbott collected widely over the island, securing additional large series of museum specimens and gaining new insights into the birds of Hispaniola's mountainous regions and satellite islands. An intensive period of field exploration occurred from 1917 to 1934 and culminated in the landmark 1931 volume *Birds of Haiti and the Dominican Republic*, by Alexander Wetmore and Bradshaw Swales. This important reference synthesized the authors' extensive data on distribution, relative abundance, systematics, and natural history, based on their field observations, specimen collections, and examination of fossilized bone deposits. Parallel work by James Bond in the 1920s, early 1930s, and 1941 added valuable information and was incorporated into his classic publication *Birds of the West Indies*, which was first published in 1936 and updated with later editions.

Few ornithological studies were conducted during the post-U.S occupation of Haiti from the late 1920s through the 1970s, or during the Dominican Republic's 30-year Trujillo dictatorship, which ended in 1961. The arrival of Donald and Annabelle "Tudy" Dod to the Dominican Republic in 1964 launched a new era in Hispaniolan field ornithology. Tudy Dod worked tirelessly to study birds, popularize them, and promote their conservation. Her illustrated 1978 book *Aves de la República Dominicana* was the first ornithological account of general interest to Dominicans. Dod's efforts helped usher in a wave of intensive field studies in the Dominican Republic during the 1960s and 1970s, primarily on specific species or groups of resident birds. Principle investigators included David B. Wingate (studying Black-capped Petrels), Wesley E. Lanyon (studying *Myiarchus* flycatchers), Angela Kay Kepler (studying todies), R. K. Selander (studying Hispaniolan Woodpeckers), and James W. Wiley (studying various species). These studies were followed in the 1980s and 1990s by a diversity of projects targeting wintering and transient bird communities which were conducted by Wayne Arendt, John Terborgh, John Faaborg, Joseph Wunderle, Jr., Chris Rimmer, and Steven Latta. This work has in turn spawned studies in nearly all major habitat types of the island by Latta and Rimmer. Ornithological field surveys in Haiti have not kept pace, mainly because of the political constraints of working in the country, but extensive fieldwork by Charles Woods, José Ottenwalder, and Florence Sergile in the 1970s and 1980s provided a noteworthy exception. Follow-up surveys by Chris Rimmer and colleagues were carried out during the winters of 2004 and 2005.

Figure 3. The basic parts of a passerine bird as illustrated by the Hispaniolan Spindalis.
Figure 4. The basic parts of a wing as illustrated by the underwing of the Black-bellied Plover. The basic parts of the waterfowl wing as illustrated by the Blue-winged Teal.

DESCRIPTIVE PARTS OF A BIRD

Body Topography

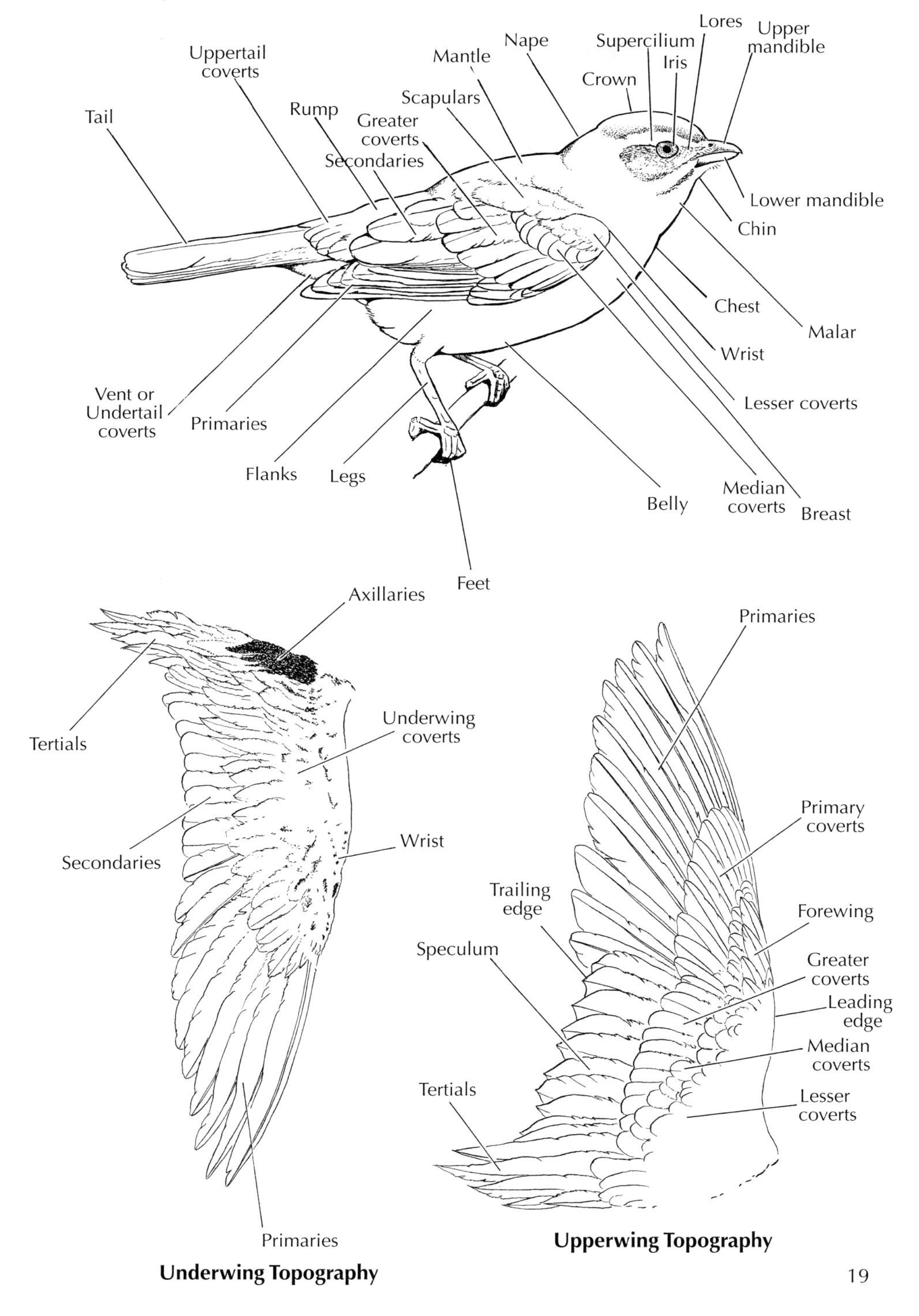

Underwing Topography

Upperwing Topography

PLATE 1: WHISTLING-DUCKS AND WIGEON

1 White-faced Whistling-Duck (*Dendrocygna viduata*) **Page 21**
1a Adult: Mostly brown above with white face; black nape, neck, belly, and tail; chest maroon; flanks barred black and white.
1b In flight: Wings dark above and below.

2 Black-bellied Whistling-Duck (*Dendrocygna autumnalis*) **Page 21**
2a Adult: Dark body, gray face with whitish eye-ring, black belly, reddish bill.
2b In flight: Bold white stripe on upperwing; dark underwing.

3 West Indian Whistling-Duck (*Dendrocygna arborea*) **Page 22**
3a Adult: Deep brown coloration overall; whitish chin and throat; white lower belly with black markings.
3b In flight: Very dark overall; lower belly mottled black and white; gray upperwing coverts.

4 Fulvous Whistling-Duck (*Dendrocygna bicolor*) **Page 22**
4a Adult: Two-toned: blackish brown above, uniform pale yellowish brown below, with thin white stripe along flanks.
4b In flight: Conspicuous white uppertail coverts; contrasting blackish wings and buffy underparts.

5 American Wigeon (*Anas americana*) **Page 24**
5a Male: White crown, large green eye patch, buffy white cheeks and neck. Light blue bill.
5b Female: Gray head, light blue bill.
5c Male in flight: White patch on upper forewing and underwing, green speculum, white belly.
5d Female in flight: White patch on upper forewing and underwing, green speculum, white belly.

6 Eurasian Wigeon (*Anas penelope*) **Page 24**
6a Male: Dark reddish brown head with cream-colored forecrown; pale gray back and sides; chest pinkish.
6b Female: Mottled brownish overall with gray head and light blue bill; may show reddish tint to head and neck.
6c Male in flight: Large white patch on forewing, green speculum, white belly, blackish flecks on axillaries.
6d Female in flight: Green speculum, white belly, blackish flecks on axillaries.

1b
2b
1a
2a
3b
4b
3a
4a
5b
5d
5a
5c
6b
6d
6a
6c

PLATE 2: DUCKS 1

1 Wood Duck (*Aix sponsa*) **Page 23**

1a Male: Distinctive facial pattern of green, purple, and white, with red eye.
1b Female: Head crested; large, white, asymmetrical eye-ring.
1c Male in flight: Long, squared tail; bill tilted down; large head; white throat.
1d Female in flight: Long, squared tail; bill tilted down; large head; white eye patch.

2 Mallard (*Anas platyrhynchos*) **Page 25**

2a Breeding male: Green head, maroon breast, yellow bill.
2b Adult female: Mottled brown overall; dark line through eye; bill orange with black markings.
2c Non-breeding male: Similar to female, but bill olive.
2d Male in flight: Blue speculum bordered with white; green head; maroon breast.
2e Female in flight: Blue speculum bordered with white; mottled brown.

3 Blue-winged Teal (*Anas discors*) **Page 25**

3a Breeding male: Mottled brown with grayish head and distinct white crescent on face.
3b Female and non-breeding male: Small size; mottled brown; light spot on lores.
3c Breeding male in flight: Conspicuous pale blue forewing and green speculum; white crescent on face.
3d Female and non-breeding male in flight: Conspicuous pale blue forewing and green speculum.

4 Northern Shoveler (*Anas clypeata*) **Page 26**

Note strikingly large, spoonlike bill.
4a Male: Green head, white breast, and reddish brown sides and belly.
4b Female: Mottled brown overall.
4c Male in flight: Pale blue forewing; green speculum; green head; white breast; reddish brown sides and belly.
4d Female in flight: Forewing gray; speculum green; mottled brown overall.

5 White-cheeked Pintail (*Anas bahamensis*) **Page 26**

5a Adult: Warm brown overall with prominent white cheek and red mark at base of bill.
5b In flight: Green speculum with buff-colored borders.
Insert bottom of Plate 2

Gadwall *(Anas strepera)* **Page 218**

Not Illustrated:
Breeding male: Gray-brown overall, with black rump and undertail coverts, white speculum, chestnut patch on wing. Slate-gray bill. Yellow legs.
Female and non-breeding male: Mottled brownish overall; white speculum; orange-yellow bill with dark longitudinal midsection.
Breeding male in flight: Black rump and undertail coverts; white speculum bordered by black and chestnut above; underwing coverts white.
Female and non-breeding male in flight: White speculum; white underwing coverts.

1b
1a
1c
1d
2c
2b
2a
2d
2e
3b
3a
3c
3d
4b
4a
4c
4d
5a
5b

PLATE 3: DUCKS 2

1 Northern Pintail (*Anas acuta*) **Page 27**
Note long, slender neck and long, pointed tail.
1a Breeding male: Brown head with long, white neck stripe; white breast.
1b Female and non-breeding male: Mottled brown overall. Bill narrow and gray.
1c Breeding male in flight: Greenish speculum with buff-colored inner border and white trailing edge.
1d Female and non-breeding male in flight: Brown speculum bordered by white bars; gray underwing contrasts with white belly.

2 Green-winged Teal (*Anas crecca*) **Page 27**
Small, compact size; dark overall coloration.
2a Breeding male: Reddish brown head, green eye patch, and white vertical bar in front of wing.
2b Female and non-breeding male: Mottled brown with dark lores and whitish belly.
2c Breeding male in flight: Green speculum edged with white or buff; no blue in forewing; green eye patch on reddish brown head.
2d Female and non-breeding male in flight: Green speculum edged with white or buff; no blue in forewing; mottled brown; whitish belly.

3 Canvasback (*Aythya valisineria*) **Page 28**
Note distinctive sloping forehead profile.
3a Male: Reddish brown head and neck, black chest, whitish back and flanks; rump brownish black.
3b Female: Pale overall; light brown head and neck, darker rump and tail.
3c Male in flight: Black breast and tail contrast with light belly and underwings.
3d Female in flight: Dark breast and tail contrast with light belly and underwings.

4 Redhead (*Aythya americana*) **Page 28**
4a Male: Pale gray back, black breast, and rounded, bright reddish head and neck.
4b Female: Uniformly dull brown; white eye-ring; blue bill tipped black.
4c Male in flight: Gray back and black neck contrast with reddish head; gray flight feathers contrast with dark forewing.
4d Female in flight: Dull brown with white eye-ring; gray flight feathers contrast with dark forewing.

5 Ring-necked Duck (*Aythya collaris*) **Page 28**
Short crest results in head's angular, peaked profile.
5a Male: Black back, white vertical bar in front of wing, and white bill-ring.
5b Female: Brown overall with light eye-ring and white bill-ring.
5c Male in flight: Dark upperwing coverts contrast with pale gray secondaries; underparts black and white.
5d Female in flight: Dark upperwing coverts contrast with pale gray secondaries; underparts brown and white.

1b
1a
1c
1d
2b
2a
2c
2d
3b
3a
3c
3d
4b
4a
4c
4d
5b
5a
5c
5d

PLATE 4: DUCKS AND GEESE

1 Lesser Scaup (*Aythya affinis*) **Page 29**
1a Male: Black head, breast, and tail; whitish back flecked gray; white flanks.
1b Female: Brown with large white mark around base of bill.
1c Male in flight: White secondaries and darker primaries; black breast.
1d Female in flight: White secondaries and darker primaries; brown breast.

2 Hooded Merganser (*Lophodytes cucullatus*) **Page 29**
2a Male: Large crest contains white, fan-shaped patch, especially striking when raised. Sides buffy cinammon.
2b Female: Grayish brown, darker above, with bushy crest.
2c Male in flight: Small white patch on secondaries; pale forewing.
2d Female in flight: Brown above with small white patch on secondaries.

3 Red-breasted Merganser (*Mergus serrator*) **Page 30**
3a Breeding male: Green head with shaggy crest, white collar, reddish brown breast.
3b Female and non-breeding male: Grayish brown with shaggy-crested, orange-brown head; chin, foreneck, and breast whitish. Bill reddish.
3c Male in flight: Secondaries and forewing white, crossed by two narrow dark bars.
3d Female in flight: Secondaries white and crossed by one dark bar.

4 Masked Duck (*Nomonyx dominicus*) **Page 30**
Small, chunky duck with conspicuous, erect tail.
4a Breeding male: Reddish brown with black face, blue bill.
4b Female, non-breeding male, and immature: Mottled buffy brown with two dark brown facial stripes.
4c Breeding male in flight: Prominent white patch on secondaries and part of forewing; reddish brown with black face.
4d Female, non-breeding male, and immature in flight: Prominent white patch on secondaries and part of forewing; brown overall with dark brown facial stripes.

5 Ruddy Duck (*Oxyura jamaicensis*) **Page 31**
5a Breeding male: Reddish brown with white cheek patch, blue bill.
5b Non-breeding male: Grayish brown overall with dark cap, white cheeks.
5c Female and immature: Grayish brown with whitish cheeks cut by single brown stripe.
5d Breeding male in flight: Chunky with long tail and dark upperwings; white cheek.
5e Female and immature in flight: Chunky with long tail and dark upperwings; cheek stripe.

6 Canada Goose (*Branta canadensis*) **Page 23**
Grayish brown with black head and neck; conspicuous white band on cheeks and throat.

1b
1a
1d
1c
2b
2a
2d
2c
3a
3b
3d
3c
4b
4a
4d
4c
5c
5b
5a
5e
5d
6

PLATE 5: GREBES AND SEABIRDS

1 Least Grebe (*Tachybaptus dominicus*) **Page 34**
Small size, blackish coloration, thin bill, yellow-orange eye.
1a Breeding: Black throat.
1b Non-breeding: White throat.

2 Pied-billed Grebe (*Podilymbus podiceps*) **Page 35**
Stocky, brownish overall, with short, conical bill.
2a Breeding: Black throat; bill whitish with black band.
2b Non-breeding: Whitish throat; bill lacks black band.
2c Juvenile: Mottled brown and buffy white markings on head.

3 Black-capped Petrel (*Pterodroma hasitata*) **Page 35**
Black above with white forecrown, rump, and nape; white below; wings sharply pointed; underwings with black leading edge forming distinctive bar on coverts.
3a Typical coloration above.
3b Darker, atypical coloration.
3c Typical coloration below.

4 Greater Shearwater (*Puffinus gravis*) **Page 36**
Large size. Grayish brown above with some white on rump and nape; no white on forecrown; mostly white below; underwings with considerable dark markings.

5 Manx Shearwater (*Puffinus puffinus*) **Page 37**
Medium size, short-tailed. Blackish above; pale crescent behind auriculars; white below, including undertail coverts; underwing coverts whitish.

6 Audubon's Shearwater (*Puffinus lherminieri*) **Page 37**
Relatively small; short wings; long, rounded tail. Dark above, white below; dark undertail coverts.

7 Wilson's Storm-Petrel (*Oceanites oceanicus*) **Page 38**
Small size. Blackish overall; rounded wrists; conspicuous white rump patch undivided; squared tail.

8 Leach's Storm-Petrel (*Oceanodroma leucorhoa*) **Page 38**
Small size. Blackish overall; sharply angled wrists; pale brown wing band; white rump patch divided in middle; tail notched.

1a
1b
2b
2c
2a
3a
3b
3c
4
5
6
7
8

PLATE 6: BOOBIES, FRIGATEBIRD, TROPICBIRDS, AND PELICAN

1 Red-footed Booby (*Sula sula*) **Page 41**
Feet orange-red to red; bill grayish.
1a Adult, brown phase: Brown with white hindparts.
1b Adult, white phase: Entirely white, including tail, with black primaries and secondaries.

2 Brown Booby (*Sula leucogaster*) **Page 41**
Dark brown head and upperparts with sharply demarcated white belly; white underwing coverts.

3 Masked Booby (*Sula dactylatra*) **Page 40**
White with blackish tail, primaries, and secondaries.

4 Magnificent Frigatebird (*Fregata magnificens*) **Page 45**
Long, forked tail; long, slender wings with pronounced crook at wrist.
4a Breeding male: Entirely black; bright red inflatable throat pouch.
4b Breeding female: Blackish overall; white breast.
4c Immature: Brownish black; white head and breast.

5 White-tailed Tropicbird (*Phaethon lepturus*) **Page 39**
5a Adult: White with heavy black bar on upperwing coverts; black on outermost primaries; long tail plumes. Bill orange to red-orange.
5b Immature: White with coarse black barring on upperparts; short tail. Bill yellow.

6 Red-billed Tropicbird (*Phaethon aethereus*) **Page 40**
White with fine black barring on back; black eye-line; long tail plumes. Bill red.

7 Brown Pelican (*Pelecanus occidentalis*) **Page 42**
Large, dark; unmistakable long, pouched bill.
7a Breeding: Back of head and nape reddish brown.
7b Non-breeding: Back of head and nape white.
7c Immature: Grayish brown plumage; lighter below.

1a
1b
2
3
4c
4a
4b
5b
5a
6
7b
7a
7c

PLATE 7: ANHINGA, CORMORANTS, BITTERNS, AND NIGHT-HERONS

1 Anhinga (*Anhinga anhinga*) **Page 44**
Long-necked; long-tailed; pointed bill.
1a Male: Glossy black; silvery white patches on upperwing and back.
1b Female: Head, neck, and breast pale brown.
1c Female swimming.

2 Neotropic Cormorant (*Phalacrocorax brasilianus*) **Page 44**
Smaller than Double-crested Cormorant; thinner bill; longer tail.
2a Breeding adult: Blackish; small, yellowish patch on face and throat is edged whitish.
2b Non-breeding adult: Similar; throat patch reduced or absent.
2c Immature: Brown above, paler below.

3 Double-crested Cormorant (*Phalacrocorax auritus*) **Page 43**
Dark waterbird with hooked bill.
3a Breeding adult: Blackish; large, orange-yellow patch on face and throat; small ear tufts.
3b Non-breeding adult: Similar; no ear tufts.
3c Immature: Brown above, paler below.

4 Least Bittern (*Ixobrychus exilis*) **Page 46**
4a Adult: Small; dark above, buffy below; cream-colored patch on upperwing.
4b Immature: Similar, but heavily streaked on back and breast.

5 American Bittern (*Botaurus lentiginosus*) **Page 46**
5a Adult: Cryptic brown plumage; black stripe on side of neck.
5b In flight: Blackish wing tips.

6 Yellow-crowned Night-Heron (*Nyctanassa violacea*) **Page 53**
Stocky; large head and short neck; red eye.
6a Adult: Gray underparts; black and white striped head markings.
6b Immature: Small white flecks on gray-brown back; narrow, distinct streaks on underparts.
6c Adult in flight: Legs and feet extend beyond tail; gray underparts.
6d Immature in flight: Legs and feet extend beyond tail; mottled brown with small white flecks.

7 Black-crowned Night-Heron (*Nycticorax nycticorax*) **Page 52**
Stocky; large head and short neck; red eye.
7a Adult: Black crown and back; white face, underparts, and head plumes.
7b Immature: Large white flecks on brown back; broad, blurry streaks below.
7c Adult in flight: Only feet extend beyond tail; contrasting black and white.
7d Immature in flight: Only feet extend beyond tail; mottled brown with large white flecks.

2b
2a
1b
3a
3b
2c
3c
1a
1c
4b
4a
5a
5b
6d
6b
6c
7d
6a
7c
7b
7a

PLATE 8: HERONS AND EGRETS

1 Little Blue Heron (*Egretta caerulea*) **Page 49**
1a Adult: Uniformly dark blue-gray body; purplish brown head and neck.
1b Immature: White overall; legs greenish; bill bluish gray with black tip.
1c Immature molting: White mottled with dark feathers.

2 Tricolored Heron (*Egretta tricolor*) **Page 50**
2a Adult: Mostly bluish gray; white belly, throat, and foreneck.
2b Immature: Similar, with reddish brown nape and upperwing coverts.

3 Cattle Egret (*Bubulcus ibis*) **Page 51**
3a Breeding: Legs and bill reddish. White with buffy orange wash on crown, breast, and upper back.
3b Non-breeding: Mostly white. Legs black; bill yellow.

4 Little Egret (*Egretta garzetta*) **Page 48**
Entirely white plumage; black legs with bright yellow feet; two long head plumes on breeding adult. Lores yellow-orange when breeding, gray-green on nonbreeding adults and immatures.

5 Snowy Egret (*Egretta thula*) **Page 49**
5a Adult: Entirely white, with black bill and legs, yellow feet and lores. Many elongated plumes on head, neck, and back.
5b Immature: Legs dark in front, greenish yellow in back.

6 Reddish Egret (*Egretta rufescens*) **Page 50**
Bicolored bill is pink at base, black at tip. Legs blue-gray. Shaggy head and neck plumes on adults.
6a Adult, dark phase: Grayish overall; head and neck reddish brown.
6b Adult, white phase: White overall.
6c Immature: Bill dark; neck feathers unruffled.

7 Great Blue Heron (*Ardea herodias*) **Page 47**
7a Adult, dark phase: Bluish gray; broad black supercilium and head plumes.
7b Adult, white phase: White with yellow bill and legs.
7c Immature: Grayish brown; entire crown blackish.

8 Great Egret (*Ardea alba*) **Page 48**
Entirely white, with yellow bill and black legs.

9 Green Heron (*Butorides virescens*) **Page 52**
9a Adult: Small; dark overall. Short, dark chestnut-colored neck. Greenish yellow to orange legs.
9b Immature: Dull brownish overall; heavily streaked below.

1a
1c
1b
2b
2a
3a
3b
4
5a
5b
6c
6b
6a
7b
7c
7a
8
9a
9b

PLATE 9: MORE LARGE WADERS

1 White Ibis (*Eudocimus albus*) **Page 54**
1a Adult: White overall; decurved, reddish bill; legs reddish.
1b Immature: Brownish, white belly and rump; decurved, orangish bill.
1c Adult in flight: Neck outstretched; black wing tips.

2 Glossy Ibis (*Plegadis falcinellus*) **Page 55**
2a Adult: Dark metallic bronze and green; long, decurved bill.
2b Immature: Paler than adult.

3 Roseate Spoonbill (*Platalea ajaja*) **Page 55**
3a Adult: Pink overall with white neck; spatulalike bill.
3b Immature: Mostly white with pink wash; spatulalike bill.
3c Adult in flight: Pink overall with spatulalike bill.

4 Wood Stork (*Mycteria americana*) **Page 56**
4a Adult: Large size. White with dark head and upper neck; heavy, black bill, decurved at tip.
4b Immature in flight: White with dusky gray head and yellowish bill; black flight feathers and tail.

5 Greater Flamingo (*Phoenicopterus ruber*) **Page 58**
5a Adult: Orangish pink coloration; long neck and legs; unique curved bill with black tip.
5b Immature: Similar to adult but white to pinkish with paler bill.
5c Adult in flight: Head and neck outstretched and drooping; flight feathers black.

6 Limpkin (*Aramus guarauna*) **Page 71**
Brown overall with white streaking and triangular spotting; long legs and neck; long bill slightly down-curved.

1a
1c
2a
1b
2b
4b
3b
4a
3a
3c
5a
5c
6
5b

PLATE 10: RAPTORS 1

1 Peregrine Falcon (*Falco peregrinus*) **Page 65**
Largest falcon. Dark mask on face; tail faintly barred with more than 5 light bands.
1a Adult: Dark gray above; white to cream colored with dark bars below.
1b Immature: Brown; heavily streaked below.

2 Merlin (*Falco columbarius*) **Page 64**
Heavily streaked underparts; pale tan supercilium; tail with 2–5 light bands.
2a Male: Dark above; rufous feathering on legs.
2b Female in flight: Dark brown above; long, pointed wings; long, narrow tail; fast and agile flight.

3 Swallow-tailed Kite (*Elanoides forficatus*) **Page 60**
3a Adult: Bicolored; white head and underparts; black back, wings, and deeply forked tail.
3b In flight: Black flight feathers and deeply forked tail.

4 American Kestrel (*Falco sparverius*) **Page 64**
Reddish brown back; tail reddish brown with terminal black band. Frequently hovers.
4a Adult male: Blue-gray wings.
4b Adult female: Reddish brown wings.
4c Female in flight.

5 Turkey Vulture (*Cathartes aura*) **Page 57**
5a Adult: Large, dark overall; bare head is red.
5b Immature: Similar, but bare head is blackish.
5c In flight: Two-toned wings held above horizontal in a broad V.

6 Northern Harrier (*Circus cyaneus*) **Page 60**
6a Adult male: Grayish blue above; lighter below.
6b Adult female: Brown above; streaked below.
6c Male in flight: Long wings and tail; white rump. Wings held above horizontal.

7 Osprey (*Pandion haliaetus*) **Page 59**
Dark brown above, white below.
7a Resident race: White head; trace of eye stripe.
7b Migratory race: White head; dark bar through eye.
7c In flight: Long, narrow wings bent at wrist; black wrist patch.

1b
2b
3b
1a
2a
3a
4c
4b
4a
5c
6c
7c
6a
5b
6b
7b
7a
5a

PLATE 11: RAPTORS 2

1 Broad-winged Hawk (*Buteo platypterus*) **Page 62**

1a Adult: Dark above; barred reddish brown below; fan-shaped tail boldly banded black and white.
1b Immature: Underparts white, heavily streaked with dark brown; tail bands more numerous and less distinct than on adult.
1c Adult in flight: Broad, rounded wings; boldly banded tail; alternates soaring and flapping.
1d Immature in flight: Broad, rounded wings; tail with finer bars; alternates soaring and flapping.

2 Sharp-shinned Hawk (*Accipiter striatus*) **Page 60**

Small forest hawk; short, rounded wings; long, narrow, squared-off tail.
2a Male: Dark, steel blue upperparts; finely barred, reddish underparts.
2b Female: Brown above; white below with coarse brown streaking.
2c In flight: Short, rounded wings; long, narrow, squared-off tail; alternately flapping and gliding.

3 Swainson's Hawk (*Buteo swainsoni*) **Page 63**

Soaring hawk with long tail and wings.
3a Adult, dark phase: Dark brown above, whitish below with rufous to dark brown bib.
3b Adult, light phase: Similar but lighter overall.
3c Immature, light phase: Similar, but underparts streaked and spotted brown.

4 Red-tailed Hawk (*Buteo jamaicensis*) **Page 63**

4a Adult: Dark above, light below, with dark belly band; tail reddish.
4b Immature: Lightly barred, grayish brown tail; belly band less distinct.
4c Adult in flight: Soars on rounded wings and tail; distinctive reddish tail and dark belly band.
4d Immature in flight: Tail lightly barred; belly band less distinct.

1a
1b
1c
1d
2c
2b
2a
3c
3a
3b
4a
4b
4c
4d

PLATE 12: RIDGWAY'S HAWK

Ridgway's Hawk (*Buteo ridgwayi*) **Page 61**

(Left) Male: Dark brownish gray above; gray underparts washed with brownish red; thighs reddish brown; tail thinly barred black and white.

(Right) Immature: Underparts buffy white, streaked pale gray and tan; tail indistinctly barred.

BKM

PLATE 13: RAILS AND THEIR ALLIES

1 Sora (*Porzana carolina*) **Page 67**
Brownish gray plumage; stubby yellow bill.
1a Adult: Blackish face, throat, and breast.
1b Immature: Black mask absent; dull buffy face, throat, and breast.

2 Black Rail (*Laterallus jamaicensis*) **Page 66**
Tiny, dark grayish rail; white spots on back; reddish brown nape; short black bill.

3 Clapper Rail (*Rallus longirostris*) **Page 66**
Primarily gray; flanks and lower belly barred black and white; long bill.

4 Spotted Rail (*Pardirallus maculatus*) **Page 68**
Spotted and barred black and white; reddish legs; long, greenish yellow bill with red spot at base.

5 Yellow-breasted Crake (*Porzana flaviventer*) **Page 68**
Tiny; pale yellowish brown; bold black and white barring on flanks; blackish crown and eye-line; short bill.

6 Northern Jacana (*Jacana spinosa*) **Page 79**
Note extremely long, slender, green toes.
6a Adult: Dark reddish brown overall; blackish head and neck; bill and frontal shield yellow.
6b Immature in flight: Olive-brown above, whitish below; white supercilium; flight feathers bright yellow.

7 American Coot (*Fulica americana*) **Page 70**
Grayish black overall; white bill and undertail coverts.

8 Caribbean Coot (*Fulica caribaea*) **Page 71**
8a Adult: Grayish black; white undertail coverts; white bill and frontal shield extending to crown.
8b Immature: Uniformly grayer than adult.

9 Purple Gallinule (*Porphyrula martinica*) **Page 69**
9a Adult: Bluish purple overall; yellow legs; red bill tipped yellow; bluish white frontal shield.
9b Immature: Buffy brown; bluish green wings; yellowish legs.

10 Common Moorhen (*Gallinula chloropus*) **Page 69**
Blackish with white flank stripe; red bill tipped yellow; high, red frontal shield.

1b
1a
3
2
4
5
7
6b
6a
8b
8a
9b
9a
10

PLATE 14: PLOVERS

1 Semipalmated Plover (*Charadrius semipalmatus*) **Page 76**
Brown upperparts; light orange legs; stubby bill; breast band may be partial or complete.
1a Non-breeding: Grayish brown mask; white forecrown; single, grayish brown breast band. Dark bill.
1b Breeding: Black mask; white forecrown; single black breast band. Orange bill with black tip.

2 Killdeer (*Charadrius vociferus*) **Page 77**
2a Adult: Grayish brown upperparts; whitish underparts; two black breast bands.
2b In flight: Orange-brown rump; white wing stripe.

3 American Golden-Plover (*Pluvialis dominica*) **Page 74**
Large, stocky plover with short bill.
3a Non-breeding: Mottled gray coloration; distinct contrast between dark crown and whitish supercilium.
3b Breeding: Mottled black and golden upperparts; black underparts extend to undertail coverts.
3c Non-breeding in flight: Uppertail coverts dark; no wing stripe; no black in axillaries.

4 Snowy Plover (*Charadrius alexandrinus*) **Page 75**
Very small size. Slender black bill; pale coloration; dark gray legs.
4a Breeding: Black auricular patch; white forecrown and supercilium.
4b Immature: Lacks black in plumage.

5 Piping Plover (*Charadrius melodus*) **Page 76**
Pale gray upperparts; short, stubby bill; yellow-orange legs.
5a Non-breeding: Black breast band partial or absent, usually replaced by brownish gray patches on sides of breast. Black bill.
5b Breeding: Single black breast band partial or complete. Orange bill with black tip.
5c In flight: White uppertail coverts.

6 Black-bellied Plover (*Pluvialis squatarola*) **Page 73**
Large plover with heavy body and thick bill.
6a Non-breeding: Light, mottled gray coloration; indistinct contrast between crown and supercilium; breast streaked.
6b Breeding: Mottled white and black above; black underparts except for white undertail coverts.
6c Non-breeding in flight: White uppertail coverts and tail with dark bars; white wing stripe; black axillaries.

7 Wilson's Plover (*Charadrius wilsonia*) **Page 75**
Single breast band; heavy, black bill; legs dull pinkish.
7a Male: Black breast band; white forecrown and supercilium.
7b Female: Brown breast band.

1b
1a
2a
2b
3c
3a
3b
4a
4b
5b
5c
5a
6a
6b
6c
7b
7a

PLATE 15: SHOREBIRDS 1

1 Wilson's Snipe (*Gallinago delicata*) **Page 93**

1a Adult: Stocky, with short neck and legs. Long bill; striped head and back; reddish brown tail.
1b In flight: Dark underwings contrast with white belly; zigzag flight, uttering call note.

2 Long-billed Dowitcher (*Limnodromus scolopaceus*) **Page 93**

Stocky build. Long, straight bill.
2a Non-breeding: Brownish gray above; whitish below with gray breast; white supercilium.
2b Breeding: Head and breast reddish brown, extending to belly; sides of breast spotted; flanks barred.

3 Red Knot (*Calidris canutus*) **Page 85**

Chunky build; greenish legs; short bill, neck, and legs.
3a Non-breeding: Plain gray above; white below, with breast finely streaked gray.
3b Breeding: Orangish red face and underparts.
3c In flight: Barred, pale gray rump; white wing stripe; pale gray underwing coverts.

4 Short-billed Dowitcher (*Limnodromus griseus*) **Page 92**

Stocky build. Long, straight bill.
4a Non-breeding: Brownish gray above; whitish below with gray breast; white supercilium.
4b Breeding: Breast and head reddish brown; belly white; sides of breast and flanks spotted and barred dark brown.
4c In flight: White rump patch extends as sharp V well up back.

5 Stilt Sandpiper (*Calidris himantopus*) **Page 91**

Long, greenish legs; long bill, thick at base and slightly drooping at tip; whitish supercilium.
5a Non-breeding: Pale brownish gray above; whitish below.
5b Breeding: Dark brownish above; reddish brown auriculars; breast and belly heavily barred.
5c In flight: White rump; pale gray to whitish tail; unmarked wings.

6 Solitary Sandpiper (*Tringa solitaria*) **Page 81**

6a Adult: Dark upperparts, finely spotted with white; white eye-ring; greenish legs.
6b In flight: Dark above; upperwings and underwings uniformly dark; black barring on white outer tail feathers.

1b
2b
1a
2a
4c
3c
3b
4b
3a
4a
5c
5b
6b
5a
6a

PLATE 16: SHOREBIRDS 2

1 Dunlin (*Calidris alpina*) **Page 91**
Chunky build; short-necked appearance. Long and heavy bill, drooping at tip. Black legs.
1a Non-breeding: Brownish gray wash on head, breast, and upperparts.
1b Breeding: Black belly; reddish brown back and crown.
1c In flight: White wing stripe; white rump divided by black bar.

2 Spotted Sandpiper (*Actitis macularius*) **Page 83**
2a Non-breeding: Grayish brown above, white below; brownish gray neck patch extends onto breast; bill dull fleshy.
2b Breeding: Bold dark spots on underparts; orange bill with black tip.
2c In flight: Short white wing stripe.

3 Sanderling (*Calidris alba*) **Page 86**
3a Non-breeding: Pale gray upperparts; pure white underparts; black mark on bend of wing.
3b Breeding: Reddish brown head and breast.
3c In flight: Pale gray upperparts; black leading edge of wing bordered by conspicuous white wing stripe; underwing bright white.

4 Ruddy Turnstone (*Arenaria interpres*) **Page 85**
Stocky; short, orange legs; wedge-shaped, slightly upturned black bill.
4a Non-breeding: Dark breast markings; belly white.
4b Breeding: Distinctive black and white facial markings; reddish orange back and wings.
4c In flight: Distinctive white pattern on upperparts.

5 Buff-breasted Sandpiper (*Tryngites subruficollis*) **Page 92**
5a Adult: Slender build. Large, dark eye on clean, buffy face. Upperparts appear scaled; underparts clear buffy with spots on sides. Legs yellow.
5b In flight: Pure white underwing coverts.

6 Wilson's Phalarope (*Phalaropus tricolor*) **Page 94**
Slender; thin, needlelike bill.
6a Non-breeding adult: Pale gray above; white below; thin, dark line through eye.
6b Breeding female: Dark reddish brown band from behind eye to shoulder; white supercilium.
6c Non-breeding in flight: Dark upperparts; white rump.

7 Red-necked Phalarope (*Phalaropus lobatus*) **Page 94**
Small and delicate; very thin, black, straight bill.
7a Non-breeding adult: Gray upperparts streaked white on mantle; white below; black smudge through eye; black crown and white forecrown.
7b Breeding female: Reddish brown neck contrasting with white throat; buffy stripes on back and upperwing coverts. Black eye patch.
7c Non-breeding in flight: White wing stripe; white stripes on back.

1c
1b
1a
2c
2b
2a
3c
3b
3a
4b
4c
4a
5b
5a
6c
6b
6a
7c
7b
7a

1 Semipalmated Sandpiper (*Calidris pusilla*) **Page 87**
Small size. Black legs; medium-length, straight black bill.
1a Non-breeding: Grayish brown above; whitish below with diffuse streaking on breast.
1b Breeding: Reddish brown tints to upperparts; fine breast streaks.
1c In flight: Narrow, white wing stripe; white rump divided by black bar.

2 Least Sandpiper (*Calidris minutilla*) **Page 88**
Very small size; brown overall; yellowish green legs; thin, drooping bill.
2a Non-breeding: Brown above; breast finely streaked and washed brown.
2b Breeding: More mottled with reddish brown tints above.
2c In flight: Dark above; faint wing stripe.

3 Western Sandpiper (*Calidris mauri*) **Page 87**
Bill black, heavy at base, narrower and slightly drooping at tip.
3a Non-breeding: Grayish above; whitish below with grayish smudge across breast.
3b Breeding: Reddish brown crown, auriculars, and scapulars.

4 Baird's Sandpiper (*Calidris bairdii*) **Page 89**
Wings project well beyond tail at rest.
4a Non-breeding: Scaled appearance: buffy brown with pale edges to feathers; breast buffy and finely streaked.
4b Breeding: Browner; faint reddish brown tints; scaling more obvious.

5 Pectoral Sandpiper (*Calidris melanotos*) **Page 90**
Sharp demarcation between streaked breast and white belly. Yellowish green legs.
5a Non-breeding: Gray-brown upperparts, head, and breast.
5b Breeding: More mottled above; breast densely streaked with black.
5c In flight: Narrow, white wing stripe; white rump divided by black bar.

6 White-rumped Sandpiper (*Calidris fuscicollis*) **Page 89**
Wings extend beyond tail at rest; white rump; distinct whitish supercilium.
6a Non-breeding: Darker gray than other small sandpipers; gray on breast gives "hooded" appearance.
6b Breeding: Darker brown than other small sandpipers; extensive fine streaking on breast and flanks.
6c In flight: White rump. Narrow, white wing stripe.

1c
2c
1b
2b
1a
2a
4b
3b
3a
4a
5c
6c
5b
6b
5a
6a

PLATE 18: SHOREBIRDS 4

1 Hudsonian Godwit (*Limosa haemastica*) **Page 84**
Long, slightly upturned bill; black tail, white at base.
1a Non-breeding: Gray overall; paler below; white supercilium.
1b Non-breeding in flight: White stripe on upperwing and underwing; white base of black tail; blackish underwing coverts.
1c Breeding in flight: Dark reddish brown below, with dark barring; white stripe on upperwing and underwing; white base of black tail.

2 Marbled Godwit (*Limosa fedoa*) **Page 84**
Long, slightly upturned bill; no white on rump.
2a Non-breeding: Buff-colored underparts.
2b Breeding in flight: Buffy below, barred with black. From above: blackish primary coverts, no white on rump. From below: light primaries, cinnamon underwing coverts.

3 Willet (*Catoptrophorus semipalmatus*) **Page 82**
Large and stocky. Light gray overall; gray legs; long, thick bill.
3a Breeding: Finely striped head, neck, and breast.
3b Non-breeding in flight: Bold black and white wing pattern.

4 Whimbrel (*Numenius phaeopus*) **Page 83**
4a Adult: Large and sturdy. Long, down-curved bill; striped crown.
4b In flight: Underwings barred.

5 American Oystercatcher (*Haematopus palliatus*) **Page 78**
Pied shorebird; black hood and back; long, heavy bill.
5a Adult: Orange-red bill, pinkish legs.
5b In flight: White uppertail coverts, broad white wing stripe.

1b
1c
2b
1a
2a
3b
3a
4b
5b
4a
5a

PLATE 19: SHOREBIRDS 5

1 Greater Yellowlegs (*Tringa melanoleuca*) **Page 80**
Large size. Long, orange-yellow legs; long, straight bill, slightly upturned.

2 Lesser Yellowlegs (*Tringa flavipes*) **Page 81**
2a Adult: Medium size. Long, orangish yellow legs; bill relatively thin and short compared with Greater Yellowlegs.
2b In flight: Dark above; white uppertail coverts.

3 Black-necked Stilt (*Himantopus mexicanus*) **Page 78**
3a Adult: Slender, with black upperparts and white underparts; long, pink legs; needlelike bill.
3b In flight: Black wings; white tail extending as V on lower back; long, trailing legs.

4 Double-striped Thick-knee (*Burhinus bistriatus*) **Page 72**
4a Adult: Large, ploverlike. Mostly brown streaked tawny and white; well camouflaged. Large, yellow eye; whitish supercilium.
4b In flight: Conspicuous white wing patches.

1
2a
2b
3a
3b
4a
4b

PLATE 20: TERNS AND JAEGERS

1 Black Tern (*Chlidonias niger*) **Page 108**

1a Non-breeding: Gray above; white below and on forecrown and nape; dark patch behind eye and on hindcrown; smudgy sides of breast.
1b Breeding: Almost entirely black head and underparts; gray mantle and wings; white undertail coverts.

2 Parasitic Jaeger (*Stercorarius parasiticus*) **Page 96**

White patches at base of primaries visible in flight. Pointed tips to elongated central tail feathers.
2a Light-phase adult: Dark brownish gray above; grayish brown cap; yellowish wash on nape and sides of neck; narrow, dark breast band.
2b Dark-phase adult: Dark brown overall.
2c Subadult and juvenile: Finely barred below; usually with reddish brown cast to plumage.

3 Long-tailed Jaeger (*Stercorarius longicaudus*) **Page 97**

Smallest, most slender jaeger.
3a Adult: Long, pointed central tail feathers; only outer primaries have white bases visible in flight; cap grayish brown; no breast band; underwing uniformly dark.
3b Subadult and juvenile (light phase): Finely barred below; fine white barring on back. Sometimes pale head and nape. Central tail feathers blunt and usually extended.

4 Pomarine Jaeger (*Stercorarius pomarinus*) **Page 95**

Large and heavy-bodied. White at base of primaries. Adult has elongated, blunt, twisted central tail feathers.
4a Light-phase adult: Blackish cap; yellowish wash on nape; broad dark band across breast.
4b Dark-phase adult: Entirely dark brown.
4c Subadult and juvenile: Heavily barred below, especially sides and underwings; rounded central tail feathers barely project.

5 Brown Noddy (*Anous stolidus*) **Page 109**

5a Adult: Dark brown overall; silvery white cap blending to gray on nape.
5b Immature: Dark brown overall; white on forecrown only.
5c Adult in flight: Dark brown overall; silvery white cap blending to gray on nape.

6 Sooty Tern (*Sterna fuscata*) **Page 108**

6a Adult in flight: Blackish above; white below. Deeply forked tail; white outer tail feathers. White of forecrown extends only to eye. Dark primaries contrast with white underwing coverts.
6b Immature in flight: Entirely dark brown flecked with white; whitish undertail coverts and underwing coverts.
6c Juvenile: Entirely dark brown flecked with white.

7 Bridled Tern (*Sterna anaethetus*) **Page 107**

7a Adult: Grayish brown above; white below; white forecrown forms V above and behind eye; narrow white collar on nape. Bill and legs black.
7b Immature: Similar to adult but upperparts flecked with pale gray.
7c Adult in flight: Grayish brown above; white below; narrow white collar on nape.

1a
1b
2a
2b
2c
3a
3b
4a
4b
4c
5a
5b
5c
6a
6b
6c
7a
7b
7c

PLATE 21: TERNS

1 Sandwich Tern (*Sterna sandvicensis*) **Page 103**
1a Breeding adult: Shaggy black crest; bill long, slender, and black with conspicuous yellow tip; moderately forked tail does not extend beyond folded wings at rest.
1b Non-breeding adult: Forecrown to behind eye white.
1c Yellow-billed form.

2 Roseate Tern (*Sterna dougallii*) **Page 104**
Long, deeply forked tail; pale gray mantle and primaries; tail extends beyond wing tips at rest.
2a Breeding adult: Bill black with some red at base; black cap.
2b Non-breeding adult: Bill blackish; forecrown white past eye.
2c Immature: Dark forecrown and crown; bill and legs blackish; back mottled; shoulder with indistinct dark markings.

3 Gull-billed Tern (*Sterna nilotica*) **Page 102**
Fairly stocky and gull-like; thick black bill; shallow fork in tail; relatively long, black legs.
3a Breeding adult: Black crown and nape.
3b Non-breeding adult: Crown whitish with gray flecks; gray spot behind eye.

4 Common Tern (*Sterna hirundo*) **Page 105**
4a Breeding adult: Bill red with black tip; black cap and trailing edge of outer primaries; moderately forked tail does not extend beyond wings at rest; legs red.
4b Non-breeding adult: Bill and legs blackish; forecrown white; shoulder with dark bar.
4c Immature: Dark forecrown and crown; bill blackish; back mottled pale brownish; shoulder with indistinct marks.

5 Forster's Tern (*Sterna forsteri*) **Page 106**
Immaculate white; long and deeply forked tail projects beyond wings at rest; thick bill.
5a Breeding adult: Orange bill with black tip.
5b Breeding adult in flight: Deeply forked tail. White body and silvery white primaries.
5c Non-breeding adult: Head and nape white; large black spot forms mask through eye; silvery white primaries.

6 Least Tern (*Sterna antillarum*) **Page 106**
Small size; short, moderately forked tail; yellow legs.
6a Breeding adult: Pale yellow bill with black tip; black cap with white forecrown.
6b Non-breeding adult: Blackish bill; head pattern less distinct.

7 Royal Tern (*Sterna maxima*) **Page 103**
Large size. Orange-yellow bill; shaggy crest.
7a Breeding adult: Black crown.
7b Non-breeding adult and immature: Crest streaked white; white forecrown and lores.

8 Caspian Tern (*Sterna caspia*) **Page 102**
Large and heavily built. Large, dark red bill; black crest; dark gray undersides of outer primaries.
8a Breeding adult: Crest entirely black.
8b Non-breeding adult: White flecks on crest.

1c
1b
1a
2a
3b
3a
2b
2c
4a
5a
4b
4c
5b
5c
6b
6a
7b
7a
8b
8a

1 Great Black-backed Gull (*Larus marinus*) **Page 101**
1a Breeding adult in flight: Large size, massive head and bill; black mantle; yellow bill with red spot.
1b First year in flight: Mottled grayish brown above and below; whitish head; black bill; broad black tail band.
1c Non-breeding adult: Large size, massive head and bill; head flecked with pale brown; black mantle; yellow bill with red spot; pinkish legs.
1d Second year: Grayish mantle mottled black; bill pinkish with broad black band near tip.

2 Lesser Black-backed Gull (*Larus fuscus*) **Page 100**
2a Breeding adult in flight: Large but slender; dark gray mantle; heavy yellow bill with red spot near tip.
2b First year in flight: Mottled grayish brown, paler on head; underparts streaked; bill black; broad black tail band.
2c Non-breeding adult: Large but slender; dark gray mantle; pale yellow legs; heavy yellow bill with red spot near tip; head and nape flecked with pale brown.
2d Second year: Bill pinkish with large black band near tip; tail with broad black band.

1a
1b
2a
2b

1c
1d
2c
2d

PLATE 23: GULLS 2

1 Herring Gull (*Larus argentatus*) **Page 100**

1a Breeding adult in flight: Gray mantle; white head and underparts; red spot on yellow bill.
1b First year in flight: Overall mottled grayish brown; black-tipped bill, pinkish at base; broad but diffuse blackish band on tail.
1c Non-breeding adult: Gray mantle; pinkish legs; red spot on yellow bill; white underparts; head and nape flecked with pale brown.

2 Ring-billed Gull (*Larus delawarensis*) **Page 99**

2a Breeding adult in flight: Black ring on yellowish bill; gray mantle; black wing tips; white head and underparts.
2b First year in flight: Mottled grayish brown upperwings with dark tips; gray back; pinkish bill with black tip; narrow black tail band.
2c Non-breeding adult: Black ring on yellowish bill; gray mantle; black wing tips; white underparts; yellowish green legs; head and nape flecked with pale brown.

1a
1b
2a
2b
2c
1c

PLATE 24: GULLS 3 AND SKIMMER

1 Laughing Gull (*Larus atricilla*) **Page 97**

1a Non-breeding adult: Head white with diffuse dark markings; dark gray mantle; black wing tips. Bill and legs dark.
1b Immature: Back and upperwings mottled brown and gray; whitish belly.
1c Breeding adult in flight: Black head; dark gray mantle; black wing tips. Bill and legs reddish.
1d First-year in flight: Back and upperwings mottled brown and gray; whitish belly; white rump; broad black tail band.

2 Black-legged Kittiwake (*Rissa tridactyla*) **Page 101**

2a Non-breeding adult: White head, dark spot behind eye; gray wash on nape; yellow bill.
2b Breeding adult in flight: All-yellow bill; gray mantle; entirely black wing tips; black legs.
2c First year in flight: Black spot behind eye; black tail band; gray mantle marked with contrasting black M.

3 Bonaparte's Gull (*Larus philadelphia*) **Page 98**

Small and ternlike; slender, black bill.
3a Non-breeding adult: Black ear spot; mantle pale gray; tail and outer primaries white; legs pink.
3b Breeding adult in flight: Black head; gray mantle; undersides of primaries white. Legs pink.
3c First year in flight: Black ear spot; black primaries and coverts show narrow M across upperwings in flight; whitish undersides to primaries; narrow black tail band.

4 Black Skimmer (*Rynchops niger*) **Page 110**

Black above; white below. Unique bill long and orange with black tip; lower mandible much longer than upper.

5 Franklin's Gull (*Larus pipixcan*) **Page 98**

Non-breeding adult in flight: White breast and underparts; gray back; partial blackish hood; white forecrown; narrow black tail band.

1a
1b
1c
1d
2a
2b
2c
3a
3b
3c
4
5

PLATE 25: PIGEONS AND DOVES

1 Ruddy Quail-Dove (*Geotrygon montana*) **Page 117**
1a Adult male: Reddish brown overall; buffy stripe below eye.
1b Immature: Browner than adult male; less distinct facial stripe.

2 Key West Quail-Dove (*Geotrygon chrysia*) **Page 116**
Reddish brown upperparts; tawny underparts; bold white stripe below eye.

3 White-winged Dove (*Zenaida asiatica*) **Page 113**
Grayish brown with large, white central wing patch; tail square, tipped white.

4 Plain Pigeon (*Patagioenas inornata*) **Page 113**
Pale gray-brown; reddish brown on wings and breast; upperwing coverts edged white; red eye-ring; white iris.

5 Common Ground-Dove (*Columbina passerina*) **Page 115**
Tiny and stocky. Sandy brown overall; stubby wings with rufous primaries; mostly dark tail with white-tipped corners.

6 White-crowned Pigeon (*Patagioenas leucocephala*) **Page 112**
6a Adult male: Slate gray body; brilliant white crown.
6b Immature: White crown of male reduced in immatures and females.

7 Scaly-naped Pigeon (*Patagioenas squamosa*) **Page 111**
Slate gray body, purplish red tint to upperparts; bare skin around eye yellow to reddish orange.

8 Mourning Dove (*Zenaida macroura*) **Page 114**
Grayish brown above, buffy below; long, wedge-shaped tail tipped with white; no white in wing.

9 Zenaida Dove (*Zenaida aurita*) **Page 114**
Cinnamon brown above, paler below; white band on trailing edge of secondaries; rounded tail with black subterminal band and white tips.

1a
1b
2
3
4
5
6b
6a
7
8
9

PLATE 26: WHITE-FRONTED QUAIL-DOVE

White-fronted Quail-Dove (*Geotrygon leucometopia*) **Page 116**
(Top) Immature: Similar to adults, but browner, lacking purplish sheen.
(Bottom) Adult: Mostly slate gray with purplish blue sheen on back; prominent white forecrown; bill reddish, paler at tip.

PLATE 27: PARROT AND PARAKEETS

Hispaniolan Parrot (*Amazona ventralis*) **Page 119**
(Top right) Large and chunky; bright green overall; short, squared tail; white forecrown; dark ear spot; maroon belly.

Hispaniolan Parakeet (*Aratinga chloroptera*) **Page 118**
(Middle left) Bright green overall; long, pointed tail; white eye-ring; red edge along bend of wing.

Olive-throated Parakeet (*Aratinga nana*) **Page 119**
(Bottom right) Bright green above; dark brownish olive underparts; long, pointed tail; blue primaries; creamy yellow eye-ring and bill.

Barry Kent MacKay

PLATE 28: ENDEMIC CUCKOOS

Bay-breasted Cuckoo (*Hyetornis rufigularis*) **Page 123**
(Left) Large cuckoo distinguished by dark reddish brown throat and breast and thick, somewhat decurved bill. Reddish brown wing patch; long tail with white tips.

Hispaniolan Lizard-Cuckoo (*Saurothera longirostris*) **Page 122**
(Right) Large cuckoo with pale gray breast; long tail with white tips; fairly straight and slender bill prominently hooked at tip. Reddish brown wing patch. Throat color varies from whitish to dull orange.

PLATE 29: CUCKOOS, ANIS, AND OWLS

1 Yellow-billed Cuckoo (*Coccyzus americanus*) **Page 121**
Slender with white underparts, no black on cheek. Long, down-curved bill is yellow at base. Tail long and graduated with broad white tips

2 Black-billed Cuckoo (*Coccyzus erythropthalmus*) **Page 120**
Slender; long tail with small but distinct white tips; long, dark, down-curved bill; reddish or yellowish eye-ring; upperwing coverts entirely gray-brown.

3 Mangrove Cuckoo (*Coccyzus minor*) **Page 122**
Slender with long, down-curved bill, yellow at base. Note black eye mask and buff-colored underparts. Long, blackish tail broadly tipped white. Wings lack reddish brown in flight.

4 Smooth-billed Ani (*Crotophaga ani*) **Page 123**
Entirely black with heavy, parrotlike bill and conspicuously long, flat tail.

5 Burrowing Owl (*Athene cunicularia*) **Page 125**
Small, long-legged owl; brown profusely spotted white above; underparts white with broad brown barring. Active in daylight; conspicuously bobs when approached. Eyes yellow.

6 Stygian Owl (*Asio stygius*) **Page 126**
Large owl with dark coloration and conspicuous ear tufts. Dark facial disk with yellowish orange eyes.

7 Barn Owl (*Tyto alba*) **Page 124**
Large, pale owl with flat, white, heart-shaped face; dark eyes.

8 Short-eared Owl (*Asio flammeus*) **Page 127**
8a Adult: Large owl. Buffy below with heavy, dark streaking on breast, reduced on lower belly. Yellow eyes set in round, whitish facial disk.
8b In flight: Conspicuous black wrist patches on whitish underwings; large buff patch on upperwings. Flight buoyant and erratic.

1
2
3
4
5
6
7
8a
8b

Ashy-faced Owl (*Tyto glaucops*) **Page 125**
Reddish brown owl with silver-gray, heart-shaped face; dark eyes.

PLATE 31: SWIFTS, POTOO, NIGHTHAWKS, AND NIGHTJARS

1 Chimney Swift (*Chaetura pelagica*) **Page 133**
Medium-sized, entirely dark swift, paler on throat, chin, and auriculars. Tail very short and rounded, barely visible in flight.

2 Black Swift (*Cypseloides niger*) **Page 132**
Fairly large, black swift with slightly notched tail. At close range, white visible on forecrown and supercilium. In flight, holds wings below horizontal.

3 White-collared Swift (*Streptoprocne zonaris*) **Page 133**
Large, black swift with distinctive white collar and slightly forked tail.

4 Antillean Palm-Swift (*Tachornis phoenicobia*) **Page 134**
Very small swift; dark brown above, whitish below with narrow dark breast band. Rump conspicuously white.

5 Northern Potoo (*Nyctibius jamaicensis*) **Page 131**
5a Large, cryptically patterned, brownish, nocturnal bird. Often perches nearly upright and motionless on end of a stump or post. Yellow eyes appear reddish in beam of light.
5b Long tail and rounded wings noticeable in flight.

6 Common Nighthawk (*Chordeiles minor*) **Page 128**
Slender, pointed wings with blackish underwing coverts and white patch across primaries. Wings project beyond tail tip at rest. Distinguished with certainty from Antillean Nighthawk only by voice.

7 Antillean Nighthawk (*Chordeiles gundlachii*) **Page 128**
Slender wings appearing somewhat rounded; underwing coverts tan; white patch across primaries. Wings same length as tail tip at rest. Distinguished with certainty from Common Nighthawk only by voice.

8 Chuck-will's-widow (*Caprimulgus carolinensis*) **Page 130**
Cinnamon brown overall; upperparts mottled brown, buff, and black. Narrow throat collar whitish or pale buff. Buffy tips to outer tail feathers; males also have white on inner webs of outer tail feathers.

1
2
3
5a
5b
4
6/7
8

PLATE 32: ENDEMIC PAURAQUE AND NIGHTJAR

Least Pauraque (*Siphonorhis brewsteri*) **Page 129**
Very small, dark nightjar with floppy, mothlike flight.
(Top) Immature: Similar to adult but less pronounced white bands.
(Second from top) Adult: Distinct white throat band and narrow white terminal band on tail.

Hispaniolan Nightjar (*Caprimulgus eckmani*) **Page 130**
Mottled dark grayish brown overall; narrow buffy band below blackish throat; breast and belly irregularly spotted with white.
(Second from bottom) Male: Outer tail feathers broadly tipped with white.
(Bottom) Female: Outer tail feathers broadly tipped with buff.

PLATE 33: EMERALD AND TODIES

Hispaniolan Emerald (*Chlorostilbon swainsonii*) **Page 135**
Small size; straight bill.
(Top right) Male: Green overall; dull black breast patch; deeply forked tail; lower mandible mostly pinkish.
(Top left) Female: Green above; dull grayish below with metallic green sides; conspicuous white spot behind eye; whitish outer tail tips.

Broad-billed Tody (*Todus subulatus*) **Page 138**
(Bottom left) Bright green above with red throat. Underparts dirty grayish white washed with yellow and pink; sides reddish pink. Underside of lower mandible entirely reddish. Iris brown.

Narrow-billed Tody (*Todus angustirostris*) **Page 139**
(Bottom right) Brilliant green upperparts with red throat. Underparts whitish, lightly tinted with yellow; sides reddish pink. Iris pale blue. Underside of lower mandible reddish with black tip.

BRM

PLATE 34: HUMMINGBIRDS, KINGFISHER, AND SAPSUCKER

1 Vervain Hummingbird (*Mellisuga minima*) **Page 136**
Tiny size; green above, whitish below; straight black bill; chin and throat often flecked greenish.
1a Male: Tail deeply notched.
1b Female: Tail rounded and tipped white.

2 Antillean Mango (*Anthracothorax dominicus*) **Page 135**
Large size; long, black, down-curved bill; pale yellowish green upperparts.
2a Adult male: Primarily black below with green throat.
2b Adult female: Whitish below, reddish brown outer tail feathers with whitish tips.
2c Immature male: Similar to female, with black stripe down center of whitish underparts.

3 Ruby-throated Hummingbird (*Archilochus colubris*) **Page 136**
Small size; metallic bronze-green upperparts; small white spot behind eye.
3a Adult male: Brilliant red throat, whitish underparts with dull greenish sides, moderately forked tail.
3b Adult female: Whitish throat, rest of underparts dull grayish white, tail rounded and broadly tipped white.

4 Belted Kingfisher (*Ceryle alcyon*) **Page 140**
Large head; bluish gray crest and upperparts; nearly complete white collar; very short legs.
4a Male: Single blue-gray breast band.
4b Female: One blue-gray and one rufous breast band.
4c Female in flight.

5 Yellow-bellied Sapsucker (*Sphyrapicus varius*) **Page 143**
Prominent black and white striped facial pattern and large white wing covert patch. Adult has red forecrown and crown with black border at rear; broad black breast band. Immature has pale brown head and underparts, less distinct facial pattern.
5a Adult male: Chin and throat red.
5b Adult female: Chin and throat white.
5c Immature: Head, back, and breast washed brown.

Images above and below line are not to the same scale.

1a
1b
2a
2b
2c
3a
3b

4c
4a
4b
5b
5a
5c

PLATE 35: ENDEMIC WOODPECKERS

Hispaniolan Woodpecker (*Melanerpes striatus*) **Page 142**

Barred blackish and greenish yellow upperparts; plain buffy olive below. Forecrown and face grayish; white and black patches on nape. Uppertail coverts red, tail black.
(Top right) Male: Crown entirely red.
(Top left) Female: Crown black, with red limited to nape.

Antillean Piculet (*Nesoctites micromegas*) **Page 141**

Small and chunky. Olive above, pale yellowish below with heavy dark spots and streaks. Pale auriculars are streaked.
(Bottom right) Male: Bright red patch in center of yellow crown.
(Bottom left) Female: Lacks red in yellow crown patch.

PLATE 36: HISPANIOLAN TROGON

Hispaniolan Trogon (*Priotelus roseigaster*) **Page 137**

Glossy green upperparts; bright red belly; gray chin, throat, and breast; blackish facial mask; yellow bill. Long, dark blue tail conspicuously tipped with white below.
(Left) Male: Wings with fine black and white barring.
(Right) Female: Lacks fine white barring on wings.

PLATE 37: FLYCATCHERS

1 Great Crested Flycatcher (*Myiarchus crinitus*) **Page 145**
Upperparts olive-brown, wings reddish brown with whitish wing bars; tail reddish brown. Throat and breast gray; belly bright yellow.

2 Hispaniolan Pewee (*Contopus hispaniolensis*) **Page 144; see also Plate 41**
Drab grayish olive upperparts; underparts lighter gray with dusky wash. Wing bars inconspicuous or absent. Bill relatively long and broad, blackish above, pinkish yellow or pale orangish below.

3 Gray Kingbird (*Tyrannus dominicensis*) **Page 146**
Gray above, pale grayish white below with distinct blackish crown and facial mask. Orangish crown patch usually concealed. Tail slightly notched.

4 Greater Antillean Elaenia (*Elaenia fallax*) **Page 144**
Upperparts grayish olive; underparts pale gray, washed with yellow. Note faint dark eye-line, two distinct wing bars, and small bill with pinkish base. White crown patch usually concealed.

5 Loggerhead Kingbird (*Tyrannus caudifasciatus*) **Page 147**
Dark brown back and tail contrasting with blackish crown and facial mask; wings and tail edged rufous. Underparts whitish washed gray. Tail squared.

6 Stolid Flycatcher (*Myiarchus stolidus*) **Page 145**
Upperparts olive-brown, darker on head, tail brownish with reddish brown inner webs. Throat and breast grayish white, belly pale yellow. Note two pale white wing bars; primaries strongly fringed with white.

7 Scissor-tailed Flycatcher (*Tyrannus forficatus*) **Page 148**
Pale grayish above with conspicuously long, forked tail. Underparts whitish with salmon pink wash on sides, flanks, and undertail coverts.

8 Fork-tailed Flycatcher (*Tyrannus savana*) **Page 148**
Grayish above with dark crown and nape, and conspicuously long, forked tail. Underparts white.
8a Adult male: Crown and nape black.
8b Adult female and immature: Duller than male; shorter tail.

Eastern Wood-Pewee (*Contopus virens*) **Page 218**
Not Illustrated: Grayish olive above, paler below with dusky wash on breast and sides; relatively broad, whitish wing bars. Head and face slighty darker than rest of upperparts; may have partial eye-ring. Legs relatively short. Bill dark above, pale yellowish orange below.

1
2
3
4
5
6
8a
8b
7

PLATE 38: VIREOS

1 Yellow-throated Vireo (*Vireo flavifrons*) **Page 150**
Upperparts olive green, belly and undertail coverts white. Note dark iris and lores; bold yellow spectacles; bright yellow chin, throat, and breast; two white wing bars; gray rump.

2 Flat-billed Vireo (*Vireo nanus*) **Page 150; see also Plate 43**
Grayish green above; dull grayish white below, washed pale yellow. Note narrow white tips to outer tail feathers, two white wing bars, white iris.

3 White-eyed Vireo (*Vireo griseus*) **Page 149**
Grayish green tinged yellow above, whitish below with yellowish sides and flanks. Nape more grayish. Note broad yellow spectacles, dusky lores, and two white to yellowish white wing bars. Adult has white iris.

4 Thick-billed Vireo (*Vireo crassirostris*) **Page 149**
Olive upperparts, dull whitish to buffy yellow underparts. Crown and nape grayish green; lores blackish. Note two white wing bars, bright yellow spectacles.

5 Black-whiskered Vireo (*Vireo altiloquus*) **Page 152**
Olive brown upperparts; whitish underparts, washed pale olive on sides and flanks. White supercilium contrasts with dusky eye-line and grayish crown. Note narrow black malar stripe. Adult has red iris.

6 Red-eyed Vireo (*Vireo olivaceus*) **Page 151**
Olive green upperparts contrast with gray crown; underparts white. Note prominent white supercilium, bordered by blackish eye-line below and distinct median crown stripe above. Adult has red iris.

7 Warbling Vireo (*Vireo gilvus*) **Page 151**
Unmarked, pale vireo. Upperparts grayish green; underparts dingy white. Sides of neck and flanks washed yellow, but buffier on neck. Note whitish supercilium, pale lores, dusky eye-line.
7a Adult: Whitish underparts.
7b Immature: Sides more yellowish or greenish.

1
2
3
4
5
6
7a
7b

PLATE 39: ENDEMIC CROWS

Hispaniolan Palm Crow (*Corvus palmarum*) **Page 153**
(Top left and right) Completely black with purplish blue sheen on back and upperwing coverts.

White-necked Crow (*Corvus leucognaphalus*) **Page 154**
(Bottom) Large crow, entirely black, with violet sheen. Note red-orange eye and large bill. White feathers at base of neck rarely visible.

Barry Kent

PLATE 40: SWALLOWS AND MARTINS

1 Barn Swallow (*Hirundo rustica*) **Page 159**
Tail long and deeply forked.
1a Adult: Dark blue-black upperparts and uniformly tan or orangish underparts; darker chestnut throat and forecrown.
1b Immature: Throat and forecrown paler; rest of underparts white.

2 Tree Swallow (*Tachycineta bicolor*) **Page 156**
2a Adult: Adult male and older females iridescent greenish blue above with entirely white underparts.
2b Immature: Yearling females and immatures have brownish upperparts. Tail moderately forked.

3 Cliff Swallow (*Petrochelidon pyrrhonota*) **Page 158**
Upperparts metallic bluish black; underparts dull white. Note reddish brown chin, throat, and auriculars; buffy forecrown, collar, and rump. Tail short and squared.

4 Cave Swallow (*Petrochelidon fulva*) **Page 158**
Stocky swallow with dark, rufous buff rump and extensive forecrown patch. Auriculars, throat, breast, and sides pale reddish brown. Tail short and square to slightly notched.

5 Bank Swallow (*Riparia riparia*) **Page 157**
Grayish brown upperparts; white underparts, with dark band across breast. Forecrown pale; tail slightly notched.

6 Purple Martin (*Progne subis*) **Page 155**
6a Adult male: Entirely glossy bluish black.
6b Female and immature: Bluish black above; scaled pattern on grayish brown breast; light gray forecrown and collar.

7 Northern Rough-winged Swallow (*Stelgidopteryx serripennis*) **Page 157**
Entirely warm brown above, white below, with pale brownish wash on chin, throat, and chest. Tail square.

8 Caribbean Martin (*Progne dominicensis*) **Page 155**
Upperparts, head, and throat metallic blue; belly white; dark band across vent.

9 Golden Swallow (*Tachycineta euchrysea*) **Page 156**
9a Male in flight: Iridescent bluish green upperparts with golden sheen; white underparts. Tail moderately forked, more deeply than that of Tree Swallow.
9b Adult male: Iridescent bluish green upperparts with golden sheen; white underparts.
9c Adult female: Duller above than males, with grayish wash on breast.
9d Immature: Duller above than females, with grayish band across breast.

1a
1b
2a
2b
3
4
5
6a
6b
7
8
9a
9b
9c
9d

PLATE 41: ENDEMIC PEWEE AND THRUSH

Hispaniolan Pewee (*Contopus hispaniolensis*) **Page 144; see also Plate 37**
(Top left) Drab grayish olive upperparts; underparts lighter gray with dusky wash, sometimes tinged yellow. Wing bars inconspicuous or absent. Bill relatively long and broad, blackish above, pinkish yellow or pale orangish below.

Rufous-throated Solitaire (*Myadestes genibarbis*) **Page 161**
(Top right) Gray above with white chin, reddish brown throat and undertail coverts; rest of underparts pale gray. Note white crescent below eye; white outer tail feathers.

La Selle Thrush (*Turdus swalesi*) **Page 164**
(Bottom) Slaty black head and upperparts, rich rufous lower breast and sides, belly and undertail coverts white. Auriculars have silvery cast. Reddish orange eye-ring; yellow bill; dusky legs.

PLATE 42: THRUSHES, MIMIDS, WAXWING, AND PIPIT

1 American Robin (*Turdus migratorius*) **Page 163**
Slaty grayish upperparts; dull orangish red underparts. Throat white streaked with black; undertail coverts white. Bill yellow.

2 Red-legged Thrush (*Turdus plumbeus*) **Page 164**
Slaty gray upperparts; breast paler gray, throat white with black stripes; rest of underparts white. Note reddish legs, bill, and eye-ring; conspicuous white tail tips.

3 Bicknell's Thrush (*Catharus bicknelli*) **Page 162**
Olive-brown upperparts and contrasting chestnut-tinged tail; whitish underparts; sides of throat and breast creamy buff prominently spotted with black. Note grayish auriculars and lores.

4 Wood Thrush (*Hylocichla mustelina*) **Page 163**
Bright rufous brown crown and nape, slightly duller on back, wings, and tail; conspicuous white eye-ring; white underparts with heavy blackish spots on breast, sides, and flanks.

5 Veery (*Catharus fuscescens*) **Page 162**
Uniformly reddish brown upperparts; underparts whitish; throat and chest buffy with indistinct brownish spots. Inconspicuous pale eye-ring.

6 Gray Catbird (*Dumetella carolinensis*) **Page 165**
Entirely dark gray with black cap, reddish brown undertail coverts, and long tail often cocked upward.

7 Cedar Waxwing (*Bombycilla cedrorum*) **Page 168**
7a Adult: Sleek and round-bodied. Overall grayish brown with warm cinnamon brown crown and back. Underparts tan, fading to pale yellowish on lower belly. Note black facial mask edged with white; black chin patch; pointed crest; yellow-tipped tail.
7b Immature: Underparts streaked.

8 Northern Mockingbird (*Mimus polyglottos*) **Page 166**
8a Adult: Pale gray above; grayish white below. Wings and tail with large white patches showing clearly in flight. Long tail is often cocked upward.
8b Immature: Brownish gray upperparts; buffier underparts.

9 American Pipit (*Anthus rubescens*) **Page 167**
Slender bodied, with thin bill and long legs; regularly bobs long tail while walking. Non-breeding adult has buffy supercilium, two faint wing bars, and underparts that vary from pinkish buff to pale gray, with blackish stripes concentrated on breast. Note conspicuous white outer tail feathers in flight.

10 Pearly-eyed Thrasher (*Margarops fuscatus*) **Page 166**
Brown upperparts; dull white underparts streaked with brown. Iris white; bill large and yellowish. Tail long and conspicuously white-tipped.
Insert bottom of Plate 42

Swainson's Thrush (*Catharus ustulatus*) **Page 218**
Not Illustrated: Upperparts uniformly olive brown; underparts white, with sides of face, throat and breast washed pale buffy; dark spotting on sides of throat and breast. Distinct buffy eye-ring and lores.

1
2
4
3
5
6
7a
7b
9
8b
8a
10

PLATE 43: SOME ENDEMIC PASSERINES

Palmchat (*Dulus dominicus*) **Page 169**
(Top) Upperparts dark brown; underparts whitish, heavily streaked brown. Bill yellow; eye reddish. Occurs in conspicuous, noisy flocks.

Hispaniolan Oriole (*Icterus dominicensis*) **Page 210; see also Plate 56**
(Second from top) Adult: Black overall with distinctive yellow scapulars, rump, and undertail coverts.
(Third from top) Immature: Mainly olive upperparts and dull olive-yellow underparts. Wings blackish; throat sometimes black or reddish brown.

Flat-billed Vireo (*Vireo nanus*) **Page 150; see also Plate 38**
(Bottom left) Grayish green above; dull grayish white below, washed pale yellow. Note narrow white tips to outer tail feathers, two white wing bars, white iris.

Barry Kent MacKay
'03

PLATE 44: KINGLET, GNATCATCHER, AND WARBLERS 1

1 Black-and-white Warbler (*Mniotilta varia*) — Page 182
1a Male: Upperparts and sides boldly streaked black and white, rest of underparts white; auriculars blackish.
1b Female: Boldly striped crown, but less distinct streaking elsewhere; buffy white underparts, grayer auriculars.

2 Tennessee Warbler (*Vermivora peregrina*) — Page 171
2a Breeding: Blue-gray crown contrasts with greener back; most pronounced in males.
2b Immature: Grayish green above, dull grayish white below, with white undertail coverts. Note dusky eye-line, narrow whitish supercilium. Breast with variable yellowish wash in females.

3 Black-throated Blue Warbler (*Dendroica caerulescens*) — Page 175
3a Male: Dark blue above with black face and throat; underparts white with black band along flanks; small but distinct white patch on wing at base of primaries. Immature male similar but upperparts washed greenish.
3b Female: Olivaceous gray above, buffy white below; small white spot at base of primaries; narrow whitish supercilium. Immature female similar but white wing spot much reduced.

4 American Redstart (*Setophaga ruticilla*) — Page 183
4a Adult male: Black with large, orange patches on wings, tail, and sides; belly and undertail coverts white.
4b Female: Olive-gray above, grayer on head, dull white below, with large, pale yellow patches on wings, tail, and sides.

5 Blue-gray Gnatcatcher (*Polioptila caerulea*) — Page 160
Tiny. Bluish gray upperparts; white underparts. Long, thin tail, often cocked or fanned, with white outer feathers; prominent white eye-ring.
5a Breeding male: Fine black eyebrow stripe.
5b Female and non-breeding male: Lighter grayer than male; lacks eyebrow stripe.

6 Ruby-crowned Kinglet (*Regulus calendula*) — Page 160
Tiny. Olive green to gray upperparts, dusky white underparts. Note bold white eye-ring, two whitish wing bars.
6a Male: Red crown patch, often concealed.
6b Female: Lacks crown patch.

7 Blackpoll Warbler (*Dendroica striata*) — Page 182
In all plumages, legs and feet yellowish orange.
7a Breeding male: White auriculars bordered by black cap and black malar stripe.
7b Breeding female: Similar to non-breeding male.
7c Non-breeding: Male has grayish olive upperparts, variably streaked with black; whitish underparts washed with yellow and faintly streaked; diffuse olive streaking on sides; undertail coverts pure white. Female similar but with reduced streaking.

1a
1b
2a
2b
3a
3b
4a
4b
5a
5b
6a
6b
7a
7b
7c

PLATE 45: WARBLERS 2

1 Worm-eating Warbler (*Helmitheros vermivorum*) **Page 184**
Large-billed, flat-headed appearance. Unmarked olive-brown above, buffy below; head buffy with bold, black crown and eye stripes.

2 Palm Warbler (*Dendroica palmarum*) **Page 180**
2a Breeding: Dull brown to cinnamon brown above, pale buff below, faint streaking on breast and flanks. Note rufous crown patch, yellow chin and throat, indistinct buffy wing bars. Rump yellow-olive; undertail coverts yellow.
2b Non-breeding: Similar, but less pronounced or nearly absent reddish brown crown, yellow chin and throat.

3 Swainson's Warbler (*Limnothlypis swainsonii*) **Page 185**
Unmarked olive-brown above, yellowish white below; chestnut brown crown; long, creamy buff supercilium; large, sharply pointed bill.

4 Northern Waterthrush (*Seiurus noveboracensis*) **Page 186**
Unmarked olive-brown upperparts, buffy white underparts with dark brown streaks, more finely streaked on throat. Buffy white supercilium narrows behind eye.

5 Louisiana Waterthrush (*Seiurus motacilla*) **Page 187**
Uniform dark olive-brown above, white below, with dark brown streaks on breast and sides. Throat white and unstreaked. White supercilium flares and broadens behind eye.

6 Ovenbird (*Seiurus aurocapilla*) **Page 186**
Uniform olive-brown above, white below, boldly streaked with blackish on throat, breast, and sides. Crown orangish bordered by blackish stripes; bold white eye-ring.

1
2a
2b
3
4
5
6

1 Yellow-throated Warbler (*Dendroica dominica*) **Page 178**

1a Breeding: Upperparts gray, throat and upper breast yellow, belly white with black streaks on sides. Note triangular black cheek patch bordered by white supercilium and white patch on side of neck; white crescent below eye.

1b Non-breeding: Similar but duller; black markings slightly reduced.

2 Common Yellowthroat (*Geothlypis trichas*) **Page 189**

2a Male: Olive green upperparts, with conspicuous black facial mask, edged above by white or grayish; throat, breast, and undertail coverts bright yellow; belly and sides dusky.

2b Female and immature: Olive green upperparts, pale buffy yellow underparts; throat, breast, and undertail coverts bright yellow; belly and sides dusky. Immature male usually shows traces of black mask.

3 Yellow-rumped Warbler (*Dendroica coronata*) **Page 176**

3a Breeding male: Bluish gray above, white below, with distinct black breast band and auriculars, bright yellow patches on sides and crown.

3b Breeding female and non-breeding male: Duller than breeding male with smaller yellow patches.

3c Immature: Grayish brown above with indistinct black streaking on back; dull white below with blurry streaks on breast, and small yellow patches on sides. Note small, yellow crown patch; bright yellow rump; white tail spots; broken white eye-ring. Female is duller and browner than male, with smaller patches of yellow; may lack yellow crown patch.

4 Bay-breasted Warbler (*Dendroica castanea*) **Page 181**

4a Breeding male: Dark chestnut cap and band on chin, throat, and sides; black mask; buff patch on sides of neck.

4b Breeding female: Duller than male; crown, breast, and sides only washed with chestnut.

4c Non-breeding: Greenish gray upperparts faintly streaked blackish; whitish underparts; undertail coverts buffy. Adult male has variable amount of chestnut on sides; females and immatures have yellowish wash to sides. Legs and feet grayish.

5 Golden-winged Warbler (*Vermivora chrysoptera*) **Page 170**

Yellow crown, grayish nape and back, and distinctive, large yellow wing patch. Head boldly patterned with broad black eye patch bordered above and below by white; black throat. Rest of underparts grayish white.

6 Chestnut-sided Warbler (*Dendroica pensylvanica*) **Page 173**

6a Breeding male: Yellow forecrown; black eye-line and malar stripe; chestnut band along sides; white underparts.

6b Non-breeding: Bright yellowish green above with variable black spotting, pale gray below; two yellowish white wing bars; conspicuous white eye-ring on gray face. Males may show some chestnut on sides.

1a
1b
2a
2b
3a
3b
3c
4a
4b
4c
5
6a
6b

1 Canada Warbler (*Wilsonia canadensis*) **Page 191**

1a Male: Bluish gray upperparts; black forecrown contrasting with complete white eye-ring and yellow lores that form conspicuous spectacles; underparts yellow with bold black breast streaks forming "necklace"; undertail coverts white.
1b Female: Similar to male but duller overall, with paler, less distinct markings on head and breast.

2 Magnolia Warbler (*Dendroica magnolia*) **Page 174**

2a Breeding male: Black cheek, prominent white supercilium, bold white wing patch, black "necklace" across breast.
2b Breeding female: Much paler than male, with gray cheeks and underparts only moderately streaked.
2c Non-breeding adult and immature: Upperparts grayish olive streaked black; head and nape gray; underparts bright yellow with variable black streaking on sides; rump conspicuously yellow. Note white eye-ring; thin, pale white supercilium; large white central spots in outer tail feathers.

3 Prairie Warbler (*Dendroica discolor*) **Page 180**

3a Breeding male: Olivaceous above with chestnut streaks on back, bright yellow below with pronounced black streaks on sides. Note distinctive facial pattern.
3b Female and immature: Female similar to breeding male, but black and chestnut markings much less pronounced. Immature similar to female but duller still; lacks chestnut streaking on back; facial contrasts reduced; auriculars grayish.

4 Black-throated Green Warbler (*Dendroica virens*) **Page 177**

4a Male: Bright olive green above with bright yellow face and dull olive-gray auriculars. Note black chin and throat patch extending onto upper breast; two prominent white wing bars; white tail spots. Underparts whitish, sides streaked with black, yellowish wash across undertail coverts.
4b Female and immature: Duller, with reduced black; yellowish or whitish chin and throat.

5 Blackburnian Warbler (*Dendroica fusca*) **Page 178**

5a Breeding male: Blackish above with black auriculars; brilliant orange throat and facial markings; prominent white wing patch.
5b Adult female and non-breeding male: Dusky olive to blackish above with pale streaks on back; black triangular auriculars bordered by yellowish orange; two bold white wing bars. Chin, throat, and breast yellowish orange; belly buffy white; sides streaked black.
5c Immature female: Similar but duller, with yellow replacing orange, and less prominent streaking on back and sides.

6 Cape May Warbler (*Dendroica tigrina*) **Page 175**

6a Male: Grayish green back with dark streaks; underparts yellow, heavily streaked black. Note narrow gray eye-line; chestnut auriculars bordered by yellow, including yellow on sides of neck; conspicuous white wing patch; yellow rump; thin, decurved bill.
6b Female: Similar to male but duller, with gray auriculars, greenish rump, and wing patch reduced to two whitish wing bars.
6c Immature: Striped breast; yellowish rump; distinctive buffy patch behind cheek.

1a
1b
2a
2b
2c
3a
3b
4a
4b
5a
5b
5c
6a
6b
6c

PLATE 48: WARBLERS 5

1 Wilson's Warbler (*Wilsonia pusilla*) **Page 191**
1a Male: Uniformly yellowish olive green upperparts, glossy black crown, bright yellow forecrown and supercilium. Underparts entirely yellow.
1b Female and immature: Similar to adult male but duller, with little or no black in crown.

2 Nashville Warbler (*Vermivora ruficapilla*) **Page 171**
2a Breeding: Unmarked olive green upperparts contrasting with gray head; underparts, including throat and undertail coverts, yellow; belly white. Distinguished in all plumages by white eye-ring. Adults may have chestnut crown patch, usually concealed.
2b Non-breeding: Similar, but head less distinctly gray, sometimes washed olive-brown, especially in female.

3 Kentucky Warbler (*Oporornis formosus*) **Page 188**
3a Breeding male: Plain olive green upperparts, solid bright yellow underparts. Crown, forecrown, and sides of face black; yellow supercilium and partial eye-ring form bold spectacles. Legs pinkish. Breeding female similar but has less extensive black on face and crown.
3b Immature female: Similar, but black on face and crown even less pronounced.

4 Hooded Warbler (*Wilsonia citrina*) **Page 190**
4a Breeding male: Olive upperparts, bright yellow underparts. Note large, dark eye and large white spots in outer tail feathers. Black hood and throat frame bright yellow face.
4b Breeding female: Black hood varies among females, ranging from almost complete to black markings only on crown.
4c Immature female: Similar to breeding female, but crown and nape usually olive green.

5 Mourning Warbler (*Oporornis philadelphia*) **Page 189**
5a Breeding male: Bluish gray hood, including broad black bib on lower edge; lacks eye-ring.
5b Breeding female: Similar to male but has pale gray or brownish gray hood without black on lower edge; often has broken eye-ring.
5c Non-breeding: Olive green upperparts, bright yellow underparts. Diffuse gray to brownish gray hood extends to lower throat, but throat often yellow. Legs and feet pinkish; often has thin white arcs above and below eye.

6 Connecticut Warbler (*Oporornis agilis*) **Page 188**
6a Breeding male: Olive to olive-brown above, with bluish gray hood, complete white eye-ring. Underparts pale yellow, with long undertail coverts extending nearly to tip of tail.
6b Breeding female: Similar to male but with pale gray to grayish brownish hood.
6c Immature: Hood even paler than that of female; throat whitish.

1a
1b
2a
2b
3a
3b
4a
4b
4c
5a
5b
5c
6a
6b
6c

1 Northern Parula (*Parula americana*) **Page 172**

1a Breeding male: Upperparts bluish gray with greenish yellow back patch. Chin, throat, and breast yellow; belly and undertail coverts white. Note two white wing bars, prominent broken white eye-ring. Male marked across chest with narrow black and chestnut band; has black lores. Female and immatures duller overall, with breast band indistinct or lacking.
1b Non-breeding: Similar, but upperparts less bright blue-gray and washed green, and breast bands of male less distinct.

2 Yellow-breasted Chat (*Icteria virens*) **Page 192**

Large and robust. Upperparts, wings, and long tail olive green. Throat, breast, and upper belly yellow; lower belly and undertail coverts white. White spectacles; lores black or gray, bordered above and below by white.

3 Pine Warbler (*Dendroica pinus*) **Page 179**

3a Male: Unstreaked greenish olive upperparts; bright yellow throat and breast with variable amounts of diffuse dark streaking. Greenish olive auriculars bordered sharply below and toward rear by yellow. Note indistinct yellow eye-ring and supercilium, white wing bars, white lower belly and undertail coverts.
3b Female: Similar, but duller overall with more grayish or brownish wash above, less streaking below.

4 Blue-winged Warbler (*Vermivora pinus*) **Page 170**

Bright yellow on crown and underparts, with bold black eye-line, greenish yellow back and nape, bluish gray wings and tail, and two distinct white wing bars. Female slightly duller than male; eye-line thin and dusky; wing bars narrower.

5 Prothonotary Warbler (*Protonotaria citrea*) **Page 184**

5a Male: Golden yellow head, breast, and belly; greenish yellow back; unmarked bluish gray wings and tail. Undertail coverts white.
5b Female: Duller and washed greenish, especially on crown and nape. Tail with conspicuous white spots. Immatures similar.

6 Yellow Warbler (*Dendroica petechia*) **Page 173**

6a Male: Upperparts olive-yellow, head yellow; underparts bright yellow. Wing bars, inner webs of tail, and tail spots yellow. Male usually shows pale chestnut patch on crown, and light to prominent chestnut red streaking on underparts.
6b Female: Similar to male but less bright, with little or no streaking below; may also show variable amounts of gray on nape.
6c Immature: Similar to female, without chestnut streaking. Gray on nape.

1a
1b
2
3b
3a
4
5a
5b
6a
6b
6c

PLATE 50: ENDEMIC TANAGERS 1

Hispaniolan Highland-Tanager (*Xenoligea montana*) **Page 194**

(Top) Back and rump olive green; underparts white washed gray on sides. Head and nape gray, with blackish spot bordered by white between eye and bill. White crescents above and below eye. Wings and tail blackish; white patch on outer primaries visible in closed wing.

Green-tailed Ground-Tanager (*Microligea palustris*) **Page 193**

(Bottom) Unmarked olive green above, except head, face, and nape grayish. Underparts uniformly dull white. Note red eye, incomplete white eye-ring.

Gray-crowned Palm-Tanager (*Phaenicophilus poliocephalus*) **Page 196**
(Top left) Adult: Yellow-green above, light gray below, with gray crown and facial mask, gray nape, and sharply defined white chin and throat. Three white spots contrast sharply with black mask.
(Top right) Immature: Duller than adult.

Black-crowned Palm-Tanager (*Phaenicophilus palmarum*) **Page 195**
(Third from top) Yellow-green upperparts, gray nape, and black crown and facial mask contrast sharply with three white spots around eye. Underparts pale gray, with diffuse white chin and throat. Immature duller than adult; crown variable, gray to dark gray or black.

Hispaniolan Spindalis (*Spindalis dominicensis*) **Page 199**
(Third from bottom) Adult male: Black head boldly striped with white. Nape yellow washed orange; yellow back broken by darker mantle. Underparts yellow with reddish brown wash on breast. Tail and wings black, edged whitish, with chestnut patch at bend of wing.
(Second from bottom) Female: Olive-brown above with yellowish rump. Underparts dull whitish with fine dusky streaks.
(Bottom) Immature male: Similar to female, but male more heavily streaked and mottled yellowish tinged chestnut especially on rump.

Barry Kent MacKay

Western Chat-Tanager (*Calyptophilus tertius*) **Page 197**
(Top) Dark chocolate brown above, mostly white below. Bright yellow spot in front of eye, yellow fringe on bend of wing. Tail long and rounded; legs and feet robust.

Eastern Chat-Tanager (*Calyptophilus frugivorus*) **Page 197**
(Bottom) Dark chocolate brown above, whitish below with dusky brownish wash on sides. Bright yellow spot in front of eye, yellow fringe on bend of wing, and broken, yellow eye-ring. Tail long and rounded; legs and feet robust.

PLATE 53: TANAGERS, EUPHONIA, BANANAQUIT, AND GRASSQUITS

1 Western Tanager (*Piranga ludoviciana*) **Page 199**

1a Breeding male: Blackish back, wings, and tail; yellow nape, rump, uppertail coverts, and underparts; most or all of head reddish.
1b Female: Olive green above; underparts variable, from bright yellow to grayish white. Wings and tail grayish brown with two yellowish white wing bars.
1c Non-breeding male: Greenish yellow head, sometimes with reddish wash on face; blackish back, wings, and tail; yellow nape, rump, and underparts. Note two distinctive wing bars, the upper yellow, the lower white.

2 Summer Tanager (*Piranga rubra*) **Page 198**

2a Breeding male: Entirely rose red.
2b Adult female and non-breeding male: Yellowish olive green above, orangish yellow below, with faint dusky eye-line. Non-breeding male may have scattered red feathers, becoming mottled red and yellow by spring.

3 Antillean Euphonia (*Euphonia musica*) **Page 211**

Small, compact bird with distinctive sky blue crown and nape.
3a Male: Blackish above, orangish yellow below and on rump and forecrown; chin and throat dark violet.
3b Female: Greenish above and yellowish green below; rump and forecrown yellowish.
3c Immature: Similar to female but washed grayish.

4 Bananaquit (*Coereba flaveola*) **Page 192**

4a Adult: Grayish black above with bright yellow rump; underparts yellow, except for gray throat. Note bold white supercilium and small white spot on wing. Slender, decurved bill has reddish pink spot at base. Tail noticeably short.
4b Immature: Similar to adult but supercilium yellowish.

5 Black-faced Grassquit (*Tiaris bicolor*) **Page 201**

5a Male: Olive-brown above, with black head and underparts.
5b Female: Uniformly drab brownish olive.

6 Yellow-faced Grassquit (*Tiaris olivaceus*) **Page 200**

6a Male: Yellowish olive overall, somewhat grayer below; distinctive yellow throat, supercilium, and crescent below eye; black on upper breast.
6b Female: Faint yellowish supercilium, crescent, and chin; lacks black on breast.

7 Scarlet Tanager (*Piranga olivacea*) **Page 218**

Not Illustrated:
Breeding male: Brilliant red with black wings and tail, dark grayish bill.
Female and non-breeding male: Uppersides uniformly greenish yellow; undersides yellowish. Wings dark and lacking wingbars; black in males.

1a
1b
1c
2a
2b
3a
3b
3c
4a
4b
5a
5b
6a
6b

PLATE 54: BUNTING, GROSBEAKS, BULLFINCH, AND SPARROWS

1 Indigo Bunting (*Passerina cyanea*) **Page 206**
1a Breeding male: Entirely bright blue.
1b Female: Entirely dull brown with pale breast stripes and wing bars.
1c Non-breeding male: Brown overall with traces of blue in wings and tail.

2 Lincoln's Sparrow (*Melospiza lincolnii*) **Page 204**
Upperparts grayish brown finely streaked with black; breast and sides washed buffy, finely streaked with black. Crown brown streaked black, with narrow grayish median stripe; broad gray supercilium; buffy malar stripes.

3 Rose-breasted Grosbeak (*Pheucticus ludovicianus*) **Page 205**
3a Breeding male: Distinguished by bold pinkish red breast; black head and back; lower breast and belly white; rump white with gray barring. Underwing coverts rosy.
3b Female: Upperparts brown and streaked; head boldly striped whitish and brown; underparts whitish, streaked brown. Underwing coverts yellow.
3c Non-breeding male: Similar to breeding male but not as bright, breast buffy to pinkish, head and back brownish to brownish black.

4 Blue Grosbeak (*Guiraca caerulea*) **Page 206**
4a Breeding male: Deep blue with reddish brown wing bars, dull black flight feathers.
4b Female and immature male: Paler gray-brown to warm rufous brown throughout, with occasional blue feathers and buffy brown wing patch.

5 Song Sparrow (*Melospiza melodia*) **Page 203**
Upperparts grayish with coarse brown steaks; underparts whitish with coarse brown streaks converging into central breast spot. Crown with broad brown stripes bordering paler central stripe; broad gray supercilium; conspicuous brown malar stripes.

6 Greater Antillean Bullfinch (*Loxigilla violacea*) **Page 202**
6a Adult: Chunky, slate gray to black bird, with heavy bill and diagnostic orange-red supercilium, throat, and undertail coverts.
6b Immature: Similar to adult but gray is washed with brown, and orange-red markings are less bright.

7 Rufous-collared Sparrow (*Zonotrichia capensis*) **Page 204**
7a Adult: Upperparts brown with coarse dark streaks; underparts grayish white. Identified by black neck band, reddish brown nape, and gray crown with black stripes.
7b Immature: Duller, somewhat spotted below, lacking black or reddish brown markings.

8 Grasshopper Sparrow (*Ammodramus savannarum*) **Page 203**
8a Adult: Intricately patterned with rufous, buffy, and gray above; plain buffy below. Note small, golden mark at forward portion of supercilium, white eye-ring, and whitish median crown stripe.
8b Immature: Golden mark is paler; fine streaks present on breast and flanks.

9 Savannah Sparrow (*Passerculus sandwichensis*) **Page 202**
Upperparts and underparts heavily streaked with brown. Conspicuous yellow supercilium and lores, pale median crown stripe, dark mustache stripe, pink legs. Tail short and slightly notched.

1b
2
3b
3a
1a
1c
3c
4b
5
4a
6a
8a
8b
7b
6b
7a
9

PLATE 55: ENDEMIC FINCHES

Hispaniolan Crossbill (*Loxia megaplaga*) **Page 212**
Note unique, crossed bill tips.
(Top left) Male: Dusky brown washed pale red overall, with red concentrated on head, upper back, and breast.
(Top right) Female: Dusky brown with finely streaked breast, yellowish wash to foreparts, yellowish rump.
(Center top) Immature: Similar to female, but browner and more heavily streaked.

Antillean Siskin (*Carduelis dominicensis*) **Page 213**
(Center bottom; two individuals) Male: Yellow body; olive green back; black head and throat; two diffuse yellow patches on black tail.
(Bottom left) Female: Olive green above and yellowish white below, with faint pale gray streaking, pale yellowish rump, two yellowish wing bars. Bill pale yellow in both sexes.
(Bottom right) Immature: Similar to female but not as bright.

Barry Kent

PLATE 56: ICTERIDS

1 Shiny Cowbird (*Molothrus bonariensis*) **Page 209**
Medium-sized blackbird with conical bill.
1a Male: Glossy black with purplish sheen.
1b Female: Drab grayish brown above, lighter brown below, with faint supercilium.

2 Baltimore Oriole (*Icterus galbula*) **Page 210**
2a Male: Orange and black plumage, white wing bar, extensive orange patches on outer tail.
2b Female and immature: Olive brownish to brownish orange above; orange-yellow below; two whitish wing bars.

3 Bobolink (*Dolichonyx oryzivorus*) **Page 207**
3a Breeding male: Black below with distinctive yellowish buff nape and white patches on wings and rump.
3b Female and non-breeding male: Warm buffy brown overall, with streaked back, rump, and sides; buffy median crown stripe; unmarked buffy throat; pointed tail feathers. Female usually paler with whitish throat.

4 Orchard Oriole (*Icterus spurius*) **Page 211**
4a Male: Entirely black head, upper breast, back, and tail; chestnut lower breast, belly, undertail coverts, and rump. Wings mostly black with narrow white lower wing bar and broad chestnut upper wing bar.
4b Female: Olive green above, greenish yellow below; wings grayish brown with two narrow white wing bars.
4c Immature male: Similar to female, but with variable black on throat and chest.

5 Hispaniolan Oriole (*Icterus dominicensis*) **Page 210; see also Plate 43**
Adult male: Black overall with distinctive yellow scapulars, rump, and undertail coverts.

6 Greater Antillean Grackle (*Quiscalus niger*) **Page 208**
6a Male: Large blackbird with long, deep, V-shaped tail and long, sharply pointed bill.
6b Female: Duller than male, and tail with smaller V shape.

7 Tawny-shouldered Blackbird (*Agelaius humeralis*) **Page 208**
7a Adult: Black overall, with tawny shoulder patch, most conspicuous when flying.
7b Immature: Shoulder patch reduced and not as bright.

1b
1a
2a
2b
4c
4a
4b
3a
3b
5
6b
6a
7a
7b

PLATE 57: INTRODUCED SPECIES

1 House Sparrow (*Passer domesticus*) **Page 214**
1a Male: Brown above, grayish below. Note chestnut nape and gray crown, pale gray auriculars, and black bib.
1b Female: Duller than male; lacks black bib. Crown gray-brown with pale buffy supercilium and thin dusky eye-line. Underparts dingy gray-brown.

2 Village Weaver (*Ploceus cucullatus*) **Page 215**
2a Male: Distinctive orange-yellow overall, with black hood, chestnut brown nape, and red iris.
2b Female: Generally yellowish green, brightest on face and breast, with darker wings and mantle, yellow wing bars. Eye red.

3 Red Avadavat (*Amandava amandava*) **Page 216**
3a Breeding male: Primarily red overall with white spots on wings, flanks, and sides.
3b Female and non-breeding male: Brownish above, paler below, but with red uppertail coverts and bill, and white spots on wing.
3c Immature: Similar to female but lacks red, and wing spots are buff colored.

4 Chestnut Mannikin or Tricolored Munia (*Lonchura malacca*) **Page 217**
4a Adult: Cinnamon colored above with black hood; white below with black belly patch; bill pale grayish.
4b Immature: Cinnamon brown above and buffy below; bill dark.

5 Nutmeg Mannikin (*Lonchura punctulata*) **Page 216**
5a Adult: Deep brown above, whitish below, with distinctive cinnamon-colored hood and scalloped black-and-white underparts.
5b Immature: Cinnamon colored above; paler below.

6 Ring-necked Pheasant (*Phasianus colchicus*) **Page 32**
6a Male: Multicolored with very long tail. Note red face, glossy greenish head and neck, and white collar.
6b Female: Mottled brown throughout, with shorter tail.

7 Helmeted Guineafowl (*Numida meleagris*) **Page 33**
Distinguished by unusual, bulbous body shape, dark gray feathering with white spots, and nearly naked head and neck.

8 Red Junglefowl (*Gallus gallus*) **Page 32**
Variable plumage of black, brown, reddish brown, gray, and white.
8a Male: Large red comb and wattle on head; long, bushy tail.
8b Female: Smaller comb and wattle, tail much reduced.

9 Rock Pigeon (*Columba livia*) **Page 110**
Plumage variable. Usually a combination of black, gray, and white, including black tail band, two dark wing bars, and white rump. Appears falconlike in flight.

10 Northern Bobwhite (*Colinus virginianus*) **Page 33**
Chunky, brown bird, finely barred tan and black above, barred white and black below, with chestnut streaking on sides.
10a Male: White throat and supercilium.
10b Female: Tan throat and supercilium.

Images above and below line are not to the same scale.

1a
2b
1b
2a
3a
3c
3b
5b
5a
4b
4a
6a
6b
7
8a
8b
9
10b
10a

SPECIES ACCOUNTS

GEESE AND DUCKS: FAMILY ANATIDAE

The birds of this large, aquatic family include six subfamilies that occur on Hispaniola. **Whistling-ducks** are primarily nocturnal and often graze in wet, grassy meadows or dip for food in shallow ponds. In flight their long legs and feet trail behind the tail, and the head is drooped, making them easily recognizable. **Geese** are larger than ducks and have a longer neck. They are a mostly terrestrial subfamily, often feeding on grains in meadows and uplands. They occur only as vagrants to Hispaniola. **Dabbling ducks** (Wood Duck through Green-winged Teal) are the best represented subfamily on Hispaniola. Dabblers only feed in shallow waters as they cannot dive, and only tip their head beneath the surface, leaving their tail pointed upward. The forward placement of the feet on the body enables dabblers to jump directly out of water into the air when disturbed. Most have colorful iridescent patches on the secondaries. Referred to as the *speculum*, this colorful patch is an excellent aid in identification. **Diving ducks** (Canvasback through Lesser Scaup) frequent areas of deep open water, often diving and swimming to cover rather than taking flight when threatened. Their feet are set far back on the body, forcing them to run over the water's surface to take flight. **Mergansers** are rare on Hispaniola. These ducks are adapted for catching fish and have a modified bill with serrated edges and a hooked tip. Both species that have occurred on Hispaniola are crested. **Stiff-tailed ducks** (Masked and Ruddy ducks) are small, chunky ducks with a short neck and a stiff tail that is frequently held erect and is a valuable aid in identification. Expert divers, they usually dive rather than fly to escape danger.

WHITE-FACED WHISTLING-DUCK—*Dendrocygna viduata* **Plate 1**
Vagrant

Description: 41–46 cm long; 690 g. A long-legged duck with a distinctive white face. Otherwise mostly brown above with black nape, neck, and tail; maroon chest and black belly; and flanks barred black and white. Immature is paler with a beige face. In flight, wings are dark above and below; there are no white markings except on the head. As in other whistling-ducks, flight is heavy with feet extending beyond the tail.
Similar species: Fulvous, Black-bellied, and West Indian whistling-ducks are similar in profile, but all lack white on the head.
Voice: High-pitched 3-note whistle.
Hispaniola: Generally inhabits open water in fresh or brackish wetlands. One record of 3 birds collected at Haina, DR, in May 1926.
Status: Vagrant.
Range: Breeds in South America from e. Panama south to c. Peru, Bolivia, and n. Argentina; also Africa. In West Indies, recorded only as a vagrant.
Local names: DR: Yaguaza Cara Blanca, Pato Silbador Cariblanca; **Haiti:** Jenjon Figi Blanch.

BLACK-BELLIED WHISTLING-DUCK—*Dendrocygna autumnalis* **Plate 1**
Vagrant

Description: 46–53 cm long; 830 g. A gooselike duck; note the dark body and gray face with a whitish eye-ring in all plumages. In adults, the large white wing patch, black belly, reddish bill, and pink legs are good field marks. Immature is much duller than adult, primarily grayish brown above and blackish below, with gray bill and legs. In flight, underwing is all blackish, but upperwing has a broad and bold white stripe; head and feet droop, with feet trailing beyond the tail.

Similar species: The black belly is distinctive.
Voice: Characteristic shrill, chattering whistle, often heard in flight.
Hispaniola: Inhabits freshwater marshes and brackish lagoons. One reliable record of about 12 seen during an aerial survey of the Artibonite River delta, Haiti, in January 1963.
Status: Vagrant.
Range: Breeds from c. Sonora, Mexico, northeast to Arizona and c. Texas, south to w. Ecuador and n. Argentina. In West Indies, a casual to rare visitor, primarily in Greater Antilles.
Local names: DR: Yaguaza Barriga Prieta, Pato Silbador Panzanegra; **Haiti:** Jenjon Vant Nwa.

WEST INDIAN WHISTLING-DUCK—*Dendrocygna arborea* **Plate 1**
Breeding Resident—*Threatened*

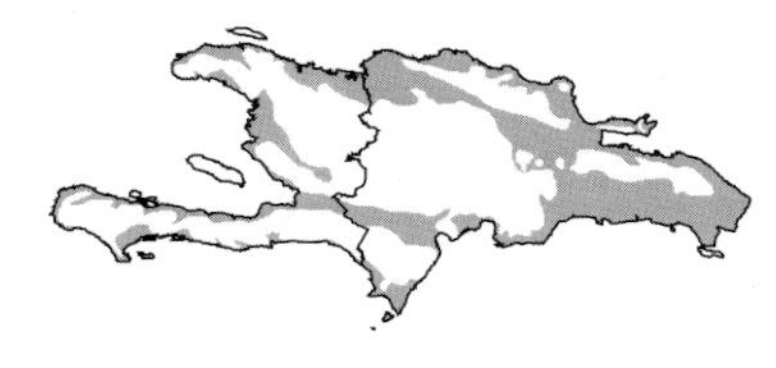

Description: 48–56 cm long; 1,150 g. Best distinguished by its deep brown coloration, white lower belly with black markings, long neck, and erect stance. Color patterns of immature are less distinct than adult's, with black on the lower belly appearing as streaks rather than splotches. In flight the head and feet droop, with feet extending beyond the tail. Very dark overall; look for the mottled belly and gray upperwing coverts.
Similar species: Fulvous Whistling-Duck is yellowish brown with white uppertail coverts and a white stripe along its side.
Voice: Shrill, whistled *chiriria*.
Hispaniola: Inhabits mangroves, palm savannas, wooded swamps, lagoons, and uplands. In Haiti, known from Étang Saumâtre, Trou Caïman, Étang de Miragoâne, swamps near Fort Liberté, and the lagoons near the mouth of the Artibonite River. In DR, most often found at Lago Enriquillo, the Samaná Peninsula area, Río Yuna and Río Yaque del Norte valleys, and especially near Monte Cristi, Laguna Saladilla and Laguna de Oviedo. Also reported from Île à Vache and Isla Beata.
Status: Local breeding resident in lowlands at favored localities. Common until about 1900, but since then greatly reduced by excessive hunting which continues. This species is threatened throughout its entire range and should be protected rigorously; hunting should be completely banned.
Comments: Feeds actively at night and roosts in deep vegetation during the day. Flocks are observed most regularly in early evening flying from mangroves or freshwater swamps, where they roost during the day, to nocturnal feeding grounds which include stands of royal palm (*Roystonea hispaniolana*), rice-growing areas, and agricultural fields.
Nesting: Nesting documented at Laguna de Rincón at Cabral and at the National Botanical Garden in Santo Domingo. Lays 10–14 white eggs in a scrape on the ground lined with fallen leaves, palm fronds, bromeliads, or other loose vegetation, or in a tree cavity. May nest at some distance from water. Recorded breeding January to March, but nesting could occur in any month of the year and may depend on rainfall. Both sexes are believed to assist in nest-building, incubating the eggs, and caring for the young. This duck does not flush from the nest until the last instant and does not sound an alarm call.
Range: Endemic to West Indies. Breeding resident in Greater Antilles, Bahamas, Barbuda, and Antigua; casual in Lesser Antilles south to Barbados.
Local names: DR: Yaguaza, Pato Silbador Caribeño; **Haiti:** Jenjon Peyi, Canard Siffleur.

FULVOUS WHISTLING-DUCK—*Dendrocygna bicolor* **Plate 1**
Breeding Resident

Description: 44–51 cm long; 710 g. A gooselike duck with long legs, long neck, and erect stance. Blackish brown above, uniform pale yellowish brown below, with a thin white stripe along the side and white uppertail coverts. In flight, note white uppertail coverts and stripe at

wing base. Underwing blackish; upperwing blackish with reddish brown coverts. Head and feet droop when flying, with feet trailing beyond tail.
Similar species: West Indian Whistling-Duck is deep brown and has dark uppertail coverts, black and white markings on sides and lower belly.
Voice: Squealing whistle, *puteow*.
Hispaniola: Prefers freshwater bodies with abundant water plants, especially rice fields. First known record in Hispaniola was at Trou Caïman, Haiti, in August 1976. Subsequently reported from DR in rice fields near Pimentel, and at Nisibón.
Status: Breeding resident. Colonized Hispaniola in the mid-1970s, presumably from eastern Cuba, but population is not known to be increasing. Often wanders widely.
Comments: Typically flocks. More of a swamp dweller than the other whistling-ducks and, unlike them, is more active during the daytime. Fairly difficult to observe because of its tendency to frequent dense aquatic vegetation.
Nesting: In DR, nesting documented only at Laguna Saladilla near Monte Cristi. Lays 11–18 dull yellowish white eggs. Breeds mostly May to June but could be any time.
Range: Breeds in s. California, c. Texas, and coastal Louisiana and Florida, south along Mexican coasts to Honduras. Also in parts of Africa, Asia, and Indian subcontinent. In West Indies, a local breeder in Cuba, Hispaniola, and Puerto Rico; elsewhere a widely dispersed vagrant.
Local names: DR: Yaguasín, Pato Silbador Fulvo; **Haiti:** Jenjon Miami, Canard Siffleur.

CANADA GOOSE—*Branta canadensis* **Plate 4**
Vagrant

Description: 64–110 cm long; 2,500 g. Mostly grayish brown with a black head and neck; distinctive white band forming a "chinstrap" on cheeks and throat; white lower belly, uppertail coverts, and undertail coverts; and black tail. In flight, note dark wings with a white band across uppertail coverts.
Similar species: Unmistakable.
Voice: Loud, resonant, honking, *h-ronk* and *h-lenk*.
Hispaniola: Inhabits borders of wetlands from saltwater lagoons to freshwater swamps, including flooded uplands. Only 1 record for Hispaniola of a bird shot at an unspecified location in DR in November 1985.
Status: Vagrant.
Range: Breeds from arctic Alaska coast and Aleutian Islands east to Baffin Island, w. Greenland, and Labrador south to c. California, s. Kansas, w. Tennessee, c. Ohio, and New Jersey. In West Indies, vagrant to casual on Andros and New Providence in Bahamas, Cuba, Cayman Islands, Jamaica, DR, and Puerto Rico.
Local names: DR: Ganso Canadiense; **Haiti:** Zwa Kanada.

WOOD DUCK—*Aix sponsa* **Plate 2**
Vagrant

Description: 47–54 cm long; 650 g. A small duck with a crested head. The male's distinctive facial pattern of green, purple, and white with a red eye and burgundy breast, is unmistakable. Female is brownish gray overall, identified by crest and large, white, asymmetrical eyering. In flight, note the long, squared tail and large head with bill tilted down.
Similar species: Unmistakable.
Voice: Males, when alarmed, emit a thin, squeaky *jeeb* call 3–4 times. Females issue a peculiar wavering note, *oo-eek, oo-eek,* reminiscent of a woodpecker, especially when taking flight.
Hispaniola: Inhabits canals, lagoons, and impoundments. Two definite records from DR and 2 unconfirmed sight records from Haiti. Occasional wanderers from Cuban population are to be expected.
Status: Vagrant.

Range: Breeds from s. British Colombia, c. Saskatchewan, s. Ontario, and Quebec through maritime provinces, south to c. California, s. Texas, s. Florida, and Cuba. Occurs in non-breeding season in southern part of breeding range. In West Indies, locally common in Cuba; rare in Bahamas; vagrant at Cayman Islands, Jamaica, DR, Puerto Rico, St. Martin, and Saba.
Local names: DR: Huyuyo; **Haiti:** Kanna Bwòdè, Canard Branchu.

EURASIAN WIGEON—*Anas penelope* **Plate 1**
Vagrant

Description: 42–52 cm long; 770 g. The male's dark reddish brown head with a golden cream-colored forecrown distinguishes it from all other red-headed ducks. Breast is pinkish. Golden forecrown is much reduced in non-breeding male, which is more rufous overall. Females occur in two color phases: gray-phase is brownish overall with a gray head and light blue bill; red-phase is similar but with a reddish tint to the head and neck. In flight, note large white patch on forewing (male), green speculum, white belly, and blackish flecks on axillaries.
Similar species: Gray-phase female is very similar to female American Wigeon but often has a darker gray head. Red-phase female is decidedly redder on the head. Both color phases of Eurasian Wigeon have axillaries with blackish flecks, whereas those of American Wigeon are whitish.
Hispaniola: Occurs in freshwater ponds and lagoons. Individuals may occur in flocks of American Wigeon. DR records of males from Laguna de Don Gregorio, Nizao, and Laguna de Rincón at Cabral.
Status: Vagrant.
Range: Breeds from Iceland, British Isles, and Scandinavia east to e. Siberia, south to c. Europe, c. Russia, and Transcaucasia. Occurs in non-breeding season on Atlantic and Gulf coasts south to Florida and west to s. Texas. In West Indies, vagrant to casual in DR, Puerto Rico, Barbuda, and Barbados.
Local names: DR: Silbon Euroasiatico; **Haiti:** Faldam Etranje.

AMERICAN WIGEON—*Anas americana* **Plate 1**
Non-breeding Visitor

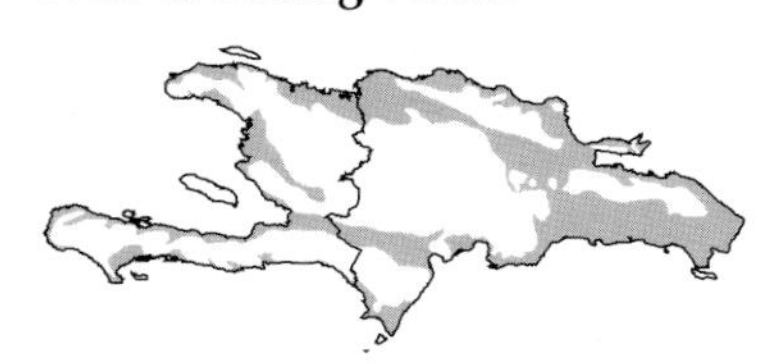

Description: 45–58 cm long; 755 g. Male is generally pinkish brown overall with a white crown, buffy white cheeks and neck, light blue bill, and green eye patch. Female is brownish with a gray head and light blue bill. In flight, note the white patch on upper forewing (broader in male), green speculum, white belly, and prominent white patch on underwing.
Similar species: See Eurasian Wigeon.
Hispaniola: To be looked for primarily in freshwater bodies, also saltwater ponds at low elevations islandwide.
Status: Now a moderately common to uncommon non-breeding visitor but numbering only a small fraction of its former abundance. In the 1950s and 1960s tens of thousands occurred on the northeastern coast of Haiti, the coastal lagoons of Monte Cristi, and Puerto Plata provinces, DR, and on the Artibonite River delta, Haiti. Since then, numbers have fallen dramatically, with reports from both DR and Haiti of only 2–20 per day, rarely as many as 100.
Range: Breeds from s. Alaska east to n. Ontario and Nova Scotia, south to s. Idaho, South Dakota, and New York. Occurs in non-breeding season from coastal s. Alaska, s. Great Lakes region, and n. Massachusetts south to Panama, casually to n. Colombia and n.Venezuela. In West Indies, regular in non-breeding season nearly throughout, but casual to rare in Lesser Antilles.

Local names: DR: Pato Cabecilargo, Silbon Americano; **Haiti:** Faldam, Kanna Zèl Blanch, Siffleur Américain.

MALLARD—*Anas platyrhynchos* **Plate 2**
Vagrant

Description: 50–65 cm long; 1,082 g. A large dabbling duck with a bright blue speculum bordered by white. Adult female is mottled brown overall, and bill is orange with black markings. Non-breeding male and immature are similar to adult female but have an olive-colored bill. Breeding male has a distinctive rounded green head, maroon breast, pale gray back and flanks, black tail-curl, and yellow bill.
Similar species: Northern Shoveler has a green head like male Mallard but has a white breast, reddish brown sides, and noticeably larger bill. Some mergansers have green heads, but they have crests and slender, hooked bills.
Voice: Among the most vocal of ducks. Female often gives loud, descending quacking calls, sometimes singly; male's call is a short, raspy quack, *quehp* or *rab*.
Hispaniola: To be looked for on bodies of calm, shallow water. One bird banded as a recently fledged duckling in New Brunswick was shot in DR, and a female was seen near Enriquillo, DR.
Status: Vagrant.
Range: Breeds from n. Alaska, nw. Mackenzie, n. Ontario, s. Quebec, and New Brunswick south to s. California and s. New Mexico east to Virginia, locally to Florida. Occurs in non-breeding season from s. coastal Alaska across extreme s. Canada south through U.S. to c. Mexico. Also widespread in Eurasia. In West Indies, regular in non-breeding season in Cuba, vagrant in Bahamas, DR, Puerto Rico, and St. Croix.
Local names: DR: Pato Inglés, Anade Azulon; **Haiti:** Kanna Kolvèt.

BLUE-WINGED TEAL—*Anas discors* **Plate 2**
Non-breeding Visitor

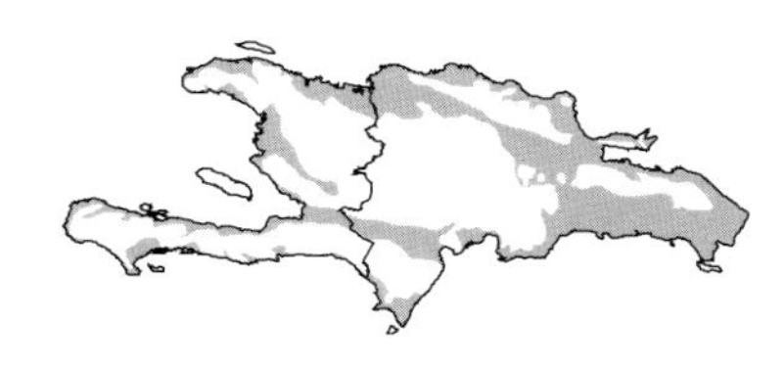

Description: 38–40 cm long; 385 g. In all plumages, this is a small, brownish duck with a pale blue forewing noticeable in flight. Female and non-breeding male are mottled brown overall with a light spot on the lores. Breeding male is mottled brown with a grayish head, distinct white crescent on the face, and rounded white patch on the flank behind the legs. In flight, blue forewing and green speculum are conspicuous.
Similar species: Female and non-breeding male Blue-winged and Green-winged teal are very similar. Blue-winged Teal is distinguished by the light spot on its lores, darker belly, and lack of a pale buffy streak beneath the tail.
Hispaniola: Found islandwide on lakes, ponds, and coastal lagoons, primarily on freshwater. Still the most common migrant duck, though greatly reduced from its earlier abundance. Birds rarely summer, but no breeding has been recorded.
Status: Now a fairly common non-breeding visitor. From the 1950s to 1980s, numbers dropped precipitously to the point where daily totals of 15–20 at favored spots were usual but 100 was exceptional. Excessive shooting was responsible for much of the decline, as evidenced by the large number of band returns. More birds have been recorded since the 1990s, but many may be transients on their way to non-breeding grounds farther south. Numbers appear to have been maintained better in western Haiti where there has been less hunting pressure.
Range: Breeds from c. Alaska east to n. Saskatchewan and Newfoundland south to n. California, w. Texas, and North Carolina. Occurs in non-breeding season from n. California, sw. Tennessee, s. Mississippi, and ne. North Carolina south to c. Peru and Argentina. Occurs throughout West Indies.
Local names: DR: Pato de la Florida, Cerceta Aliazul; **Haiti:** Sasèl.

NORTHERN SHOVELER—*Anas clypeata* **Plate 2**
Non-breeding Visitor

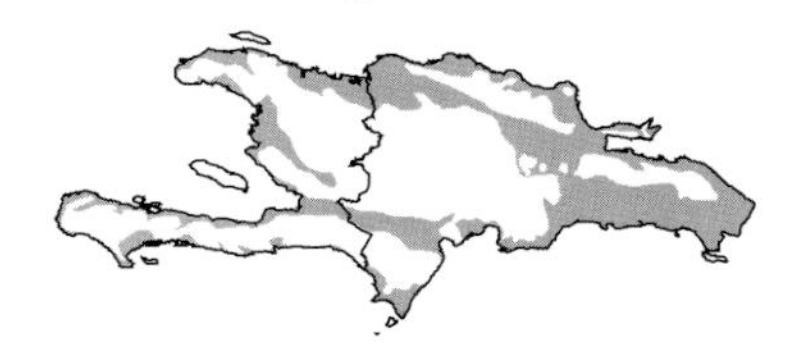

Description: 45–50 cm long; 615 g. A medium-sized dabbling duck with an unusually large bill, widened and spoonlike toward the tip. Male has a green head, white breast, and reddish brown sides and belly. Female is mottled brown overall. In flight, note the blue forewing and green speculum.
Similar species: Male Mallard also has a green head, but its breast is maroon and bill is smaller and yellow. Blue-winged Teal also has a blue forewing and green speculum, but it is considerably smaller.
Hispaniola: Inhabits fresh and brackish water bodies, rarely saline lagoons. Since the 1970s, reports include small numbers of birds from near Barahona, Puerto Viejo, Laguna Saladilla, Cabo Rojo, Laguna de Rincón at Cabral, and Gonaïves, Haiti.
Status: Now an uncommon and local non-breeding visitor, apparently much reduced from earlier years. Historically, numbers highly variable, ranging from none to 2,000, but since the 1970s only scattered reports of generally 5–25 birds, rarely more.
Range: Breeds from n. Alaska east to Manitoba and New Brunswick, south to s. California, n. Illinois, and Delaware. Occurs in non-breeding season from coastal sw. British Columbia, s. Kansas, s. Illinois, and se. Massachusetts south to w. Costa Rica, rarely to n. Venezuela and n. Colombia. Also widespread in Eurasia. In West Indies, regular in the non-breeding season in Greater Antilles, casual to rare elsewhere.
Local names: DR: Pato Cuchareta, Cuchara Norteño; **Haiti:** Jeneral, Kanna Souchè.

WHITE-CHEEKED PINTAIL—*Anas bahamensis* **Plate 2**
Breeding Resident—*Threatened*

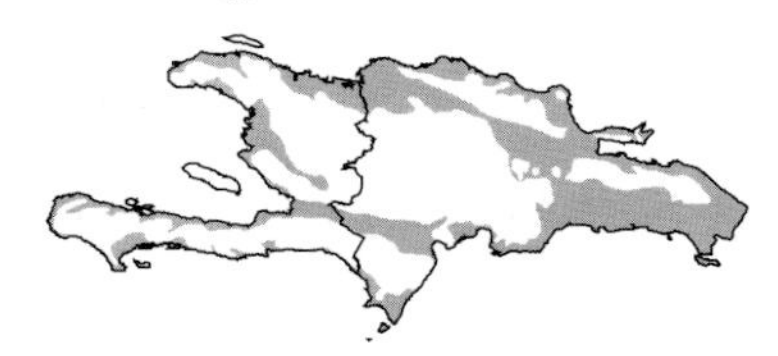

Description: 41–46 cm long; 535 g. Warm brown overall with a prominent white cheek and a red mark at base of bill. Buffy tail is decidedly pointed. Speculum is green with buff-colored borders.
Similar species: Ruddy Duck also has a prominent white cheek, but it is smaller and more compact, has a blue bill, and short tail is often cocked upright.
Voice: Male gives a squeaky call. Female's call is a *quack*.
Hispaniola: Resident of lowlands, usually on freshwater but regular also on brackish ponds, mangrove marshes, and saltwater lagoons. In Haiti, occurs at Étang Saumâtre, the lower Artibonite River and lagoons near its mouth, Trou Caïman, and Étang de Miragoâne. In DR, occurs at Lago Enriquillo, Laguna de Oviedo, Laguna de Rincón at Cabral, Puerto Alejandro, and the northwestern coastal lagoons from Monte Cristi to Puerto Plata. Also reported from Islas Beata, Catalina, and Saona.
Status: Now uncommon. Formerly fairly common and widespread, the most numerous of the resident ducks, but has declined sharply because of excessive hunting. Now expected only at favored localities.
Comments: Because the timing and length of the breeding season are somewhat unpredictable, reproductive strategies of males and females vary within this sedentary species. Both monogamy and polygyny have been recorded, and there is considerable variation in length of the pair bond and how long the female is attended by the male. There appears to be intense competition for high-quality mates in comparison with that seen in *Anas* species of the Northern Hemisphere.
Nesting: A scrape is made on dry land concealed under a clump of vegetation, sometimes at a great distance from water. Lays 5–12 pale brown eggs. Generally breeds February to June, but the season varies depending on rainfall and availability of invertebrates.

Range: Breeding resident in Bahamas, Greater Antilles, Lesser Antilles south to Guadeloupe, n. Venezuela, and Colombia south to n. Chile and c. Argentina.
Local names: DR: Pato de la Orilla, Rabudo Cariblanco; **Haiti:** Kanna Tèt Blanch, Canard des Bahamas.

NORTHERN PINTAIL—*Anas acuta* **Plate 3**
Non-breeding Visitor

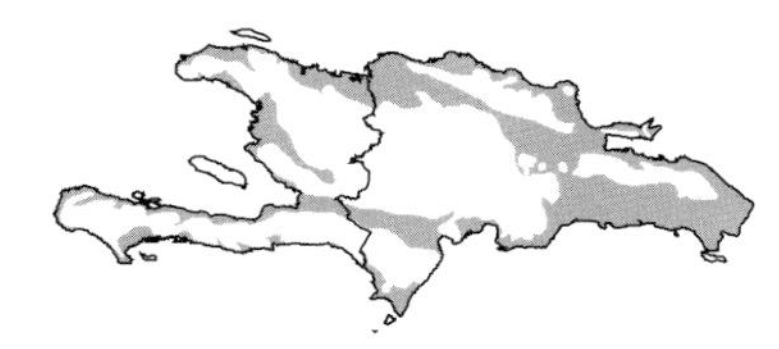

Description: Male: 57–76 cm long; 1,035 g; female: 51–63 cm long; 986 g. A slender, elegant duck with a long neck and long, pointed tail. The bill is narrow and gray. Female and non-breeding male are mottled brown. Breeding male has a brown head and white breast and neck stripe. In flight, female and non-breeding male show a brown speculum bordered by a white trailing edge. Gray underwing contrasts with white belly. Breeding male has a greenish speculum with buff-colored inner border and white trailing edge.
Similar species: Male is unmistakable. Female may be confused with Fulvous Whistling-Duck, but pintail lacks the thin white stripe along the side and shows a white trailing edge to speculum. West Indian Whistling-Duck has darker wings and less pointed tail.
Hispaniola: Occurs primarily at low elevations in freshwater bodies, but also on salt ponds.
Status: Now a rare non-breeding visitor islandwide, more often reported for DR than Haiti. Thousands reported in the 1940s and 1950s in DR, but very few recorded since then. A February 2000 report of 20–40 near Gonaïves was the first Haiti report in 25 years. The striking decline from previous abundance levels suggests excessive hunting pressure.
Comments: This has traditionally been one of the first migrant ducks to arrive in the West Indies, occurring as early as September and usually leaving by April.
Range: Breeds from n. Alaska east to Nova Scotia and south to c. California, Kansas, and Massachusetts. Occurs in non-breeding season from se. Alaska, sw. Nebraska, s. Illinois, and se. Massachsetts south to nw. Costa Rica. Also widespread in Eurasia. In West Indies, regular in non-breeding period in Greater Antilles, casual elsewhere.
Local names: DR: Pato Guineo, Pato Pescuecilargo, Rabudo Norteño; **Haiti:** Kanna Pilè, Pilet.

GREEN-WINGED TEAL—*Anas crecca* **Plate 3**
Vagrant

Description: 33–39 cm long; 340 g. A small duck with a bright green speculum and no blue in the forewing. Female and non-breeding male are mottled brown with dark lores and a whitish belly. Breeding males have a reddish brown head, large green eye patch, and white vertical bar in front of the wing. All plumages show a buffy streak beneath the tail. In flight, note the green speculum edged with white or buff.
Similar species: Female and non-breeding male Green-winged and Blue-winged teal are very similar, but Green-winged Teal lack a distinctive whitish spot on the lores and have a smaller bill, whitish belly, and pale patch beneath the tail. In flight, they lack the blue forewing of Blue-winged Teal.
Hispaniola: To be looked for in shallow freshwater bodies. Three known records: 30 in DR seen in an aerial waterfowl survey in January 1959; 1 female in a hunter's bag near Port-au-Prince, Haiti; and 2 males at Laguna Saladilla, DR.
Status: Vagrant.
Range: Breeds from Aleutian Islands east through n. Alaska, Quebec, and Labrador south to c. California, Kansas, and Pennsylvania. Occurs in non-breeding season from se. Alaska, s. Kansas, s. Illinois, and e. Massachusetts south to s. Florida and s. Mexico, rarely to Costa

Rica. Also widespread in Eurasia. In West Indies, uncommon in Bahamas and Cuba; rare or vagrant in rest of Greater and Lesser Antilles.
Local names: DR: Pato de la Carolina, Pato Serrano, Cerceta Aliverde; **Haiti:** Sasèl Zèl Vèt.

CANVASBACK—*Aythya valisineria* — Plate 3

Vagrant

Description: 48–52 cm long; 1,220 g. A large diving duck with a distinctive sloping forecrown in profile. Male has a rich reddish brown head and neck, black chest, whitish back and flanks, and brownish black rump. Female and immature are pale overall with light brown head, neck, and chest; throat is buffy brown, back and flanks are light gray, and rump is brownish. Sloping bill is black in all plumages. In flight, the long head and neck give the bird an elongated appearance. Dark breast and undertail coverts contrast with the very light belly.
Similar species: Similarly patterned Redhead lacks the sloping forecrown of Canvasback.
Hispaniola: Occurs in large, relatively deep lagoons and canals with well-vegetated edges. One DR record of 2 males at Laguna Saladilla.
Status: Vagrant.
Range: Breeds from c. Alaska east to n. Yukon, ne. Manitoba, and w. Ontario, south to ne. California, c. New Mexico, nw. Iowa, and s. Ontario. Occurs in non-breeding season from s. British Columbia, s. Ontario, and coastal Massachusetts south to Gulf Coast and s. Mexico, rarely to Guatemala and Honduras. In West Indies, vagrant in Cuba, DR, and Puerto Rico.
Local names: DR: Pato Lomo Blanco; **Haiti:** Kanna do Blanch.

REDHEAD—*Aythya americana* — Plate 3

Vagrant

Description: 42–54 cm long; 1,045 g. A diving duck with a rounded head and a steep forecrown. Male has pale gray back and black breast contrasting with bright reddish head and neck; rump and tail coverts are black; bill is blue-gray with black tip. Female is uniformly dull brownish overall. White eye-ring and blue band around a black-tipped bill are diagnostic.
Similar species: Canvasback is the only other duck with a rufous red head and neck, but it has a distinctly sloping forecrown. Female Redhead is similar to female Lesser Scaup and Ring-necked Duck, both of which have a different head shape, white belly, and white patch at the base of the bill.
Voice: Generally silent during the non-breeding season but may give catlike *meow* and wheezy *whee-ough* calls.
Hispaniola: Occurs on ponds and lagoons. The only DR record is of 100 found on a non-breeding-season waterfowl survey in 1953.
Status: Vagrant.
Range: Breeds from sc. Alaska and ne. British Columbia east to n. Saskatchewan and nw. Minnesota, south to s. California, n. Texas, and n. Iowa. Occurs in non-breeding season from s. British Columbia, s. Great Lakes region, and s. New England south to sw. Guatemala. In West Indies, vagrant in Bahamas, Turks and Caicos Islands, Cuba, Jamaica, and DR.
Local names: DR: Pato Cabeza Roja; **Haiti:** Kanna Tèt Wouj.

RING-NECKED DUCK—*Aythya collaris* — Plate 3

Non-breeding Visitor

Description: 39–46 cm long; 705 g. A medium-sized diving duck with a ringed bill and a short crest resulting in an angular, peaked profile. Male has a white bill-ring; black head, chest, back, and tail; gray sides; and white vertical bar in front of wing. Female is uniformly dark grayish brown overall with a light bill-ring and eye-ring, sometimes with a trailing white

streak between the auricular and crown. In flight, dark upperwing coverts contrast with pale gray secondaries in both sexes.

Similar species: Male is distinguished from Lesser Scaup by black back, white bill-ring, and white vertical bar on the side of the breast. Female is best distinguished by the wide gray, rather than white, wing stripe.

Hispaniola: Most often found on lowland freshwater lakes and ponds. Sight records since the 1960s include flocks of up to 20 birds at a small lake south of St. Marc, Haiti, Laguna del Rincón at Cabral, Lago Enriquillo, and a lagoon near Enriquillo, DR. The species occurs primarily from October to March.

Status: A rare non-breeding visitor. Substantial numbers of up to 5,500 were reported in the 1950s, but since then the species has been seldom reported.

Range: Breeds from c. Alaska east to c. Manitoba and Nova Scotia, south to n. California, Illinois, and Massachusetts. Occurs in non-breeding season from s. British Columbia, s. Great Plains, and s. New England south to s. Mexico, occasionally to e. Panama. In West Indies, regular in non-breeding period in Greater Antilles, rare to casual elsewhere.

Local names: DR: Pato Negro, Cabezón, Porron Acollarado; **Haiti:** Kanna Tèt Nwa, Canard Tête-noire.

LESSER SCAUP—*Aythya affinis* **Plate 4**

Non-breeding Visitor

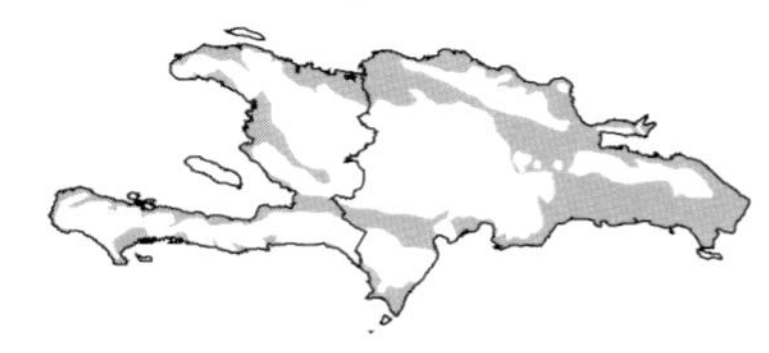

Description: 39–45 cm long; 820 g. The male's black head (glossed purple in good light), black breast and tail, white flanks and belly, and whitish back flecked gray are good field marks. Female is brown with a large white mark behind the bill. In flight, secondaries are white and primaries are dark in both sexes.

Similar species: Very similar Greater Scaup has not yet been recorded from Hispaniola but has a more rounded head with a green gloss sometimes visible in good light. Male Ring-necked Duck has a black back, white bill-ring, and white vertical bar on the side of the breast. Female Lesser Scaup is best distinguished from female Ring-necked Duck by the bold white, rather than gray, stripe on secondaries, and by the broad white patch at the base of the bill.

Hispaniola: Principally found on freshwater ponds and lakes. Since 1960 reported infrequently and usually in small numbers from such locations as a small lake south of St. Marc, and at Savane Desolée near Gonaïves, Haiti; and in DR at Laguna del Rincón at Cabral, near Baní, near San Pedro de Macorís, and Monte Cristi. Occurs primarily November to March.

Status: Uncommon non-breeding visitor. Formerly considered to be common to abundant, this species declined sharply by the 1950s and 1960s, and its numbers remain low today.

Range: Breeds from c. Alaska to w. Quebec and south to ne. California, c. Nebraska, and n. Ohio. Occurs in non-breeding season from s. British Columbia, n. Colorado, s. Michigan, and nw. Vermont south to Colombia and Venezuela. In West Indies, regular in non-breeding period in Greater Antilles, rare farther east and south in Lesser Antilles.

Local names: DR: Pato Turco, Porron Menor; **Haiti:** Kanna Zombi.

HOODED MERGANSER—*Lophodytes cucullatus* **Plate 4**

Vagrant

Description: 40–49 cm long; 610 g. Crest and slender, serrated bill, hooked at the tip, identify this species as a merganser. Male is distinctive, with a black head and neck and a large crest containing a white, fan-shaped patch when raised. Bill is black, sides are buffy cinam-

mon. Female has generally dark plumage, and the grayish brown crest is bushy, giving a "hammer-headed" appearance. Dark bill is dull orange near its base. In flight, male is dark above with a small white patch on secondaries and a pale grayish forewing; female is brown above with only a small white patch on secondaries. Wing beats of Hooded Merganser are very fast and shallow, and the flight is more horizontal than in most ducks.
Similar species: Female is similar to female Red-breasted Merganser but is much smaller with a darker face, back, and bill.
Hispaniola: One record of 2 birds seen at Cabo Rojo, Pedernales Province, DR.
Status: Vagrant.
Range: Breeds from se. Alaska east to sw. Alberta, c. Ontario, New Brunswick, and s. Nova Scotia, south to e. North Dakota, c. Arkansas, n. Louisiana, and Alabama to c. Florida. Non-breeding range is disjunct, from se. Alaska south to s. California and New Mexico, and from c. Wisconsin, s. Ontario, and c. Maine south to Texas and Gulf Coast. In West Indies, occurs in non-breeding season in Bahamas and Cuba, vagrant in DR, Puerto Rico, and Virgin Islands.
Local names: DR: Merganse de Caperuza; **Haiti:** Kanna Kouwonnen.

RED-BREASTED MERGANSER—*Mergus serrator* **Plate 4**
Vagrant

Description: 51–64 cm long; 1,020 g. A large diving duck with a shaggy crest and slender, hooked bill. Non-breeding male and female are mostly grayish brown with a reddish brown head and whitish chin, foreneck, and breast. The bill is reddish. Breeding male has a green head, white collar, and reddish brown breast. The back is dark, and the sides and flanks are grayish. There are white patches on the sides of the breast and on the wings. In flight, male's secondaries and forewing are white crossed by two narrow, dark bars; female's secondaries are white and crossed by one dark bar. Like other mergansers, flies more horizontally than most ducks.
Similar species: Female differs from female Hooded Merganser by her larger size, lighter face and back, and reddish bill.
Hispaniola: Occurs primarily in open water of bays, the ocean near shore, and inland lagoons. Two DR records of a female at Puerto Plata in December 1995 and a male at Cabo Rojo in January 1999.
Status: Vagrant.
Range: Breeds from n. Alaska east to Labrador and Newfoundland, south to n. Alberta, c. Minnesota, s. Ontario, Maine, and Nova Scotia. Occurs in non-breeding season along Pacific coast from Alaska to s. Baja California, along the Atlantic coast south from Nova Scotia, around the Florida peninsula to ne. Mexico, and in Great Lakes. Also widespread in Eurasia. In West Indies, rare to vagrant in Bahamas, Cuba, DR, and Puerto Rico.
Local names: DR: Merganse Pechiroja; **Haiti:** Kanna Fal Wouj.

MASKED DUCK—*Nomonyx dominicus* **Plate 4**
Breeding Resident

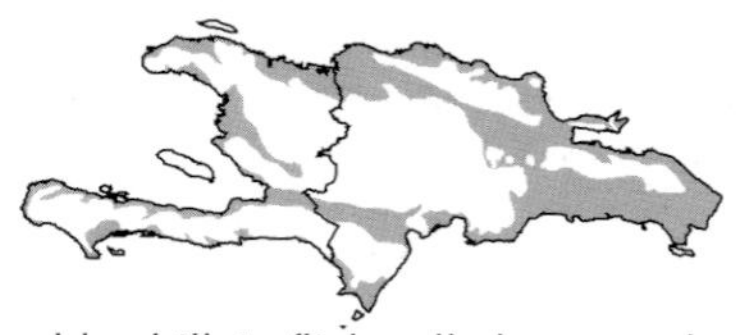

Description: 30–36 cm long; 365 g. A small, chunky duck with a conspicuous, erect tail. Non-breeding male, female, and immature are mottled buffy brown with two dark brown facial stripes and a white wing patch. Breeding male is reddish brown overall with blackish barring on back and sides, a black face, and a blue bill. In flight, all plumages show a long tail and a conspicuous white patch on the secondaries and part of the forewing.
Similar species: Ruddy Duck is the only other duck that typically cocks up its tail. Female Ruddy Duck has only one dark facial stripe and lacks a white wing patch.
Voice: Generally silent, but displaying males give a series of short clucking calls, *coo-coo-coo*.

Hispaniola: Occurs in lowland freshwater ponds and lakes with dense emergent vegetation. A retiring, skulking species, easily overlooked and thus infrequently reported. Reports from the 1930s exist from scattered localities on DR northern coast (Samaná Peninsula, Gaspar Hernández) and at Trou Caïman and the Artibonite River valley, Haiti. Since the 1960s the species has been recorded at Bayaguana, Villa Riva, Laguna Limón, Laguna de Engombe, Laguna El Derramadero, Laguna de Rincón at Cabral, Laguna de Nogales, Río Yásica, Río Haina, Pilancón, Guerra, and Laguna Saladilla.
Status: Rare to uncommon and local breeding resident. Although reliable data are not available, its numbers must have declined significantly since the mid-1950s because of drainage of marshes for rice fields and excessive hunting.
Comments: Frequents thick aquatic vegetation and is easily overlooked.
Nesting: Lays 8–18 relatively large, white eggs in a nest built in swamp vegetation over or near water. Breeding confirmed at Río Haina west of Santo Domingo, Villa Mella, and in a wetland near Río Nigua at El Tablazo north of San Cristóbal, DR. Breeds May to August.
Range: Resident locally from Nayarit east to coastal Texas, south to s. Bolivia and n. Argentina. In West Indies, regular but local in Greater Antilles, rare but local breeder in most of Lesser Antilles.
Local names: DR: Pato Criollo, Malvasia Enmascarada; **Haiti:** Kanna Maske, Kroube, Canard Masqué.

RUDDY DUCK—*Oxyura jamaicensis* **Plate 4**
Breeding Resident

Description: 35–43 cm long; 545 g. A small-bodied, compact, stiff-tailed duck. Breeding male is reddish brown overall with a black cap, contrasting white cheek patch, and blue bill. Non-breeding male is similar to female but with white cheeks. Females and immatures are grayish brown with whitish cheeks cut by a single brown stripe below the eye. In flight, appears chunky with a relatively long tail and dark upperwings.
Similar species: Female Masked Duck has two dark facial stripes rather than one.
Voice: Silent most of the year, but males give muffled staccato clucking calls during courtship displays.
Hispaniola: Occurs principally in lowland freshwater ponds and lakes; apparently never abundant. Specific records from favored localities such as Laguna de Rincón at Cabral, Laguna El Derramadero, Laguna Mala Punta, and La Altagracia Province, DR, and at Trou Caïman, Haiti. The lack of band recoveries implies that non-breeding visitors from North America may not come often or occur only in small numbers.
Status: Uncommon and local resident breeder. The West Indian subspecies has declined substantially in Bahamas, Puerto Rico, and Virgin Islands to the point of being threatened on these islands. However, populations on Hispaniola, as well as on Cuba and Jamaica, appear to be relatively stable.
Nesting: Lays 4–12 whitish eggs in a nest built over water in emergent swamp vegetation. Has nested in small numbers fairly regularly at Laguna de Rincón at Cabral. Breeds June to August.
Range: Breeds from c. Alaska east to se. Manitoba, south to s. California, s. Arizona, New Mexico, and Florida. Occurs in non-breeding season from Washington, c. Kansas, s. Michigan, and se. Maine south to Honduras. In West Indies, a breeding resident in Bahamas and Greater Antilles; casual at Cayman Islands and in Lesser Antilles, where it has not been proven to breed.
Local names: DR: Pato Espinoso, Pato Rojizo, Malvasia Rojiza; **Haiti:** Kanna Plonjon, Koukourèm, Canard Plongeur.

PARTRIDGES AND GUINEAFOWL: FAMILY PHASIANIDAE

This diverse family consists primarily of gregarious, terrestrial birds with short, stout bills and large, sturdy feet. Locomotion is typically via walking or running. Wings are short, broad, and rounded; members of this family fly low and strongly, but only for short distances, and usually only after being disturbed. Junglefowl, pheasants, and guineafowl are large birds, the males larger than females and possessing very long tails and brightly colored plumage. Food is mostly fruit and seeds gathered from the ground. All species are either domesticated or considered game birds on Hispaniola and were introduced from the Old World by European colonizers after 1492.

RED JUNGLEFOWL—*Gallus gallus* **Plate 57**
Introduced Breeding Resident

Description: Male: 71 cm long; female: 43 cm long; mass variable. Extremely variable plumage of black, brown, reddish brown, gray, and white. Male, or rooster, is resplendently plumaged with a red comb and wattle on its head and a long, bushy tail. Female, or hen, has a smaller comb and wattle, and tail is much reduced.
Similar species: Familiar and unmistakable.
Voice: Universally recognized *cock-a-doodle-doo* or *chi-qui-ri-qui*. Also a variety of clucks and other notes. Chicks give a soft, characteristic call note, *pee-o*.
Hispaniola: Feral populations are found locally in Los Haitises National Park and in Sierra de Bahoruco, DR, and perhaps elsewhere.
Status: Well-known, introduced bird, found in a wild state only very locally in the DR. Domesticated birds are common on farms and in villages throughout the island, and are often left to run free.
Nesting: Lays 1–9 white to light brown eggs in a scrape on the ground, sometimes lined with twigs.
Range: Resident from Himalayas and s. China south to c. India, se. Asia, Sumatra, and Java; domesticated and introduced worldwide. In West Indies, feral populations are found in Bahamas, Hispaniola, Mona Island, and Grenadines.
Local names: DR: Female: Gallina; male: Gallo; **Haiti:** Female: Poul; male: Kòk.

RING-NECKED PHEASANT—*Phasianus colchicus* **Plate 57**
Introduced Breeding (?) Resident

Description: Male 60–70 cm long; 1,317 g; female 50–60 cm long; 953 g. Male is a spectacularly multicolored, chickenlike bird with a very long tail. Note the red face, glossy purple head and neck, and white collar. Mantle and flanks are golden brown and spotted black, blending to gray on rump. Tail is chestnut brown and buff, prominently barred black. Breast and belly are dark brown. Female is mottled brown throughout, with a shorter, pointed tail.
Similar species: Unmistakable.
Voice: Harsh, loud crowing *uurk-iik*, not unlike male junglefowl, and various clucking sounds.
Hispaniola: Escapees from a private hunting reserve are known to occur in the area of Rio Chavón, La Romana Province, and may be possible at El Limón between La Romana and Bayahibe.
Status: Introduced to a private hunting reserve in the mid-1990s; a small feral population now exists.
Nesting: Not known to breed, but likely to do so.
Range: From s. British Columbia, sw. Manitoba, c. Minnesota, and maritime provinces, south locally to c. California, se. Texas, and c. Virginia. Also found in Hawaiian Islands, New Zealand, and much of Eurasia. In West Indies, introduced to Bahamas and DR.
Local names: DR: Faisán; **Haiti:** Fezsan, Faisan.

HELMETED GUINEAFOWL—*Numida meleagris* **Plate 57**
Introduced Breeding Resident

Description: 53 cm long; 1,299 g. Distinguished by its unusual, bulbous body shape, dark gray feathering with white spots, and nearly naked head and neck.
Similar species: Unmistakable.
Voice: Raucous, cackling calls in rhythmic series.
Hispaniola: Most common in semiarid thorn scrub, dry forest, and savanna habitat. Has been recorded as high as 760 m on Morne La Selle, Haiti.
Status: Uncommon and local resident breeder. Possibly introduced as early as 1508, but at least by the early 1700s, by which time it was widespread and common in the lowlands of both countries and actively hunted as a game species. Has declined sharply from being very common nearly islandwide in the 1930s in both countries, especially in Haiti. Now rather local and increasingly confined to sparsely populated regions, private property, or other protected areas where hunting is controlled or not allowed. Population increased noticeably from 1982 to 1996 in Sierra de Bahoruco National Park, especially on the northern side.
Comments: Ground-dwelling in small flocks; flushes with a burst of whirring wings. Roosts in trees and bushes. Most birds receive handouts from local farmers and are somewhat domesticated, but truly wild populations also exist.
Nesting: Lays 12–15 light brown or buff-colored eggs with spots in a scrape on the ground hidden in vegetation.
Range: Resident in Arabia and much of Africa. In West Indies, introduced and now feral in Cuba, Hispaniola, Puerto Rico, Virgin Islands, St. Martin, and Barbuda.
Local names: DR: Guinea; **Haiti:** Pentad Mawon, Pintade Sauvage.

BOBWHITES: FAMILY ODONTOPHORIDAE

Formerly included within the Phasianidae, this family consists of a variety of terrestrial birds from the New World. Bobwhites are characterized by their small size and short tail. They tend to be shy and secretive, foraging on the ground in thick brush and preferring to run for cover rather than fly. Bobwhites are more often heard than seen, and they are often observed bursting into short flights from just below one's feet. Like members of the Phasianidae, the bobwhite is introduced to Hispaniola.

NORTHERN BOBWHITE—*Colinus virginianus* **Plate 57**
Introduced Breeding Resident

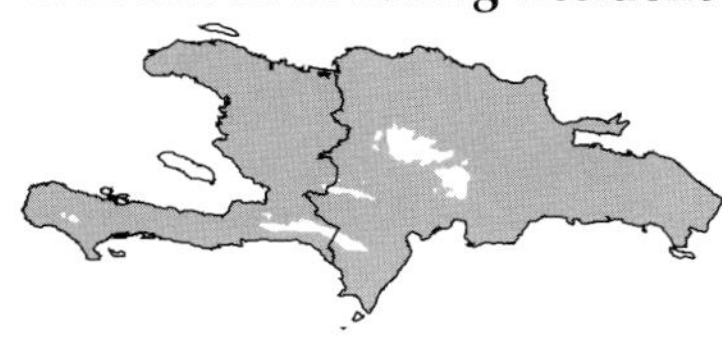

Description: 25 cm long; 178 g. A chunky, brown bird, finely barred tan and black above, barred white and black below, with chestnut streaking on sides. Resembles a small, rounded chicken quickly scampering in small groups about the underbrush. Often does not flush until nearly underfoot, when it bursts from cover. Male has a white throat and supercilium and shaggy head crest. Female is similar but with a tan throat and supercilium.
Similar species: Unmistakable.
Voice: Clear whistled rendition of its name, *bob-white* or *bob, bob-white*.
Hispaniola: Found in scrubland and pasture with ample cover, and especially in pine forest. Introduced into Haiti and DR at different times and from different sources. Introduced to Haiti before 1796. Now common in La Visite National Park and uncommon in Macaya Biosphere Reserve, Haiti. Apparently introduced in DR near Santo Domingo from Cuba about 1890. By the 1930s it had become widespread and relatively common in lowland savanna habitat.

Subsequently spread to the pine zone in Sierra de Bahoruco and Cordillera Central, where it is found regularly to 1,930 m. The species is generally, though sparsely, distributed throughout eastern DR.
Status: Uncommon and increasingly local breeder, much reduced from its earlier abundance during the 20th century, primarily because of excessive hunting.
Comments: Birds introduced to Haiti were of the race *C. v. virginianus*, and those introduced to DR were of the endemic Cuban race, *C. v. cubanensis*. The latter have an almost entirely black breast in the male and heavy black barring below in the female. It is not known whether populations from Haiti and DR have come into contact along the Massif de la Selle-Sierra de Bahoruco ridge and, if so, whether they have interbred.
Nesting: Lays 12–18 dull white eggs on the ground amidst a clump of grass. Breeds May to July.
Range: Resident from se. Wyoming, c. South Dakota, c. Michigan, s. New York, and Massachusetts south to e. New Mexico, w. Texas, and s. Florida, south to Chiapas, Mexico. In West Indies, occurs in Cuba; introduced to Hispaniola, Puerto Rico, and Bahamas.
Local names: DR: Cordoníz; **Haiti:** Kay, Caille.

GREBES: FAMILY PODICIPEDIDAE

Entirely aquatic, grebes have a slender, pointed bill, an extremely short tail, and flattened lobes on each toe. They are excellent swimmers and divers, rarely taking flight, which has led to the local belief that the birds are flightless. Legs are placed far back on the body, and grebes must run some distance over the water before taking off. On land they cannot walk, but must push themselves along on their breast. Characteristic behavior includes gradually sinking out of sight, or submerging until only the head remains above the surface. Floating nests are constructed of aquatic vegetation. The downy young are sometimes carried on a parent's back and may remain there while the adult dives.

LEAST GREBE—*Tachybaptus dominicus* **Plate 5**
Breeding Resident

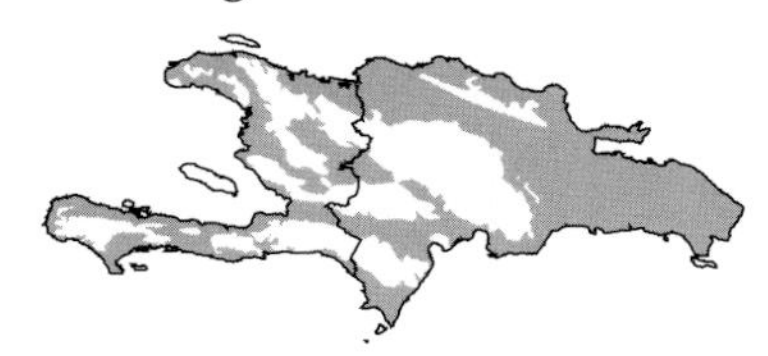

Description: 22–27 cm long; 122 g. Small size, blackish coloration, thin bill, and yellow-orange iris are diagnostic. Breeding birds have a black throat; non-breeding adults and immatures have a white throat. The white wing patch is a good field mark but is not always visible.
Similar species: Pied-billed Grebe is larger and bulkier, with a stouter bill, dark eye, and all-dark wing.
Voice: Descending chatter, *te-te-te-te-te-te*; also a rising, reedlike *week* or *beep*.
Hispaniola: Breeding resident in many freshwater localities in both countries, principally in the lowlands but occasionally at elevations of 1,000 m or higher. Occurs primarily in freshwater cattail swamps and small ponds with water plants for cover. Current status in Haiti is unknown; in DR, most common in the Río Dajabón basin and the Samaná Peninsula.
Status: Decidedly uncommon and local year-round resident on Hispaniola. Generally shy and difficult to observe, so may be more common than indicated, but it has certainly declined since 1930 because of habitat loss.
Comments: Generally solitary or in small family groups. Often stays among vegetation, where it is extremely difficult to observe. When alarmed, dives rapidly and only re-emerges under cover of dense vegetation.
Nesting: Lays 1–7 whitish eggs in a floating nest among emergent vegetation. Breeds throughout the year but most often in spring and fall.

Range: Resident from s. Baja California east to s. Texas and south to s. Peru and n. Argentina. In West Indies, occurs in most of Bahamas and Greater Antilles, east to Virgin Islands.
Local names: DR: Tígua, Zaramagullón Chico; **Haiti:** Ti Plonjon, Grèbe.

PIED-BILLED GREBE—*Podilymbus podiceps* — Plate 5
Breeding Resident

Description: 30–38 cm long; 442 g. A small, stocky, ducklike bird with a brown back, buffy gray sides of neck and flanks, and a short, conical bill. Breeding adult has a black throat, and bill has a black band. Non-breeding adult has a whitish throat, and bill lacks the black band. Head of juvenile is distinguished by mottled brown and buffy-white markings.
Similar species: Least Grebe is smaller and has a thin, straight bill and yellowish eye.
Voice: Harsh cackle breaking into a distinctive *kowp, kowp, kowp,* slowing at the end.
Hispaniola: Occurs on many lakes, ponds and slow-moving rivers in both countries, primarily in the lowlands. Prefers freshwater but also inhabits brackish and hypersaline lagoons where freshwater is absent. Most commonly reported from Étang Saumâtre and near Gonaïves, Haiti, and at Lago Enriquillo, Laguna de Rincón at Cabral, Puerto Escondido, the Samaná Peninsula, and San Pedro de Macorís and La Altagracia provinces, DR. Also recorded from Île à Vache.
Status: Common breeding resident in both countries.
Comments: The typical grebe of Hispaniolan ponds and lagoons. Usually solitary or in small family groups, but can occur in loose aggregations of up to 50 individuals. Often frequents open water. If alarmed, dives instantly and swims beneath the surface to the protection of dense aquatic vegetation.
Nesting: Lays 2–6 whitish eggs in a floating nest built among emergent vegetation. Thought to breed throughout the year, but primarily March to June.
Range: Breeds from s. Alaska east to maritime Canada and south to Argentina and Chile. Occurs throughout West Indies but most common in Greater Antilles.
Local names: DR: Zaramagullón; **Haiti:** Plonjon Fran, Grand Plongeon.

SHEARWATERS AND PETRELS: FAMILY PROCELLARIIDAE

Shearwaters and petrels are highly pelagic birds with long, narrrow wings, webbed feet, and large external tubular nostrils at the base of a hooked bill, from which their colloquial family name "tubenose" is derived. Rarely coming within sight of land except to breed, these birds are not readily observed even during the breeding season, for they nest colonially on remote islets and mountain cliffs, usually in burrows, and are only active about the nest after dark. At night the braying and wailing cries of these birds are conspicuous, and the musty odor of their nesting areas often remains long after the birds have gone. Their typical pattern of flight involves stiff-winged gliding low over the waves and in the troughs, alternating with periodic series of short, rapid flaps. These birds feed on small fish, squid, and other organisms on or near the surface of the ocean. They are equipped with special salt glands that enable them to drink salt water.

BLACK-CAPPED PETREL—*Pterodroma hasitata* — Plate 5
Breeding Resident—*Critically Endangered*

Description: 35–40 cm long; 278 g. Upperparts appear black except for variable white patches on the rump, nape, and forecrown. Underparts are mostly white; leading and trailing

edges of wings are black; distinctive black bar on underwing. In flight, tail is pointed; wings are relatively long and pointed.

Similar species: Distinguished from Greater Shearwater by whiter upperparts, especially on the forecrown and rump; mantle is blacker, heightening the contrast with its more extensive white markings. In flight, wrist is more noticeably bent than in shearwaters, and flight is more erratic with faster wing beats and high, arching glides.

Voice: Heard only at night around the nesting colony. Issues three distinct calls which are repeated often: a drawn-out, wailing *aaa-aw, eek*; a drawn-out *ooow, eek*; and yelps like "a hurt puppy."

Hispaniola: Recorded at sea from Gulf of Gonâve, Bay of Port-au-Prince, Cabo Engaño, Mona Passage, Cabo Cabrón, Samaná, and 35 km southwest of Punta Las Salinas. Also recorded from near Islas Beata and Alto Velo. The only significant breeding populations are in the Massif de la Selle between Morne La Selle and Morne La Visite, and in the Massif de la Hotte on the southern face of Pic Macaya and possibly on Pic Formon in southwestern Haiti. An estimated 20–40 pairs breed on Loma de Toro in the Sierra de Bahoruco of the DR.

Status: Critically endangered rangewide. Populations are greatly diminished because of harvesting by humans and losses to introduced predators. Largest remaining nesting colonies are on Hispaniola, where it is considered a rare, very local, and declining breeding resident. A tall telecommunications mast with stay wires originally erected on Loma de Toro in 1995 poses a major hazard to the survival of birds. Historic reports of birds off the northern coast and inland at Moca, DR, suggest a possible nesting locality in the Cordillera Central, but that population, if ever present, may now be extirpated. Systematic surveys of this species' current status on Hispaniola are needed.

Comments: This seabird only comes to land via the cover of darkness to breed among remote cliffs. Historically, fires were lit on the cliff tops above a breeding colony on foggy, moonless nights. The disoriented petrels would crash into and around the fire, whereupon they would be gathered up. Non-breeding birds occur in numbers in the Gulf Stream off the southeastern coast of the U.S., primarily from April to October. These birds may migrate via the Antilles Current east of the Bahamas. It has been suggested that breeding petrels may actually commute to North Carolina waters to feed. Believed to feed on squid and small fish.

Nesting: A single white egg is laid in an excavated burrow, cavity of rock talus, or cave on steep mountain slopes from 1,500 to 2,000 m elevation. Breeding season is presumed to extend from October to June with a peak from December to February. Birds may arrive at the breeding sites as early as late September, but the main courtship period is probably late October and November with a prelaying exodus in December, egg-laying in January, hatching in late February and early March, and fledging in May to early June.

Range: Occurs in most of Caribbean, Gulf of Mexico, and off se. North America and ne. South America. Breeds at high elevations in Hispaniola and possibly still in Guadeloupe, Dominica, and Martinique. Formerly bred in Jamaica. Unproven to have ever bred in Cuba, but suspected to have done so.

Local names: DR: Diablotín; **Haiti:** Chanwan Lasèl, Canard de Montagne.

GREATER SHEARWATER—*Puffinus gravis* **Plate 5**
Vagrant

Description: 48 cm long; 840 g. One of the two largest shearwaters in the West Indies. Grayish brown upperparts, mostly light below, but underwings with considerable dark markings. Note the distinctive black cap, noticeable white collar, and narrow white band on the rump.

Similar species: Generally similar to Black-capped Petrel but larger; white of upperparts is much reduced on nape and rump and is absent from forecrown. Black-capped Petrel's mantle is blacker, heightening the contrast with its whiter upperparts. Greater Shearwater's wing beats are much slower than those of Manx and Audubon's shearwaters or Black-capped Petrel.

Hispaniola: Two reports: 1 seen in the Mona Passage and another reported off the northeastern coast of DR.
Status: Vagrant, but likely underrepresented in records because of its occurrence far offshore.
Comments: This shearwater nests only among the small Tristan da Cunha Islands in the southern Atlantic from November to April. It then migrates up the western Atlantic past the West Indies, later returning to its nesting islands primarily via the eastern Atlantic. Greater Shearwaters not yet mature enough to breed appear to be widely dispersed throughout the Atlantic Ocean.
Range: Breeds on Tristan da Cunha, Gough, and Falkland islands in s. Atlantic; common seasonally in n. Atlantic. In West Indies, casual in Gulf of Mexico and off Puerto Rico, regular off Lesser Antilles.
Local names: DR: Pardela Capirotada; **Haiti:** Gwo Cahen, Gros Puffin.

MANX SHEARWATER—*Puffinus puffinus* **Plate 5**
Vagrant

Description: 30–38 cm long; 400 g. An intermediate-sized shearwater with a short tail. Blackish above and white below, including underwing and undertail coverts. Pale crescent behind auriculars. Flight characterized by four or five distinctive snappy wing beats and a rocking glide, especially in light winds or flat seas.
Similar species: Audubon's Shearwater is slightly smaller, with a browner back and upperwings, longer tail, grayer underside of primaries, and undertail coverts usually dark. Greater Shearwater has slower wing beats, is larger, and has a white rump. Manx Shearwater has a long-winged and short-tailed appearance, whereas Audubon's Shearwater appears short winged and long tailed.
Hispaniola: Two birds found dead on a beach in La Altagracia Province had both been banded as nestlings in Wales, U.K.
Status: Vagrant, primarily from November to March, but may also be expected in other months. Scarcity of records likely results from the species occurring far offshore.
Range: Breeds in n. Atlantic from Newfoundland and Iceland south throughout British Isles, Brittany, Madeira, Azores, and most of Mediterranean Sea; formerly Bermuda. Seasonally ranges south to coast of South America from Trinidad to Argentina and off African coast from Canary Islands to South Africa. Vagrant in West Indies.
Local names: DR: Pardela Pichoneta; **Haiti:** Cahen Zangle.

AUDUBON'S SHEARWATER—*Puffinus lherminieri* **Plate 5**
Non-breeding Visitor

Description: 30 cm long; 180 g. A relatively small, short-winged and long-tailed, blackish brown and white shearwater; the only shearwater regularly encountered in the West Indies. Dark blackish brown above; white below but with dark undertail coverts.
Similar species: Rounded tail and distinctive rapid wing beats between glides distinguish this species from dark-backed pelagic terns. It has faster wing beats than the larger Greater Shearwater and lacks the white nape and rump. Manx Shearwater has longer wings, a shorter tail, and white undertail coverts.
Voice: Highly vocal at night around the nest, when adults utter mournful, catlike cries while flying. In their burrows, they issue a loud, distinctive *plimico*. The young utter plaintive, liquid, twittering notes.
Hispaniola: Surprisingly few records exist for such a relatively common and widespread species; this is most likely due to incomplete searching of nearshore waters. Although it has never been documented, the species may breed on isolated offshore islands. Reports exist for birds at sea from north of Puerto Plata, offshore of Cabo Mongó, the Mona Passage, and northeast of Cabo Francés Viejo, DR. Also reported from near Navassa Island, north of Île de la Tortue, near Isla Beata, and off Isla Saona.

Status: Generally an uncommon and local non-breeding visitor, but rare outside the breeding season. When not breeding, birds are believed to disperse to the seas around their breeding grounds, where they are not likely to be seen except from a boat far offshore.
Range: Widespread in tropical Atlantic and Gulf of Mexico and in other tropical seas. Ranges throughout West Indies, breeds locally.
Local names: DR: Pardela de Audubon, Pampero de Audubon; **Haiti:** Cahen, Puffin.

STORM-PETRELS: FAMILY HYDROBATIDAE

Storm-petrels are the smallest of pelagic birds, only approaching land to breed. Little larger than swallows, they characteristically swoop and flutter low over the water, sometimes pattering the surface with their webbed feet. Found alone or in loose flocks, they feed on zooplankton and other small marine animals. Storm-petrels nest in burrows or crevices on remote islets, visiting their colonies at night. In many species, one parent tends the nest for days without food before being relieved by its mate.

WILSON'S STORM-PETREL—*Oceanites oceanicus* **Plate 5**
Non-breeding Visitor

Description: 18–19 cm long; 34 g. A small, dark brownish black seabird with a conspicuous white rump and white patch on lower flank. Tail is short and squared; legs long with feet projecting beyond tail in flight.
Similar species: Distinguished from similar Leach's Storm-Petrel by being blacker and having shorter, broader, and more rounded wings; less angled wrists; and more direct flight with briefer glides, reminiscent of a swallow. Rump patch is entirely white, tail is square, and feet extend beyond tail in flight.
Hispaniola: One observed off the southeastern DR coast, 2 records off the northeastern DR coast, and another about 40 km off the coast from Pedernales, DR. One also seen off Isla Saona, and 2 birds seen off Isla Beata.
Status: Generally a rare non-breeding visitor to offshore waters of Hispaniola, primarily in May and June. Likely more frequent than records indicate.
Comments: Wilson's Storm-Petrel is believed by some experts to be the most numerous bird in the world.
Range: Breeds around coastal Antarctica and on subantarctic islands off South America and in s. Indian Ocean; common seasonally in n. Atlantic. Casual throughout West Indies.
Local names: DR: Golondrina del Mar, Pamperito de Wilson; **Haiti:** Oseanit.

LEACH'S STORM-PETREL—*Oceanodroma leucorhoa* **Plate 5**
Vagrant

Description: 18–22 cm long; 35 g. A small, brownish black seabird with a white rump, notched tail, and pale brown vertical bar on wings. Rump often bisected by dark median line. Flight erratic, with sudden, sharp changes of direction and deep wing beats.
Similar species: Distinguished from similar Wilson's Storm-Petrel by its longer, narrower, and more pointed wings with more sharply angled wrists; pale brown wing band; white rump patch appearing divided at close range; and forked tail. In flight, feet do not extend beyond tail.
Hispaniola: Two records: 1 collected in Samaná Bay, DR, and one found dead at Laguna de Oviedo, DR.
Status: Vagrant. Likely more frequent than records indicate. Conscientious offshore surveys would almost certainly show the species to be an uncommon but fairly regular transient, primarily November to June.

Comments: Leach's Storm-Petrel does not follow boats as does Wilson's. Diet consists of zooplankton, shrimp, and small fish.
Range: Breeds in n. Atlantic from Massachusetts north to s. Labrador and from s. Iceland south to n. Scotland; also breeds in n. Pacific; ranges south seasonally in Atlantic to e. Brazil and South Africa. Vagrant throughout West Indies.
Local names: DR: Lavapiés, Paiño Boreal; **Haiti:** Oseanit Dèyè Blanch.

TROPICBIRDS: FAMILY PHAETHONTIDAE

Elegant and spectacular, tropicbirds are confined to tropical and subtropical oceans. Two of the three living species worldwide occur in Hispaniolan waters. Adults are characterized by long, streaming central tail feathers, which barely project in young birds. The pointed tail and dorsal pattern of narrow black bars distinguish immature tropicbirds from terns. When near shore, a tropicbird's habit of making numerous approaches to the nesting cliffs before landing is distinctive. Tropicbirds feed primarily on squid and fish, which they capture by diving from substantial heights. Because their nasal openings are almost entirely blocked (as is the case with related families), excess salt filtered from the sea water they drink passes through the mouth to the tip of the bill, where the droplets are shaken off. Tropicbirds occasionally alight on the water, with their tail held cocked.

WHITE-TAILED TROPICBIRD—*Phaethon lepturus* **Plate 6**
Breeding Resident

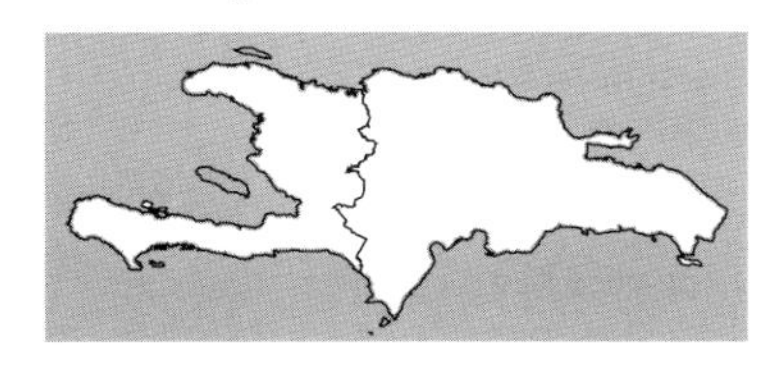

Description: 60–80 cm long (including 30- to 40-cm tail plumes); 350 g. Adult is white overall with narrow black patch through eyes, long tail feathers, a heavy black bar on upperwing coverts, and black on outermost primaries. Bill is orange to red-orange. Immature is coarsely barred black and white above, with short central tail feathers; bill is yellowish, sometimes ringed with black near the tip.
Similar species: Immature differs from immature Red-billed Tropicbird in having coarser black barring on upperparts and lacking a black band across nape.
Voice: Raspy, harsh, ternlike *crick-et*.
Hispaniola: Fairly common breeding resident present year-round at suitable coastal cliff localities on both the main and several satellite islands. Regularly observed both in offshore waters and along stretches of coastline where it does not nest. Reported from Navassa Island, Île à Vache, Île de la Gonâve, Île de la Tortue, and Islàs Alto Velo and Beata.
Status: Widespread, but only a very local breeding resident in Hispaniola. It is the commonly encountered tropicbird of the region.
Comments: Best observed flying near the vertical sea cliffs it uses to breed. Aerial courtship displays are performed, the male drooping its tail as it flies before the female. Landing on the cliffs is often difficult, so several passes are sometimes required before a successful landing is executed.
Nesting: The only documented DR records appear to be a colony of about 20 pairs at Cabo Cabrón, Samaná, in August 1996; an estimated 60 pairs between Cabo Rojo and Bahía de Las Águilas in June 1998, including a chick seen in a nest; and a small colony nesting near the National Aquarium, Santo Domingo, since 1999. However, it certainly breeds at many other places along the coastline. Strongly suspected DR breeding localities include Punta Martín García and Isla Alto Velo. Formerly nested in Haiti near Jérémie and Jean Rabel, and Île de la Tortue, but present status is not known. Lays a single tan egg, heavily splotched with dark brown. Breeds primarily March to July.

Range: Widespread in tropical oceans. Found virtually throughout West Indies.
Local names: DR: Rabijunco Coliblanco; **Haiti:** Pay-an-ke Bèk Jòn.

RED-BILLED TROPICBIRD—*Phaethon aethereus* **Plate 6**
Vagrant

Description: 91–107 cm long (including tail plumes); 750 g. A large and spectacular, mostly white seabird. Adult is white overall with fine black barring on back, a black eye-line, conspicuously black wing tips, long tail plumes, and a red bill. Immature lacks long tail plumes; bill is yellowish, ringed with black near the tip; a black band extends from the eyes across the nape; and the back is barred.
Similar species: Immature White-tailed Tropicbird is more coarsely barred on back, and black eye-line does not meet across nape.
Voice: Shrill, harsh, grating *keé-arrr*.
Hispaniola: Several records from Punta Medina, Miches, Playa San Luis, and Oviedo, DR. Also reported from Cayos Siete Hermanos. The nearest confirmed breeding sites are in the U.S. Virgin Islands.
Status: Vagrant.
Range: Local in tropical Atlantic, Pacific, and Indian oceans. In West Indies, largely confined to Puerto Rico and Lesser Antilles.
Local names: DR: Rabijunco Piquirojo; **Haiti:** Pay-an-ke Bèk Wouj.

BOOBIES: FAMILY SULIDAE

Boobies are large, sleek-plumaged birds with a long, stout bill, long, pointed wings, and a tapered tail. They sometimes wander well out to sea from their colonial roosting and nesting areas, but they often frequent nearshore waters. Their leisurely flapping and gliding flight, low over the water, is characteristic, as are their spectacular plunge-dives into the sea in pursuit of fish and squid. Boobies vigorously defend a small area around their nests, usually using exaggerated and stereotyped head movements in interactions with other birds. Guano deposits from some nesting populations of boobies have been harvested as an important source of fertilizer. Fossils of boobylike birds date back at least 20 million years.

MASKED BOOBY—*Sula dactylatra* **Plate 6**
Non-breeding Visitor

Description: 74–86 cm long; 1,500 g. Adult is primarily bright white with brownish black tail, primaries, and secondaries. Subadult is similar to adult, but upperparts show brown on head and rump and brown flecks on upperwing coverts. Immature has a brown head and upperparts with a white nape; underparts are primarily white except for brownish throat, undertail, and flight feathers.
Similar species: Adult is distinguished from white-phase Red-footed Booby at a distance by its dark tail. Immature Masked Booby differs from Brown Booby by its lack of brown on upper breast and its white nape.
Hispaniola: Seen at Cabo Engaño, La Altagracia Province, DR, and in Mona Passage. Also reported from Isla Saona. It is not surprising that occasional birds from the nearest breeding colony at Mona Island should appear in waters off the eastern coast of the island.
Status: Very rare and local non-breeding visitor. Masked Booby is threatened in the West Indies because of habitat loss, disturbance from development, and predation by introduced mammals.
Comments: Masked Booby is more restricted in its nesting habitat use than Brown and Red-

footed boobies, preferring flat, unforested terrain on small islands. As a result, it is decidedly less abundant, with colonies small and confined to remote areas.
Range: Widespread in tropical Atlantic, casual north to North Carolina, and coastal Pacific Middle America north to Mexico; also in other tropical oceans. In West Indies, rare and local but ranges widely from scattered breeding localities in s. Bahamas, Puerto Rico, Virgin Islands, and Lesser Antilles.
Local names: DR: Bubí, Bubí de Cara Azul, Bubí Enmascarado; **Haiti:** Gwo Fou Maske.

BROWN BOOBY—*Sula leucogaster* **Plate 6**
Breeding Resident

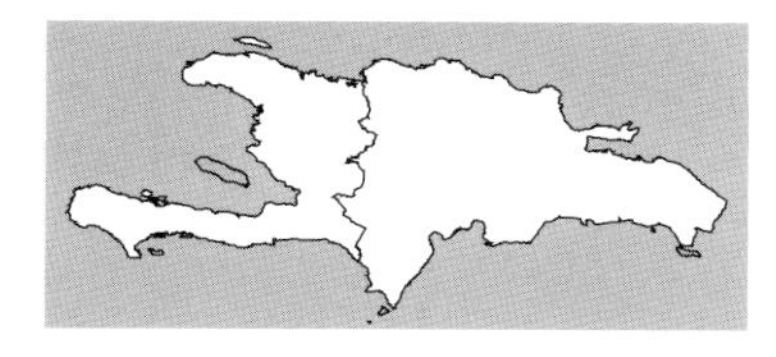

Description: 64–85 cm long; 1,200 g. A rich, dark brown booby with a whitish or dull brown belly. Adult identified by entirely brown head and upperparts which are sharply demarcated across breast from white belly; underwing coverts are white. Feet are bright yellow. Immature is uniformly dull brown above and on upper breast, sharply demarcated from mottled light brown and white belly; feet are drab yellowish. Plunge dives like other boobies, but often at lower angles and with a fanned tail.
Similar species: Immature Masked Booby has a white nape and lacks brown on upper breast. Immature Red-footed Booby is also brownish, but the demarcation between breast and belly is not well defined.
Voice: Hoarse, grunting *kak*, heard only during the breeding season.
Hispaniola: Regularly found around most of the DR coast, especially the southern and eastern portions; less often found in Haitian waters or near Haitian satellite islands. The largest Hispaniolan roost site is at Los Frailes, near Isla Alto Velo. Birds from the sizable breeding colonies on Mona and Desecheo islands regularly range into east coast DR waters. Reported from Navassa Island, Île de la Gonâve, and Islas Alto Velo, Beata, Catalina, and Saona.
Status: Uncommon breeding resident. This is the most common of the three boobies in Hispaniola and generally the most likely to be seen from shore.
Comments: Generally seen singly or in small, loose aggregations, though sometimes occurs in numbers when feeding on large fish schools. Often rests on buoys or rocky cliffs. Frequently roosts with herons or other seabirds on islands where it is thought not to breed. Flies with a series of short, unhurried flaps followed by tilting glides, often low over the sea.
Nesting: Lays 1–2 white eggs in a nest scrape on the ground. Breeding period varies but generally peaks March to June and September to October. This species has bred in varying numbers on Navassa Island since at least the early 1930s. Believed to nest on Isla Alto Velo, and a few may still breed on small islands off the northern and northeastern coasts of DR.
Range: Breeds from Bahamas south through Gulf of Mexico and Caribbean Sea to South America. Also widespread elsewhere in tropical oceans. Occurs throughout West Indies.
Local names: DR: Bubí Pardo; **Haiti:** Gwo Fou Gri, Fou Brun.

RED-FOOTED BOOBY—*Sula sula* **Plate 6**
Breeding Resident

Description: 69–79 cm long; 1,000 g. The smallest booby, with a relatively slim body, slender bill, and long, pointed tail. In all adults, feet are orange-red to red, bill grayish, and bare face patch pinkish. Two variable color phases, brown and white. Brown-phase birds are pale buffy brown on head and underparts, pale brown on back and upperwing coverts, but with contrasting white tail, rump, and undertail coverts. White-phase birds are entirely white, including tail, with black primaries and secondaries. Immature is sooty brown, including all-dark underwings; paler below, sometimes with a slightly darker band across breast. Flight is more buoyant with deeper wing beats than in other boobies.

Similar species: Distinguished from both Masked and Brown boobies at a distance by its white tail. Brown breast of immature Brown Booby is sharply demarcated from lighter belly.
Voice: Guttural *ga-ga-ga-ga*, of variable length, which trails off. Also a distinctive rattling squawk.
Hispaniola: Uncommon resident, ranging regularly into Haitian coastal waters from the colony on Navassa Island and to the southern and eastern coasts of DR and Isla Saona from the colony on Desecheo Island in the Mona Passage. This species is rarely observed elsewhere, although more intensive work at sea might yield more regular sightings.
Status: Widespread but very local year-round resident. It is abundant primarily near its remote roosting and nesting islands which are widely scattered. Away from these colonies, it is not often seen from shore.
Comments: Sometimes flies low over the sea in a V-shaped formation of several birds.
Nesting: Lays 1–2 white eggs in a loose stick nest built in a tree or bush. Breeds primarily April to June.
Range: In w. Atlantic, occurs principally in Caribbean region; also occurs widely elsewhere in tropical seas. In West Indies, resident and local breeder in Swan Islands, Little Cayman, Puerto Rico, Virgin Islands, Les Saintes, and Grenadines.
Local names: DR: Bubí de Patas Coloradas, Bubí Patas Rojas; **Haiti:** Fou Blanc.

PELICANS: FAMILY PELECANIDAE

Pelicans are among the largest of flying birds, with a long, pouched bill, short, thick legs, and long, broad wings. All pelicans are highly adapted for swimming and flying. The resident Brown Pelican prefers coastal waters. Pelicans are often seen flying low across the water in a stately formation, and they may hunt cooperatively. Wing beats are deep and powerful; flight often features gliding and soaring. The large, extensible throat pouch, used for fishing, is generally not evident. The bill typically rests on the long neck, even in flight. Pelicans are sometimes persecuted by fishermen who consider the bird a competitor for fish.

BROWN PELICAN—*Pelecanus occidentalis* — Plate 6
Breeding Resident

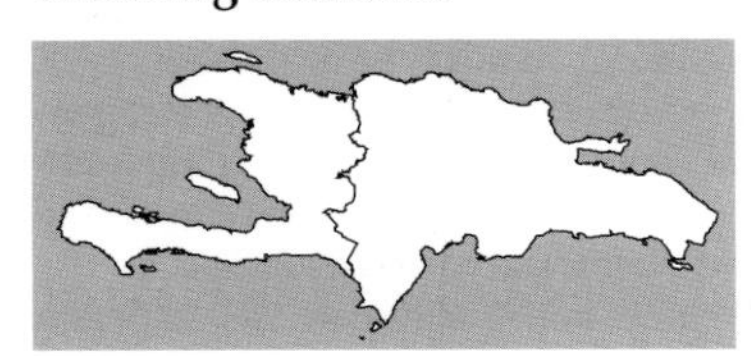

Description: 100–137 cm long; 3,400 g. A large, dark, coastal seabird with an unmistakable bill. Breeding adult has head and neck with yellowish wash on crown and a reddish brown nape and back of head, though infrequently the nape remains white. Upperparts and wings are silvery gray, underparts darker, flight feathers blackish. Nonbreeding adult has a white nape and back of head. Immature is grayish brown overall, paler below.
Similar species: Unmistakable.
Voice: Adults are generally silent. Nestlings give screeching or groaning calls.
Hispaniola: Occurs along nearly the entire coast of the main island and all of the satellite islands, occasionally gathering in flocks of more than 250 birds at favored localities such as shallow bays. Also occurs in lagoons and other protected coastal areas, and occasionally at inland freshwater reservoirs.
Status: Common year-round resident. Migrants that breed in North America augment local numbers primarily from November to February. Appears to have declined modestly since the mid-1950s, because of environmental degradation and disturbance of its nesting colonies.
Comments: A well-known denizen of bays, docks, and fishing wharfs, the Brown Pelican makes spectacular aerial dives for fish, its entire body sometimes disappearing beneath the surface. Fish are snared in its immense bill, the water drained, and the fish then swallowed

whole. Typically in small flocks, Brown Pelicans often fly in an impressively tight formation low over the wave crests with wing beats in perfect unison. Brown Pelicans fly with deliberate flaps of their huge wings and surprisingly lengthy glides. They often congregate in large mangrove roosts.
Nesting: Lays 2–4 white eggs. Nest may be on the ground or in a tree and is built of sticks, grass, leaves, rope, or other scavenged materials. Pelicans may breed at any time of year. Formerly recorded breeding at Cayos de Los Pájaros, Bahía de San Lorenzo, DR, and Isla Catalinita, DR, and Île de la Gonâve, Haiti. In the 1970s a colony of about 150 pairs nested in DR at Bayahibe, 125 at Laguna de Oviedo, and 200 at Isla Beata. In the 1980s, 25 pairs bred on a small cay near Cayo Levantado, Samaná. No breeding reported since.
Range: Breeds from Maryland south along Atlantic seaboard, and along Gulf of Mexico and Caribbean coasts to South America; also on Pacific coast of tropical Americas. In West Indies, occurs throughout, breeding in s. Bahamas, Greater Antilles east to Virgin Islands, and Antigua.
Local names: DR: Pelícano, Alcatraz; **Haiti:** Pélican, Blague-à-Diable.

CORMORANTS: FAMILY PHALACROCORACIDAE

Cormorants are large aquatic birds with a long neck, hooked bill, and long, stiff tail. Uniformly black or dark brown, they have distinctively colored bare facial skin and throat pouches. Cormorants inhabit coastal waters and inland lakes, swimming with their head angled up and their body low in the water. They perch upright, frequently with their wings horizontally stretched to dry. Expert divers, cormorants feed primarily on fish, but also crustaceans and amphibians. They are colonial breeders, often nesting with other species, and large groups often congregate to roost or feed. Flocks fly low in a straight line or V formation.

DOUBLE-CRESTED CORMORANT—*Phalacrocorax auritus* **Plate 7**
Non-breeding Visitor

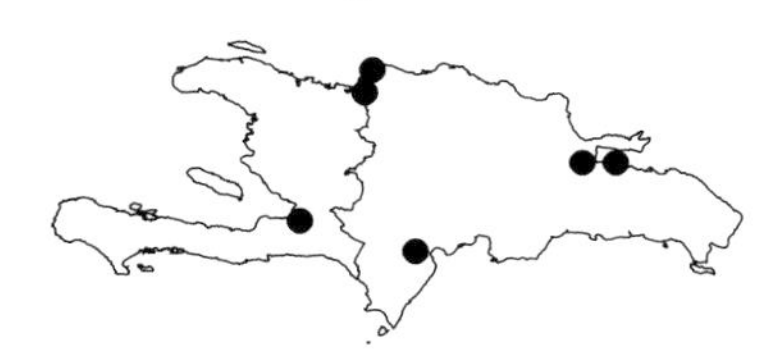

Description: 70–90 cm long; 1,700 g. A large, dark waterbird with a long neck and hooked bill; often seen perched on rocks, pilings, or trees. Breeding adult appears totally black except for bright orange-yellow skin on face and throat. Small ear tufts are sometimes visible. Non-breeding adults lack ear tufts but are otherwise similar. Immature is similar to adults but browner above and much paler.
Similar species: Neotropic Cormorant (which is a vagrant) is smaller bodied and has a longer tail, especially noticeable in flight, and smaller, more yellowish throat patch.
Voice: Generally silent away from its breeding colonies, but emits a variety of deep guttural grunts.
Hispaniola: Occurs at lakes and protected coastal waters, almost exclusively along the northern coast; to be expected August to April. Recorded at Bahía de San Lorenzo, Laguna Saladilla, Los Haitises National Park, and Monte Cristi, DR. Records from off the northern coast include 1 at Laguna de Rincón at Cabral, DR, and 1 at Lac Bois Neuf, Haiti.
Status: Uncommon non-breeding visitor. The species is presently expanding its range eastward in the West Indies.
Comments: Usually in flocks of variable size, but sometimes found singly. A strong swimmer and diver, it favors open, deeper water. Like other cormorants, it must regularly perch in the sun with wings spread out to dry its feathers, which otherwise become waterlogged. Cormorants can gradually submerge their body underwater until only the head is above the surface, or they can sink entirely out of sight.
Range: Breeds from s. Alaska east to James Bay and Newfoundland south to Baja California,

Florida, and Gulf Coast. In West Indies, breeds in Cuba and n. Bahamas; occurs in non-breeding season outside breeding range from Greater Antilles east to Virgin Islands.
Local names: DR: Corúa, Cormoran Bicrestado; **Haiti:** Kòmoran Lanmè.

NEOTROPIC CORMORANT—*Phalacrocorax brasilianus* **Plate 7**
Vagrant

Description: 63–69 cm long; 1,150 g. A dark waterbird with a long neck and hooked bill, often seen perched on rocks, pilings, or trees. Breeding adult appears totally black except for a small, yellowish patch of skin edged with white on the face and throat. Non-breeding adults are similar, but white edging of throat patch is reduced or absent. Immature is similar to adults but browner and much paler.
Similar species: Double-crested Cormorant is considerably larger and has a heavier bill, shorter tail (especially noticeable in flight), and larger and brighter yellow throat patch. Also look for feathering on the head continuing in front of the eye, and a sharper angle to the skin patch at the base of the bill on Neotropic Cormorant.
Voice: Generally silent away from its breeding colonies, but emits a guttural grunt.
Hispaniola: Recorded twice at Trou Caïman, Haiti, once in fall 2003 and once in spring 2004. Further investigation in Haiti may reveal this species to be regular in low numbers along the coast.
Status: Vagrant.
Comments: Like other cormorants, must regularly perch in the sun with wings spread out to dry its feathers, which otherwise become waterlogged.
Range: Breeds from nw. Mexico, c. Texas, and sw. Louisiana south through Middle America and all of South America. In West Indies, breeds in Cuba and Bahamas; vagrant on Jamaica, Hispaniola, Puerto Rico, and Virgin Islands, as well as St. Barthélemy and Dominica.
Local names: DR: Corúa Neotropical, Cormoran Neotropical; **Haiti:** Kòmoran Dlo Dous.

DARTERS: FAMILY ANHINGIDAE

Closely related and superficially similar to cormorants, anhingas have a longer, heron-like neck and bill and a long, fan-shaped tail. Anhingas are found mainly in freshwater lakes and rivers where they dive for fish, amphibians, and aquatic reptiles, which they skewer with their bill. On the water, they often swim with just their head and neck exposed, giving them the nickname of "snakebird." Nesting is colonial, often with other waterbirds, in trees or shrubs.

ANHINGA—*Anhinga anhinga* **Plate 7**
Vagrant

Description: 75–95 cm long; 1,200 g. A distinctive large, dark, long-necked bird with a long, white-tipped tail, pointed bill, and large silvery white patches on the upperwing and back. Adult male is mostly glossy black. Adult female's head, neck, and breast are pale brown. Immature is similar to female but more buffy brown on the head and neck, and with reduced white markings above.
Similar species: Resembles a cormorant, but neck more snakelike, tail longer, and bill longer and pointed, not hooked.
Voice: Sometimes vocal when perched, producing guttural croaking and clicking sounds.
Hispaniola: Recorded only once: 1 bird seen near Grande Saline, Haiti, on the Artibonite River. To be looked for in shallow, calm water bodies, either fresh, brackish, or saline.
Status: Vagrant.

Comments: Often soars high, hawklike, neck extended, with long tail spread. Its habit of swimming with only its slender neck and head above the surface gives the appearance of a large snake. Perches like a cormorant in a spread-wing posture to dry its feathers.
Range: Breeds from sc. U.S. south and east through Middle America and n. South America to Ecuador, e. Peru, and n. Argentina. In West Indies, resident in Cuba; vagrant elsewhere.
Local names: DR: Marbella, Corúa Real, Anhinga; **Haiti:** Aninga.

FRIGATEBIRDS: FAMILY FREGATIDAE

Large, impressive seabirds of tropical and subtropical oceans, frigatebirds have exceptionally long, narrow wings, a long, deeply forked tail, and a long, hooked bill. Highly aerial, frigatebirds have the greatest wing surface area relative to body weight of any living bird. The skeleton weighs only about 250 g. Frigatebirds cannot dive or land on the ocean as do most other seabirds, but rely on their speed and agility in the air to rob other seabirds of prey, or snatch food from the water surface with their long bill. Their piracy has earned them the name "man-o'-war" birds. During the breeding season, males develop a brilliant red throat pouch, which they can inflate to impressive proportions during displays.

MAGNIFICENT FRIGATEBIRD—*Fregata magnificens* **Plate 6**
Breeding Resident

Description: 89–114 cm long; 1,500 g. Easily identified by silhouette alone by its long, forked tail; long, slender, pointed wings sharply bent at the wrist; and habit of soaring motionless in the air. Adult males appear entirely black, but during courtship the inflatable throat pouch is bright red. Adult females are blackish overall with a white breast. Immature is dark brownish black with a white head and breast.
Similar species: Unmistakable.
Voice: Silent except for guttural noises during courtship.
Hispaniola: Occurs along the entire coast of the main island and its satellites.
Status: Regular and moderately common but somewhat local resident, perhaps somewhat less common in Haiti. Breeding grounds may serve as year-round roosts, but non-breeding birds may establish roosts some distance from the breeding grounds and sometimes wander great distances to forage.
Comments: Notorious for stealing fish from other seabirds. Frigatebirds do not rest on the water; consequently, they return each evening to a roost, often among mangroves or on an offshore islet. They occur both singly and in large flocks of 100 or more birds. Frequently seen soaring in one spot at moderate heights with minimal wing movement as they search for prey.
Nesting: A single egg is laid in a stick nest built in a low bush. The breeding season is variable but may last from August to April with a peak from November through February. It has been found nesting on Cayos de Los Pájaros, Bahía de San Lorenzo, DR; Navassa Island; and Isla Saona. The only recorded nesting in Haiti occurred on Frégate Island off the eastern end of Île de la Gonâve in 1927.
Range: Widespread on both coasts of tropical Americas. Occurs throughout West Indies.
Local names: DR: Tijereta, Fragata Magnifica; **Haiti:** Sizo, Frégate.

BITTERNS AND HERONS: FAMILY ARDEIDAE

Most of the members of this large, cosmopolitan family of graceful birds wade in shallow water in search of fish, frogs, and other prey which they spear with their long, dagger-shaped bills. All are characterized by a long neck and long legs with long, slender toes. Wings are broad and rounded, and the short tail is concealed at rest by the closed wings. All fly with the neck drawn back in the form of an S and the legs projecting beyond the tail. Plumages vary from cryptic blacks, browns, and buffs (bitterns) to entirely white (some egrets). Many species have elongated plumes on their head, neck, or back for courtship displays during the breeding season. Sexes are generally similar. Most ardeids (except bitterns) nest and roost in mixed-species colonies and they can be seen flying in long, streaming flocks to and from the colonies at dawn and dusk. In addition to aquatic prey, herons feed on terrestrial vertebrates, including small mammals and nestling birds.

AMERICAN BITTERN—*Botaurus lentiginosus* **Plate 7**
Vagrant

Description: 60–85 cm long; 706 g. A stocky, cryptically plumaged heron with a short neck. Brown above flecked finely with black, coarsely streaked brown and white below, with a black stripe along the side of the neck. This mark, and its habit of "freezing" with bill pointing upward, are diagnostic. Immatures lack the black neck mark. In flight, blackish primaries and secondaries contrast strongly with lighter brown underwings and body.
Similar species: Immature night-herons are more grayish brown, show less contrast between flight feathers and body, lack black on the neck, and have shorter bills. Immature Green Heron is smaller, darker, and lacks black neck mark.
Voice: When flushed, may give a hoarse, grunting *wok* or *coc*.
Hispaniola: To be looked for in thick, emergent vegetation of freshwater swamps. Since 1973 there are only 3 DR records, from Laguna de Rincón at Cabral and Lago Enriquillo.
Status: Known only as a vagrant, though probably underrecorded because of its secretive nature.
Comments: Usually goes undetected until flushed. Similarity to immature night-herons and propensity for remaining motionless in dense aquatic vegetation result in it being easily overlooked.
Range: Breeds from se. Alaska east to n. Manitoba and Newfoundland, south to s. California, and east through c. Missouri to Virginia, locally in Texas, Louisiana, Florida, and c. Mexico. In West Indies, ranges seasonally to Cuba, Bahamas, Cayman Islands, Greater Antilles east to Virgin Islands; casual in Lesser Antilles.
Local names: DR: Martinete, Ave Toro, Guanabó Rojo; **Haiti:** Makwali Ameriken.

LEAST BITTERN—*Ixobrychus exilis* **Plate 7**
Breeding Resident

Description: 28–36 cm; 86 g. The smallest heron, with a distinctive bright buffy overall coloration. Upperparts are greenish black; neck and underparts pale buff; belly and undertail coverts white; wings chestnut with a large, cream-colored patch on the upperwing. Bill is thin and yellow, and legs are yellowish. Adult male is identified by the black crown and back; adult female has a dark brown crown and back. Immature resembles female but is paler brown above and heavily streaked on the back and breast.

Similar species: Small size and buffy color are distinctive. Green Heron is larger and lacks chestnut-colored wing.
Voice: Breeding "song" is a series of low, descending cooing notes, *koo-koo-koo-koo*, accelerating slightly. Gives a loud, harsh *kak* or *kak-kak-kak* when flushed.
Hispaniola: Found along rivers, lake shores, and lagoons in the lowlands where rushes, cattails, or mangroves occur. Nonmigratory and sedentary, and there is no evidence of any influx of northern migrants in the non-breeding season. Apparently absent from eastern DR (i.e., east of 69°30′ W longitude), it continues to be moderately common in appropriate habitat, often more numerous than suspected because of its retiring habits.
Status: Fairly common and widely distributed breeding resident.
Comments: Solitary. Unless flushed it usually goes unnoticed among dense marsh vegetation where it often stands motionless.
Nesting: Lays 2–5 bluish white eggs in a nest of twigs and leaves built over water in reeds or mangroves. Breeds May to August.
Range: Breeds from s. Oregon east through s. Manitoba to New Brunswick, south through Middle America to coastal Peru and east to n. Argentina, but populations between Mississippi River and Pacific coast are discontinuous. In West Indies, breeds in Greater Antilles, and in Lesser Antilles at Guadeloupe; casual at Dominica and Barbados.
Local names: DR: Martinetito, Martinetico, Ave Toro Menor; **Haiti:** Makwali, Krabye Mang Lanmè.

GREAT BLUE HERON—*Ardea herodias* **Plate 8**
Breeding Resident; Non-breeding Visitor

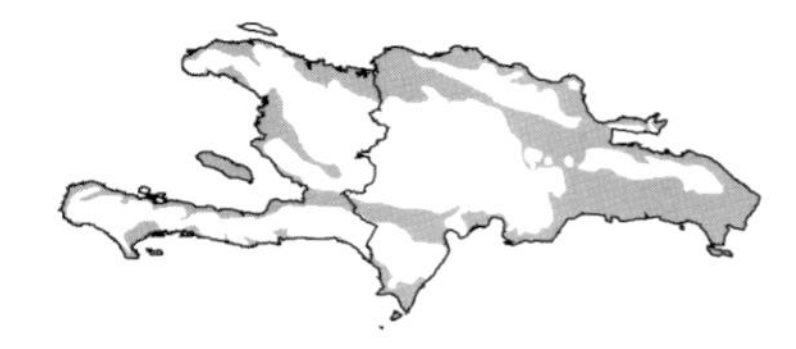

Description: 97–137 cm long; 2,400 g. The largest regularly occurring heron on Hispaniola. There are two color phases. The much more common dark phase is primarily bluish gray with a large, heavy bill and broad, black supercilium. Neck feathers are gray with a violaceous tinge, and there is black streaking on the foreneck. Thighs are feathered chestnut, and legs are brownish green. A rare white phase, considered by some to be a separate species called Great White Heron (*Ardea occidentalis*), is entirely white with yellow bill and legs. Dark-phase immatures have entire crown black and overall plumage more grayish brown. Flies with deep, slow wing beats.
Similar species: White form of Great Blue Heron is distinguished from Great Egret by its larger size and yellowish legs. Although only hypothetically recorded from Hispaniola, Gray Heron (*Ardea cinerea*) is to be expected and is slightly smaller with a pale gray neck and white feathers on thighs.
Voice: Deep, throaty croak, *guarr* or *braak*, like a large frog, often repeated 3–4 times.
Hispaniola: Occurs in both saltwater and freshwater ponds and lagoons, primarily in lowlands, breeding sparingly. Most often noted in coastal areas, especially near river mouths, but also found routinely at large inland lakes such as Étang Saumâtre, Lago Enriquillo, Laguna de Rincón at Cabral, and Laguna de Oviedo, and in rice-growing areas inland. Recorded from most major satellite islands, including Île à Vache, Île de la Gonâve, Cayos Siete Hermanos, and Islas Beata, Catalina, and Saona. Band recoveries suggest a substantial influx of migrants from the north.
Status: Fairly common breeding resident and non-breeding visitor islandwide. The islandwide population has apparently remained fairly stable since the beginning of the 20th century.
Comments: A shy, solitary species. This heron often stands motionless in search of prey.
Nesting: Lays 2–4 bluish eggs in a nest of sticks high in a tree. Usually breeds April to August. Since the 1970s, nesting is known only at Laguna de Oviedo. The few documented breeding records suggest that a significant number of summering birds may be non-breeders from North America and that the local breeding population does not usually nest colonially.
Range: Breeds from s. Alaska east to maritime Canada and south to s. Mexico. Occurs throughout West Indies but breeds only in Greater Antilles.

Local names: DR: Garzón, Garza Ceniza, Garzón Cenizo, Garcilote; **Haiti:** Gwo Krabye Ble, Lajirond, Crabier Bleu.

GREAT EGRET—*Ardea alba* **Plate 8**
Breeding Resident; Non-breeding Visitor

Description: 94–104 cm long; 870 g. A large, slender, entirely white egret with a yellow bill and black legs.
Similar species: Rare white form of Great Blue Heron is somewhat larger, with heavier bill and yellowish legs. Cattle Egret resembles Great Egret in coloration but is much smaller, stockier, and prefers drier habitat.
Voice: Deep, hoarse, throaty, drawn-out croak, *karrr* or *ahrr-rr*.
Hispaniola: Occurs principally at lower elevations in large freshwater and saltwater swamps, grassy marshes, riverbanks, and turtlegrass beds in shallows behind reefs. Recorded from most major satellite islands, including Île de la Gonâve, Île à Vache, and Islas Alto Velo, Beata, and Saona. Banding returns indicate that a significant number of migrants come from mainland North America in the non-breeding season.
Status: Originally a common and widespread breeding resident and non-breeding visitor throughout the entire island. During the plume hunting era (circa 1900–1910) its population was decimated, and it was considered rare until 1930. From then until the 1950s the population recovered, though not to its former abundance, but it began to decline again because of habitat disturbance, conversion of marshes to rice fields, and possibly pesticide poisoning. The species has again recovered and is now common to abundant at favored localities.
Nesting: Lays 2–3 bluish green eggs in a loose platform nest of sticks built on a tree branch or in a shrub. This species usually nests colonially in mangroves, as at Bucán de Base, or in mixed colonies with other herons, as at Laguna de Oviedo and in Los Haitises National Park.
Range: Breeds from s. Oregon locally east to Manitoba and Maine and in coastal and s. U.S. to s. Argentina and Chile. Also widespread in Eastern Hemisphere. In West Indies, regular in Greater Antilles, Bahamas, and most of Lesser Antilles.
Local names: DR: Garza Blanca, Garza Real; **Haiti:** Gwo Krabye Blanch, Crabier Blanc, Grande Aigrette Blanche.

LITTLE EGRET—*Egretta garzetta* **Plate 8**
Vagrant

Description: 55–65 cm long; 500 g. A small, slender egret with entirely white plumage. Breeding adult has two long head plumes, stringy breast plumes, nearly straight back plumes, black legs with bright yellow feet, and yellow-orange lores. Non-breeding adults and immatures have yellowish extending onto lower legs, and dark lores.
Similar species: Very similar to Snowy Egret, distinguishable only with great care. Head plumes of breeding adult Little Egrets are longer, back plumes are straighter; non-breeding birds have dark lores and darker legs. Cattle Egret is smaller, more stocky, with a thicker neck.
Voice: Hoarse, croaking *kark*.
Hispaniola: One record of a single bird seen near the coast near Laguna de Rincón at Cabral in April 2000.
Status: Vagrant.
Range: Breeds locally in s. Europe, Africa, s. Asia, and Japan south to Australia. Occurs in non-breeding season primarily in s. Asia, Africa, and Australia; accidental in Quebec, Newfoundland, Nova Scotia, New Hampshire, Massachusetts, Virginia, Trinidad, Suriname, and Bermuda. Regular in West Indies and has bred in Barbados; vagrant in Puerto Rico, St. Lucia, Guadeloupe, Martinique, and DR.
Local names: DR: Garza Chiquita; **Haiti:** Ti Krabye Etranje, Petite Aigrette.

SNOWY EGRET—*Egretta thula*

Plate 8

Breeding Resident; Non-breeding Visitor

Description: 56–66 cm long; 371 g. A medium-sized, slender, entirely white heron. Note its black legs, yellow feet and lores, and thin, pure black bill. Breeding-plumaged birds have elongated plumes on head, neck, and back. Legs of immature are dark in front and greenish yellow in back. Very active while foraging, running and jumping to seize prey, and often stirring bottom sediments with feet.

Similar species: Immature Little Blue Heron has a bicolored bill, lacks yellow on lores, and has greenish legs.

Voice: Hoarse, rasping *guarr* or *raarr*, higher pitched and more nasal than Great Egret.

Hispaniola: Frequents coastal and interior wetlands of all types islandwide, principally in the lowlands. Slightly less common in parts of Haiti and in the three easternmost DR provinces than in other lowland parts of Hispaniola. Has been recorded on most of the major satellite islands, including the Cayemites, Île à Vache, Île de la Gonâve, and Islas Beata and Saona. Band recoveries indicate presence of substantial numbers of non-breeding visitors from North America.

Status: Commonly occurring breeding resident, with counts of 100–300 at favored localities such as Lago Enriquillo not unusual. Like many other waterbirds, was common to locally abundant until hunting for its filamentous nuptial plumes decimated the species. Population recovery began in the 1950s, then numbers declined somewhat, but the species has since become common again, although still less numerous than Great Egret.

Nesting: Lays 2–5 greenish blue eggs in a simple stick nest. This species frequently nests in colonies, sometimes with other heron species, and often in mangroves. Nests reported in DR at Cumayasa, Laguna de Oviedo, Laguna Salada, and Laguna de la Piedra near Monte Cristi, and on islands in Los Haitises National Park. Breeds April to July.

Range: Breeds from n. California locally east to South Dakota and Maine, south along coasts to Argentina and Chile. In West Indies, regular in Bahamas, Greater Antilles, and northern Lesser Antilles south to Guadeloupe; casual elsewhere.

Local names: DR: Garza de Rizos, Garceta Nivea; **Haiti:** Zegrèt Blan, Aigrette Blanche.

LITTLE BLUE HERON—*Egretta caerulea*

Plate 8

Breeding Resident; Non-breeding Visitor

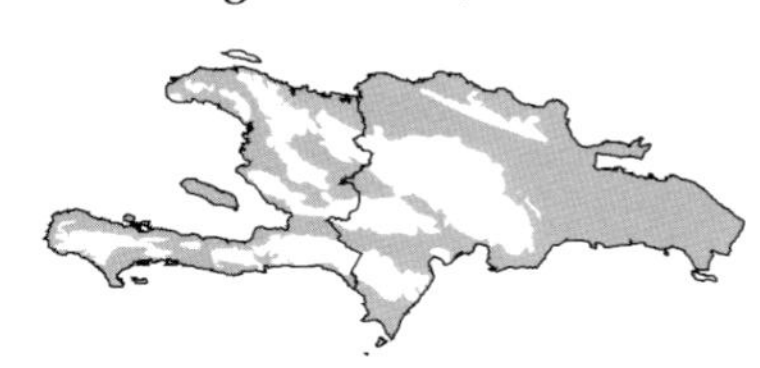

Description: 56–74 cm long; 340 g. A medium-sized, uniquely plumaged heron with all-white immatures and all-dark adults. Adult identified by its uniformly dark slaty blue-gray body and wings and purplish brown head and neck. Immature is initially entirely white; late in first year it becomes mottled with the dark feathers of adult plumage. In all plumages bill is bluish gray with a black tip, and legs are pale, dull green. A relatively inactive feeder, it moves slowly or stands motionless.

Similar species: White immature strongly resembles Snowy Egret but is distinguished by pale gray base of bill, lack of yellow on lores, and greenish legs.

Voice: Hoarse, complaining, drawn-out *rraa-aahh* or *gruuh-uhh*.

Hispaniola: Occurs at lakes, rivers, marshes, and coastal lagoons islandwide, principally in the lowlands. But this species also ranges inland to higher elevations than most other herons anywhere there is water. Has been found on most major satellite islands, including the Cayemites, Île à Vache, Île de la Gonâve, Cayos Siete Hermanos, and Isla Beata. Banding returns indicate that a substantial number of migrants from North America arrive in Hispaniola in the non-breeding season.

Status: Common and occasionally abundant breeding resident and non-breeding visitor. Although it is often secretive and solitary, numbers in the range of 10–20 can typically be found now at favored localities in a day. This species was not hunted for its plumes, so it remained common throughout the 20th century.
Nesting: Lays 3–4 greenish eggs in a platform nest of twigs constructed high in a tree. There are reports of nesting from San Lorenzo Bay, Laguna de Oviedo, and Bucán de Base. Breeds April to July.
Range: Breeds from s. California and Maine south along coasts, and in s. Missouri and se. U.S., south to Peru and Uruguay. Found throughout West Indies.
Local names: DR: Garza Azul, Garza Blanca (immature), Garza Pinta, Garceta Azul; **Haiti:** Ti Krabye Ble, Metis, Krabye Blanch, Crabier Blanc (immature), Aigrette Bleue.

TRICOLORED HERON—*Egretta tricolor* **Plate 8**
Breeding Resident; Non-breeding Visitor

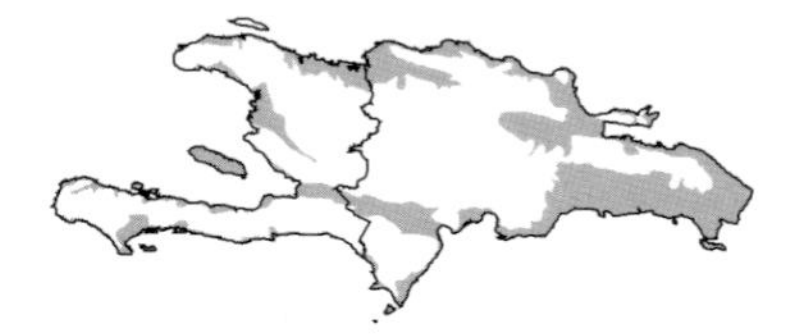

Description: 60–70 cm long; male 415 g, female 334 g. An active heron with a very long bill and neck. Adult distinguished by its slaty bluish gray head, neck, and upperparts, contrasting with a white belly, undertail coverts, and underwing coverts. Throat and foreneck are white with a narrow median stripe of chestnut flecking. Immature has a chestnut nape and upperwing coverts. Bill is generally yellowish with a dark tip but turns bluish with a dark tip in breeding birds. Legs are usually greenish yellow but turn pinkish red in breeding birds. Forages actively, singly or in loose groups, by running and chasing; occasionally forages by stealthy wading.
Similar species: Little Blue Heron and Reddish Egret have dark bellies, entirely dark neck, and more conspicuously bicolored bill.
Voice: Guttural, drawn-out *aahhrr* or *guarr*, similar to call of Snowy Egret.
Hispaniola: Occurs in mangrove swamps, lagoons, river mouths, and borders of major inland lakes and marshes, but only in the lowlands and most frequently on or near the coast. Also found on most of the major satellite islands, including the Cayemites, Île à Vache, Île de la Gonâve, and Islas Beata and Saona. Band recoveries indicate that North American migrants augment the resident population in the non-breeding season.
Status: Common breeding resident and non-breeding visitor islandwide. Counts of 10–20 in a day are not unusual.
Nesting: Lays 3–4 bluish eggs in a platform nest of sticks built on a tree limb. Typically nests in colonies, often in mangroves. Reports of nests from February to August at Laguna Salada, DR, on islands near Los Haitises National Park, and at Laguna de Oviedo and Bucán de Base, DR.
Range: Breeds on Atlantic coast from Maine to Florida, Gulf of Mexico coast, Caribbean coast to ne. Brazil, and Pacific coast from Sonora to Peru. In West Indies, breeds in Bahamas, Greater Antilles, Providencia, and San Andrés islands; birds wander south through Lesser Antilles in non-breeding season.
Local names: DR: Garza Pechiblanco, Garza Tricolor, Garceta Tricolor; **Haiti:** Krabye Vant Blanch, Aigrette Tricolore.

REDDISH EGRET—*Egretta rufescens* **Plate 8**
Breeding Resident

Description: 70–80 cm long; 450 g. A medium-sized heron with both dark and white color phases. The dark form is slate gray overall, with head and neck reddish brown. White-phase birds are entirely white. The long, heavy bill is bicolored in both forms and is pinkish with a black tip. Legs and feet are slaty blue-gray. Plumes on the head and neck on breeding birds

impart a ruffled, shaggy appearance. Dark-form immature is a unique pale, buffy gray color overall, with a pinkish cinnamon tint on the neck; white-form immature is white. Immatures of both color phases have entirely dark bills, lores, and legs; neck feathers are unruffled. The distinctive foraging behavior includes an energetic, dashing pursuit of prey, often with wings spread to form a canopy.

Similar species: Adult Little Blue Heron has smaller, black-tipped bill which is blue-gray at the base, not pink. Immature Little Blue Heron is similar to white-phase Reddish Egret but also has a gray, rather than pink, bill.

Voice: Low, nasal groaning or grunting, similar to that of Tricolored Heron.

Hispaniola: Resident of coastal lagoons, brackish ponds, salt ponds, and bays islandwide, but somewhat unevenly distributed. DR localities where it is usually found include Las Salinas, Cabo Rojo, Laguna de Oviedo, Puerto Alejandro, and Monte Cristi; also found inland at Lago Enriquillo. Status in Haiti is uncertain as few reports are available. There is no evidence of an influx of northern migrants in the non-breeding season. Among the major satellite islands it is recorded only from Isla Alto Velo.

Status: Locally common breeding resident. Appears to be more common today than in 1930s, particularly in the DR where it is not unusual to find 3–5 birds in a day at favored localities.

Comments: White-plumaged form is much more common than reddish-phase birds, based on DR records, unlike in mainland North American and Cuban populations, in which white form is much rarer.

Nesting: Lays 2–5 pale greenish blue eggs in a platform nest of sticks and leaves, usually in a colony with other herons and most often in mangroves. Breeds December to April.

Range: Breeds in coastal Baja California and Sonora south to Oaxaca and coastal Florida west to coastal Texas, Yucatán Peninsula, nw. Bahamas, Cuba, and Hispaniola; formerly Jamaica. In West Indies, occurs in non-breeding season in breeding range and east to Puerto Rico.

Local names: DR: Garza Rojiza, Garceta Rojiza; **Haiti:** Zegrèt Ble, Aigrette Roussâtre.

CATTLE EGRET—*Bubulcus ibis* **Plate 8**

Breeding Resident; Non-breeding Visitor

Description: 48–56 cm long; 338 g. A stocky, white, short-legged, and thick-necked heron. Distinguished by its relatively small size and short, thick, yellowish bill, it is the only heron regularly occurring in the uplands. Breeding individuals have reddish legs and eyes, a reddish tinted bill, and a buffy orange wash on the crown, breast, and upper back. Non-breeding birds have black legs and yellow bill; the buffy orange wash is reduced or absent. Immatures have a blackish bill.

Similar species: Much larger Great Egret is aquatic and longer necked, with a longer, more yellow bill.

Voice: Generally silent away from breeding colonies, where it emits short, grunting croaks or clucks.

Hispaniola: Found virtually throughout the island except at the highest mountain elevations. Also reported from Navassa Island, Île à Vache, Île de la Gonâve, and Isla Saona. Band recoveries indicate some migrants come from North America in the non-breeding season, but they are indistinguishable from local birds. Most non-breeding visitors are believed to arrive in September or October and to leave in March or April.

Status: Common to abundant breeding resident and non-breeding visitor. A fairly recent colonist of the West Indies, it was first recorded in Haiti and the DR in 1956 and rapidly became common and widespread.

Comments: At dawn and dusk these herons fly in formation to and from their roosts. Whereas other herons are rarely found away from water, Cattle Egret seldom visits water except to drink. This species is commonly seen in small groups foraging alongside cattle in pastures. Cattle Egret invaded South America from Africa in the 1870s and rapidly expanded through the Americas, arriving in the West Indies in the 1950s. As it colonized the West Indies, it apparently spread the African cattle tick through the islands. Cattle Egret colonies appear to attract other colonial waterbirds to nest, possibly deterring predators through the establishment of noisy, conspicuous colonies.
Nesting: Lays 2–3 pale greenish eggs in a platform of twigs built in a tree. Nests in suitable locations islandwide, especially in colonies in mangroves. Breeds primarily April to July but also in other seasons.
Range: Breeds from n. California east to s. Saskatchewan, Ontario, and Maine, south to Argentina and Chile. Widespread in Eastern Hemisphere where it originated. Found throughout West Indies.
Local names: DR: Garza Ganadera, Garceta Bueyera; **Haiti:** Krabye Gad-bèf, Pècho.

GREEN HERON—*Butorides virescens* **Plate 8**
Breeding Resident, Non-breeding Visitor

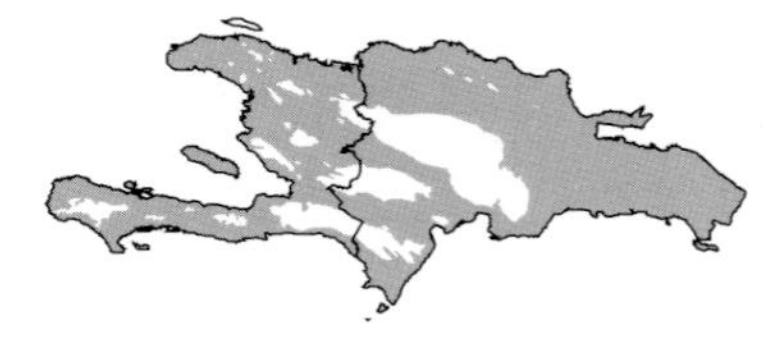

Description: 40–48 cm long; 212 g. A small, dark, compact heron. Adult is blue-gray overall with a glossy greenish black cap and back. Neck is dark chestnut with a narrow white stripe from the throat down the foreneck. Wings are blackish, and coverts are edged with buff. Belly is gray. Legs may be greenish yellow to orangish but turn bright orange in breeding adults. Immatures have a duller, brownish back and are heavily streaked below.
Similar species: Least Bittern is smaller, with a conspicuous cream-colored wing patch. Immature Green Herons differ from American Bittern in being smaller, darker, and lacking a black neck mark.
Voice: When flushed, gives a distinctive, piercing *skeow*. When undisturbed, issues an irregular series of low, clucking notes, *kek kek kek, kuk kuk kuk*.
Hispaniola: Most abundant in the coastal lowlands, but routinely found in the interior wherever there is water, and even occasionally at fairly high elevations in the Cordillera Central and Sierra de Bahoruco. Also reported from the Cayemites, Île à Vache, Île de la Gonâve, and Islas Beata and Saona. Occurrence of non-breeding visitors from North America is indicated by specimens.
Status: Common breeding resident and non-breeding visitor islandwide. Appears to have been common continuously since the 1930s.
Comments: This is the small, dark, common heron of Hispaniolan wetlands. Usually solitary, it is often discovered by its piercing call which it issues when disturbed. It typically perches motionless on a snag low over the water awaiting unwary prey. Feeds on fish, insects, and terrestrial vertebrates.
Nesting: Lays 2–3 glaucous green eggs in a platform of twigs built at variable heights in a tree or bush. Usually breeds March to September. This species does not usually nest in colonies and will use a variety of trees, including mangroves.
Range: Breeds from sw. British Colombia east to Ontario and New Brunswick, south to Panama. Found throughout West Indies.
Local names: DR: Cra-Crá, Martinete, Cuaco; **Haiti:** Krakra, Rakrak, Ti Krabye Riviè, Valet de Caïman.

BLACK-CROWNED NIGHT-HERON—*Nycticorax nycticorax* **Plate 7**
Breeding Resident; Non-breeding Visitor

Description: 58–66 cm long; 883 g. A medium-sized, stocky, black, white, and gray heron with a large head and short neck. Adult is identified by its black crown and back; gray wings,

rump, and tail; and white face, underparts, and head plumes. The eyes are red. Immature is brown with white flecks on the back; underparts have broad, blurry, pale brown streaks. Bill is black in adult, dusky yellow in immature. Legs are greenish yellow on birds of all ages. In flight, only the feet extend beyond tail. Usually stands with neck hunched.

Similar species: Immature is distinguished from similar immature Yellow-crowned Night-Heron by its browner appearance, larger white flecks on wings and upperparts, more blurred streaks on underparts, thinner and paler bill, and shorter legs. Limpkin is distinguished from immature Black-crowned Night-Heron by its larger size and longer, paler, and slightly down-curved bill.

Voice: Distinctive, barking *quok* or *quark*, heard before sunup and after sundown and often given in flight.

Hispaniola: Found almost exclusively in the lowlands but occasionally inland at higher elevations. Primarily frequents freshwater swamps, although also found at brackish lagoons and salt ponds. Among satellite islands, reported from Île à Vache, Île de la Gonâve, and Isla Saona.

Status: Uncommon breeding resident and non-breeding visitor. As the name suggests, night-herons are active primarily at night, roosting in dense vegetation during the day, which makes population estimates difficult. Resident population does not appear to be large, though the species is regularly recorded throughout the year. Band recoveries suggest a significant influx of North American birds in the non-breeding season. Considered rather rare in the 1930s, it appears to have increased since that time.

Comments: Nocturnal, usually seen at dawn or dusk. Solitary, except at the breeding colony.

Nesting: Lays 2–5 pale greenish blue eggs in a simple stick nest in colonies near water, usually well up in a tree in a swamp or on a cay. Breeding season is variable but generally January to July.

Range: Breeds from c. Washington east to s. Manitoba, Quebec, and Nova Scotia south to Tierra del Fuego. Also widespread in Eurasia and Africa. In West Indies, occurs in Bahamas and Greater Antilles; uncommon to casual in Lesser Antilles.

Local names: DR: Rey Congo, Garza Nocturna Coroninegra; **Haiti:** Kòk Lannwit, Kòk Dlo, Coq de Nuit.

YELLOW-CROWNED NIGHT-HERON—*Nyctanassa violacea* **Plate 7**
Breeding Resident

Description: 55–70 cm long; 685 g. A medium-sized, stout-billed heron with a stocky appearance. Adult is mostly blue-gray, with back feathers edged light gray. Head is boldly patterned, with a black and white striped crown and white auriculars. Immature is grayish brown with fine white to buffy spotting above; underparts are narrowly streaked brown and white. In all plumages, bill is blackish, legs are yellow, and eyes are red.

Similar species: Immature Yellow-crowned Night-Heron is similar to immature Black-crowned Night-Heron but has a heavier bill, grayer plumage with smaller white flecks on wings and upperparts, more narrow and distinct streaks on underparts, and longer legs which extend farther beyond tail in flight. Limpkin is distinguished from immature Yellow-crowned Night-Heron by its larger size and longer, paler bill which is slightly down-curved.

Voice: Distinctive, squawking *quok* or *quark*, similar but higher than that of Black-crowned Night-Heron. It is most often heard before sunup and after sundown.

Hispaniola: Occurs primarily in mangrove swamps, but also along water courses and the edges of lakes and ponds, and occasionally in dry thickets away from water. Recorded from several major satellite islands, including the Cayemites, Île à Vache, Île de la Gonâve, and Isla Beata. Since the 1960s, greatest numbers have been reported from the Artibonite River delta

of Haiti and Las Salinas and the Río Culebra mouth in the DR. No evidence of seasonal migrants from North America.
Status: Common breeding resident in appropriate habitat islandwide. Appears to have increased modestly since the 1930s.
Comments: Solitary. Active primarily at night, roosting in dense vegetation during daylight. Heavy bill is adapted for eating hard-shelled crustaceans, especially crabs, but species also feeds on fish, insects, worms, and a wide variety of terrestrial vertebrates.
Nesting: Lays 2–4 pale greenish blue eggs in a bulky twig platform nest in a tree, sometimes far from water. Breeds March to July. Successful nests have been found atop large communal Palmchat nests in royal palms (*Roystonea hispaniolana*) at the National Botanical Garden, Santo Domingo.
Range: Breeds from s. Texas northeast to s. Minnesota, Michigan, and Massachusetts, south along coasts to Peru and e. Brazil. Found throughout West Indies.
Local names: DR: Rey Congo, Yaboa, Garza Nocturna Coroniamarilla; **Haiti:** Kòk Lannwit, Crabier Gris, Coq de Nuit.

IBISES AND SPOONBILLS: FAMILY THRESKIORNITHIDAE

Ibises and spoonbills are fairly large, gregarious wading birds of shallow coastal lagoons. Ibises have a distinctive decurved bill adapted for probing shallow water, mud, and grass for small animals, whereas spoonbills have a long, flattened, spatulate bill used for catching and straining floating prey out of shallow water. All have bare facial skin. Ibises and spoonbills often fly with their neck outstretched, and in long lines in V formation, alternating flapping with short glides. Breeding occurs in colonies, often with other waterbirds.

WHITE IBIS—*Eudocimus albus* **Plate 9**
Breeding Resident

Description: 56–68 cm long; male 1,036 g, female 764 g. Entirely white with black wing tips and a long, decurved reddish bill. Face and legs also red. Immature is primarily brown with a white belly and rump, and the bill and legs are more orangish.
Similar species: Glossy Ibis is entirely dark. Immature Roseate Spoonbill has a spatulate yellowish bill and lacks black wing tips.
Voice: Series of low, hoarse grunts and a nasal *oohh-oohh*.
Hispaniola: Found near the coast islandwide. Except at saltwater Lago Enriquillo, not recorded in the interior. Most numerous in the DR at Monte Cristi, Lago Enriquillo, and Laguna de Oviedo. Also reported from Île à Vache and Islas Beata and Saona.
Status: Fairly common breeding resident. Formerly considered rare because of overhunting, but today sight records of 15–25 per day are not uncommon.
Comments: Typically occurs in flocks. Species is often nomadic, disperses widely, and may change roost sites or colonies frequently.
Nesting: Lays 2–3 whitish or grayish eggs covered with many brown spots in a nest of twigs and grasses typically built in mangroves, about 3 m above water. Because nestlings are less tolerant of salt, nesting colonies must be near freshwater feeding sites. Nesting was first confirmed at Laguna de Oviedo and Laguna Salada, DR, in 1987. Breeds April to September.
Range: Breeds from c. Baja California and c. Sinaloa east to s. Texas, Gulf of Mexico coast north to coastal North Carolina, and south along both coasts of Middle America to nw. Peru

and French Guiana. In West Indies, occurs primarily in Greater Antilles; casual in Puerto Rico and Bahamas.
Local names: DR: Coco Blanco, Ibis Blanco; **Haiti:** Ibis Blanch, Ibis Blanc.

GLOSSY IBIS—*Plegadis falcinellus* **Plate 9**
Breeding Resident; Non-breeding Visitor

Description: 48–66 cm long; 550 g. An entirely dark wader, appearing black from a distance, but metallic bronze and green in good light, with a long, brownish olive decurved bill. Legs are grayish but may become reddish in breeding season. Immature is lighter and duller than adult.
Similar species: Immature White Ibis has white, rather than dark, belly and rump.
Voice: Generally silent but occasionally gives a repeated, nasal grunt, *urnn urnn urnn*, or a quacking *waa waa waa*.
Hispaniola: Occurs in coastal lowlands nearly islandwide, including mudflats, marshy savannas, and rice fields, and sparingly inland along rivers and edges of major lakes such as Laguna de Rincón at Cabral and Lago Enriquillo, DR, and Étang Saumâtre, Haiti. Rarely found at higher elevations. Formerly more common in southern DR and central western Haiti than in northern DR, but is now found routinely in northwestern coastal DR, inland in the Río Yaque del Norte basin, and in the Río Yuna drainage. Also reported from Île à Vache and Isla Beata. Peak non-breeding-season counts reflect an influx of North American birds.
Status: Fairly common breeding resident and non-breeding visitor.
Comments: Typically flocks. Apparently wanders among islands as habitat conditions warrant.
Nesting: Lays 2–4 typically dark bluish green eggs in a simple stick platform nest in a tree over water. Nesting is usually colonial. Breeds primarily March to June.
Range: Breeds on Atlantic and Gulf of Mexico coasts from Maine to Florida and Louisiana; also in nw. Costa Rica. Widespread in Eurasia, Africa, and Australasia. In West Indies, breeds in Cuba, Hispaniola, and Puerto Rico; casual to rare elsewhere.
Local names: DR: Coco Oscuro, Coco Prieto, Ibis Lustroso; **Haiti:** Ibis Pechè, Ibis Noir, Pêcheur.

ROSEATE SPOONBILL—*Platalea ajaja* **Plate 9**
Breeding Resident

Description: 71–86 cm long; 1,500 g. Extraordinary large pink wader with carmine red upperwing coverts and white neck, upper mantle, and upper chest. Head is unfeathered and whitish, with a black nape band. Tail is orangish with reddish pink coverts. Note the unique spatulalike, grayish white bill. Legs are reddish. Immature is almost entirely white but displays a pink wash and yellowish bill.
Similar species: Unmistakable.
Voice: Generally silent. On the breeding grounds, emits low, soft grunts and a rapid, dry, rasping *rrek-ek-ek-ek-ek*.
Hispaniola: Spottily distributed in shallow saltwater lagoons and edges of mudflats in the coastal lowlands islandwide and at selected inland lakes. Most common along the northwestern DR coast near Monte Cristi, shores of Samaná Peninsula, coastal Barahona and Pedernales provinces, Lago Enriquillo, and Laguna de Rincón at Cabral, DR; and coastal marshes near the mouth of the Artibonite River and Étang Saumâtre, Haiti. Also reported from Île de la Gonâve and Isla Beata.

Status: Locally common breeding resident. Numbers of 10–30 per day are routine.
Comments: Typically occurs in small flocks. Feeds by wading through shallow water sweeping its head from side to side, the bill straining out organisms from bottom sediments; sometimes the head is entirely submerged. Feeds primarily on small fish but also crustaceans, insects, and aquatic plants.
Nesting: Lays 2–4 dull white eggs. Typically nests in colonies in mangroves or trees near shore. Breeds August to December. Reported to be breeding at Bucán de Base, Laguna Salada, and Laguna de Oviedo, DR, in the 1970s; formerly bred at Isla Beata in the 1920s but not since then.
Range: Breeds from n. Sinaloa, Gulf Coast of Texas, sw. Louisiana, and s. Florida south along both Middle America coasts to c. Chile and c. Argentina. In West Indies, breeds in Greater Antilles and on Great Inagua; casual elsewhere.
Local names: DR: Cuchareta; **Haiti:** Espatil, Spatule.

STORKS: FAMILY CICONIIDAE

The storks comprise a small, mainly Old World family of long-legged birds closely related to the ibises and spoonbills. All are large, with a long neck, broad wings, and a short tail. The long, usually straight bill is stout and pointed. Storks fly with strong, slow wing beats, usually with the neck extended and the legs trailing behind, sometimes soaring high on thermals. Breeding occurs in colonies or in scattered pairs, and nesting may vary in timing from year to year depending on water levels.

WOOD STORK—*Mycteria americana* **Plate 9**
Former Breeding Resident—*Extirpated?*

Description: 100 cm long; male 2,600 g, female 2,100 g. A large, long-legged, mostly white bird with a dark, naked head and upper neck; dark, iridescent primaries, secondaries, and tail feathers; and a large bill, decurved at the tip. Adult is distinguished by its bald and blackish head and a heavy, black bill. Legs are blackish to bluish gray, feet are pink. Immature has a feathered and paler, dusky gray head with a yellowish bill. In flight, frequently soars with feet trailing beyond tail. Note the black trailing edge of the wing contrasting with the white coverts.
Similar species: Much larger and with a much heavier bill than White Ibis. Adult plumage patterns are similar, but Wood Stork's legs are blackish, not pink; its head is dark and bare; and both primaries and secondaries are black.
Voice: Usually silent, but quite vocal on the breeding grounds, giving nasal barking and hissing sounds, as well as bill clattering.
Hispaniola: Inhabited swamps, mangroves, and coastal mudflats. Also rice fields, ponds, and inland water bodies. Formerly reported as common near Neiba, DR, and other reports exist from the mouth of the Río Yuna and Isla Saona.
Status: Formerly a locally common breeding resident. No recent records exist, so this species is generally assumed to be extirpated from Hispaniola. The last report is of 1 seen in mangroves near Monte Cristi in 1968. The population was probably never large and was apparently confined to a few localities. The disappearance of the Wood Stork is likely due to excessive hunting.
Range: Breeding resident from s. Sonora, coastal Gulf of Mexico, Atlantic coast from South Carolina to Florida, south on both coasts of Middle America to w. Ecuador and n. Argentina. In West Indies, found in Cuba; casual in Jamaica, Bahamas, and Dominica.
Local names: DR: Faisán, Cocó Cigueña Americana, Cigueña; **Haiti:** Tantal Rak Bwa.

AMERICAN VULTURES: FAMILY CATHARTIDAE

This small, Western Hemisphere family contains only seven species, all of which are large birds with long, broad wings and an unfeathered head and neck. They are excellent soarers, and all feed almost exclusively on carrion. Although superficially similar to hawks, vultures have much weaker bills and feet, are better suited for walking than for grasping prey, and are now known to be more closely related to storks than to hawks.

TURKEY VULTURE—*Cathartes aura*

Plate 10

Breeding Resident

Description: 64–81 cm long; 2,000 g. The large size, blackish brown coloration, and small bare head are diagnostic. Distinctive soaring flight with dark two-toned wings held well above the horizontal in a broad V is distinctive. Tail is fairly long. Adult is distinguished by a reddish head, but this is noticeable only at close range. Immature has a blackish head.
Voice: Usually silent, but young and even adults at the nest make some noises by expelling air, producing a soft hissing whistle, clucking, and whining.
Hispaniola: Occurs in open areas, including scrublands, open forests, cane fields, pasturelands, towns, and garbage dumps, primarily in the eastern and northeastern parts of the DR, the Cordillera Central, and the southern peninsula of Haiti. Reported only rarely on the southern slope of the Sierra de Bahoruco, DR, and once from Santo Domingo, but regular in small numbers in coastal Haiti from the Artibonite River delta north to Gonaïves. Still apparently absent from northern and northwestern Haiti and most of southwestern DR.
Status: Increasingly widespread breeding resident. Probably colonized naturally from Cuba sometime after 1900, and now well established and expanding its range.
Comments: Feeds on carrion, and sometimes assembles in numbers on cloudy days. Often sunbathes on exposed perches with wings outstretched. Has remarkable senses of sight and smell, which it uses to locate carrion.
Nesting: No data from Hispaniola available, but breeding likely. In other localities, lays 1–2 whitish or grayish eggs with brown spots. Nest is typically in a shallow depression on the ground under vegetation, in a crevice among rocks, or on a cliff ledge. May breed year-round but primarily February to April.
Range: Breeds from s. British Columbia east to s. Manitoba, n. Michigan, s. Vermont, and s. Maine south to s. Tierra del Fuego. In West Indies, common in Greater Antilles, rare in Cayman Islands, vagrant in Virgin Islands.
Local names: DR: Aura Tiñosa, Maura; **Haiti:** Malfini Karanklou, Vautour.

FLAMINGOS: FAMILY PHOENICOPTERIDAE

The flamingos form a unique and spectacular family of large, social wading birds, characterized by a very long neck and legs. The distinctively thick, decurved bill is used to filter small mollusks, crustaceans, and other organisms from shallow lagoons. Their peculiar feeding behavior involves inverting the head and swinging the bill rhythmically side to side in the water, straining tiny food particles with a specially adapted tongue and bill. Flamingos are gregarious and wary. In flight, the long, thin neck is extended and the legs trail far behind the short tail.

GREATER FLAMINGO—*Phoenicopterus ruber* **Plate 9**
Non-breeding Visitor

Description: 107–122 cm long; 2,550 g. A large, slender wading bird with orangish pink coloration overall, long legs and neck, and strangely curved bill with a pronounced black tip. Immature is much paler than adult. In flight, the head and neck are outstretched and drooping; flight feathers are black.
Similar species: Roseate Spoonbill is smaller, paler, lacks black on wings, and has a longer, spatulate bill.
Voice: Distinctive gooselike honks in flight. Feeding birds in flocks often give a low, conversational gabble.
Hispaniola: Inhabits shallow lagoons and coastal estuaries with high salinity. Historically known in DR from the shores of Lago Enriquillo, the northwestern coast near Monte Cristi (including the area west to Liberté Bay and Fort Liberté, Haiti), and several localities along the coasts of Barahona and Pedernales provinces. In Haiti, recorded from the coastal mangrove lagoons between Grande Saline and Gonaïves, Étang Saumâtre, and Trou Caïman. Also reported from the satellite islands of Île de la Gonâve, Île à Vache, Île de la Tortue, and Islas Beata and Saona. A strong flyer, this species might be encountered at almost any favorable feeding locality along the coast of the entire island. Most reports since the 1970s come from Laguna de Oviedo, Bucán de Base, Lago Enriquillo, Puerto Viejo, and Laguna Salina (Monte Cristi), DR, and marshes of the Artibonite River delta, Haiti.
Status: Locally common non-breeding visitor. Its favored localities have probably changed little since the early 1700s, though its numbers have likely fluctuated substantially during that period. Flamingos on Hispaniola are largely seasonal migrants from the large breeding colony of approximately 60,000 birds at Great Inagua, Bahamas. This species was hunted islandwide from the 1600s at least to 1900. Whereas it may have been common in the early part of that period, it was probably never abundant and was confined principally to areas rarely accessed by hunters. As the human population grew, flamingos ceased to occur at places where hunting pressure and other human disturbance became severe. On balance, the species appears to have increased in abundance from 1930 to the present day. It is currently threatened by habitat loss and collection for the pet-bird trade; flamingos are frequently captured for display at artificial ponds at tourist hotels and resorts in the DR.
Comments: Typically occurs in flocks. Sometimes travels large distances in search of adequate habitat. Forages in shallow water.
Nesting: Currently no known active breeding sites. Historically, breeding has been documented at Lago Enriquillo, Bucán de Base and Laguna Salada on the Barahona Peninsula, and Isla Beata. Unconfirmed reports of nesting have been made for Laguna de Oviedo and Laguna Limón, DR, and Gonaïves, Étang Saumâtre, Liberté Bay, and Île de la Gonâve, Haiti. Most flamingo nestings in Hispaniola should probably be viewed as satellite colonies indicative of an expanding population in Cuba and the Bahama Islands.
Range: Breeding resident on Yucatán Peninsula, Bonaire, and Greater Antilles, wandering widely. Also widespread in South America, Africa, w. Asia, and Indian subcontinent. In West Indies, breeds in Cuba and s. Bahamas; casual elsewhere.
Local names: DR: Flamenco, Flamenco Mayor; **Haiti:** Flanman, Flamant Rose.

OSPREY: FAMILY PANDIONIDAE

The single species of this family breeds worldwide, except in South America (which it visits only during the non-breeding season) and Antarctica. Osprey capture fish by hovering overhead and plunging feet first with half-closed wings, often submerging completely. The Osprey is well adapted for capturing fish, having spine-studded soles on its feet and large claws for gripping its slippery prey. Partially as a result of pesticide contamination, Osprey numbers declined to dangerously low levels in the 1960s and 1970s. Many populations have recovered strongly since several of the more lethal chemicals were banned.

OSPREY—*Pandion haliaetus* **Plate 10**
Non-breeding Visitor

Description: 53–61 cm long; male 1,403 g, female 1,568 g. The distinctive flight silhouette features long, narrow wings angled at the wrist and bowed down. In all plumages, birds appear white below with a dark wrist patch. Adults are uniformly dark brown above with white breast and belly. The more widespread migratory race (*P. h. carolinensis*) has a white crown and forecrown with a dark bar through the eye. The race resident in Bahamas and Cuba (*P. h. ridgwayi*) has a whiter head with only a trace of an eye-line.
Similar species: None.
Voice: Series of short, piercing whistles.
Hispaniola: Inhabits coastal lagoons, rivers, and major lakes and ponds islandwide. Recorded from several satellite islands, including Navassa Island, Île à Vache, Cayos Siete Hermanos, and Islas Alto Velo, Beata, and Catalina. Found in summer several times: 1 at Barrage de Péligre in central Haiti, 1 at Isla Beata, and 1 at Laguna de Rincón at Cabral.
Status: Regular and sometimes relatively common non-breeding visitor, occasionally lingering through summer but not yet proven to breed.
Comments: Both races sometimes breed in mixed pairs on Cuba. This is the only fish-eating hawk on Hispaniola.
Nesting: Nesting on Hispaniola has been suspected several times but never proven; a large nest probably built by this species was found at Bucán de Base, DR, in 1977, but there is no documentation that eggs were laid.
Range: Breeds from nw. Alaska east to Newfoundland and south to Baja California and Yucatán Peninsula. Occurs in non-breeding season from California, Texas, and Louisiana south to Argentina and Chile. Nearly cosmopolitan outside Western Hemisphere. In West Indies, breeds in Bahamas and Cuba; a non-breeding visitor elsewhere.
Local names: DR: Guincho, Aguila Pescadora; **Haiti:** Malfini Lanmè, Malfini de Mer.

KITES, HAWKS, AND ALLIES: FAMILY ACCIPITRIDAE

Members of the Accipitridae occur on all continents except Antarctica and have even reached many of the most isolated oceanic islands. These diurnal predators soar more frequently than falcons and are further distinguished by their rounded, rather than pointed, wings. Females are larger than males. Diets vary from small mammals and birds to insects and snails. Most species are persecuted because of their reputation for taking poultry or other livestock. In fact, because these raptors also prey on rats, mice, and insect pests, their economic value far outweighs any losses resulting from their occasional depredation of domestic stocks.

SWALLOW-TAILED KITE—*Elanoides forficatus* **Plate 10**
Vagrant

Description: 50–64 cm long; 442 g. An unmistakable, bicolored kite with long, pointed wings and a deeply forked tail. White head and underparts contrast with black back, wings, and tail. Flight is graceful and flowing.
Similar species: None.
Voice: High, shrill, whistled *ke-wee-wee, hewee-we,* the first note short.
Hispaniola: Occurs in coastal swamps, savannas, and river mouths. One report of 2 birds among 60 Osprey in August 1999 arriving on north winds at Gonaïves Bay, Haiti. More likely to occur migrating southward from August to October than during its northward migration from February to June, but one photographed near Punta Cana in spring 2004.
Status: Vagrant.
Comments: When hunting, glides slowly, low to the ground, with steady wings but with tail constantly balancing. Feeds mainly on reptiles, frogs, and insects. Often consumes prey in flight. In migration, often soars more than 500 m above ground, frequently in small flocks.
Range: Breeds from South Carolina south to Florida and west to Louisiana, and from se. Mexico south to e. Peru and n. Argentina. Occurs in non-breeding season in South America. In West Indies, regular passage migrant in Cuba; vagrant in Bahamas, Cayman Islands, Jamaica, and Haiti.
Local names: DR: Gavilán Cola de Tijera; **Haiti:** Malfini Ke Chankre.

NORTHERN HARRIER—*Circus cyaneus* **Plate 10**
Non-breeding Visitor; Passage Migrant?

Description: 46–53 cm long; male 358 g, female 513 g. A large, slender hawk with long wings and tail, long legs, and a distinctive white rump and facial disk. Adult male is pale gray above, whitish below, with chest and sides flecked reddish brown. Adult female is dark brown above and white below, with underparts heavily streaked with brown. Immature is brown above, entirely reddish brown below with dark brown streaks limited to upper breast. Flight behavior is characteristic: usually flies low over fields or marshes with an erratic series of heavy flaps and distinctive tilting glides, the wings uptilted. Soars only occasionally.
Similar species: Buteos have a heavier build and broader wings, and often soar at fairly high altitudes.
Voice: Generally silent except at nest.
Hispaniola: Occurs in marshes, swamps, open savannas, and rice fields. Reported from DR near Sosúa, Laguna de Rincón at Cabral, Sabana de la Mar, over rice fields near El Valle, and at Laguna Limón. Reported from Haiti at Trou Caïman, Saint-Michel de L'Atalaye, and Port-au-Prince; also from Île de la Tortue.
Status: Rare non-breeding visitor and possible spring passage migrant to both countries.
Range: Breeds from n. Alaska east to Newfoundland, south to n. Baja California, s. Missouri, and Virginia. Occurs in non-breeding season from s. Canada south to Colombia and Venezuela. Also widespread in Eurasia. In West Indies, a regular migrant and non-breeding visitor to Cuba and Bahamas, uncommon to rare in rest of Greater Antilles, rare in Lesser Antilles.
Local names: DR: Gavilán Sabanero, Gavilán de Ciénaga; **Haiti:** Gwo Malfini Savann, Busard Saint-Martin.

SHARP-SHINNED HAWK—*Accipiter striatus* **Plate 11**
Breeding Resident—*Threatened*

Description: Male 24–27 cm long, 103 g; female 29–34 cm long, 174 g. A small forest hawk with short, rounded wings; relatively small head; and long, narrow, squared-off tail, boldly barred with

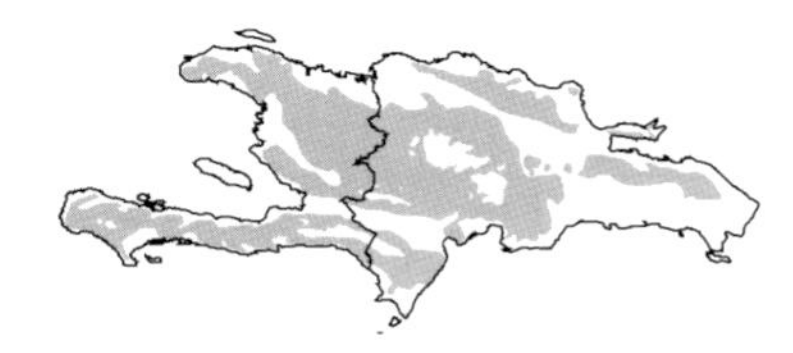

black. Adult birds are dark steel blue above with narrow reddish bars below. Females are much larger than males. Immatures are brown above; buffy below, streaked with dark brown. In flight, the short, rounded wings and long, narrow tail are characteristic. Sharp-shinned Hawks fly rapidly through the forest with alternate bursts of shallow wing flaps and glides.

Similar species: Adult Broad-winged Hawk has a similar color pattern but is much larger, chunkier, and with shorter tail and broader wings. Ridgway's Hawk has a dark, fan-shaped tail with thin whitish bars.

Voice: Series of short, sharp, high-pitched notes, *que-que-que-que*.

Hispaniola: Principally inhabits mature forests in interior hills and mountains; seldom seen near sea level. Most often confined to pine, shade coffee, and broadleaf forest from 300 to 1,465 m in Cordillera Central, Sierra de Neiba, and Sierra de Bahoruco, DR; also occurs in Macaya range of Haiti.

Status: Uncommon and increasingly local resident. The Hispaniolan race of Sharp-shinned Hawk is threatened; since the 1930s its numbers have declined noticeably because of habitat destruction and incidental shooting, despite legal protection. Current status in Haiti is poorly known. The subspecies *A. s. striatus* is endemic to Hispaniola.

Comments: In early morning, performs soaring display flights during courtship. Feeds almost exclusively on small birds.

Nesting: Lays 3–4 light brown eggs spotted with dark brown in a stick nest built high in a tree in dense vegetation. Only 2 described nests: 1 from Loma Yautia between Piedra Blanca and Rancho Arriba, DR, and 1 in pine forest in the Aceitillar sector of the Sierra de Bahoruco.

Range: Breeds from w. Alaska, n. Saskatchewan, c. Quebec, and Newfoundland south locally to c. California, n. Texas, South Carolina, and in Mexican highlands to Oaxaca. Most North American breeders pass non-breeding season in se. U.S., occasionally south to Greater Antilles. In West Indies, resident in Cuba, Hispaniola, and Puerto Rico; casual in Bahamas, Cayman Islands, Jamaica, and Virgin Islands.

Local names: DR: Guaraguaíto de Sierra; **Haiti:** Emouchet, Malfini Mouchè.

RIDGWAY'S HAWK—*Buteo ridgwayi* **Plate 12**

Breeding Resident—*Critically Endangered Endemic*

Description: 36–41 cm long; mass unavailable. Distinguished by its dark brownish gray upperparts, gray underparts washed with brownish red, reddish brown thighs, and dark tail with thin whitish bars. Females are somewhat larger than males. Males are grayer than females and have a bright reddish brown bend of wing. Females are browner overall with a drab, brown bend of wing. Also note female's lighter breast with more barring, gray belly with reddish pink tint, and more heavily barred tail. Immature is brownish above, with buffy white underparts with pale gray and tan streaks, and tail is less distinctly barred. In flight, soars on broad wings with a fan-shaped tail. Underwings display light patches.

Similar species: Broad-winged Hawk has a boldly banded tail with wide bands. Swainson's Hawk has a gray tail with many thin black bands and a wide subterminal black band. Ridgway's Hawk is considerably smaller than Red-tailed Hawk which, as an adult, has a reddish tail.

Voice: Shrill calls of three basic types: (1) *kleeah*, used in self-assertive and aggressive contexts; (2) *weeup*, given in food exchanges and displays; and (3) a whistle-squeal, given during high-intensity interactions.

Hispaniola: This species appears to prefer undisturbed forested foothills but has been found historically in a wide variety of habitat types, including rainforest, subtropical dry and moist forests, pine forest, limestone karst forest, and even rarely in second-growth woodlands and

agricultural areas. It has been recorded from sea level to 2,000 m. Historically considered local but fairly widely distributed in much of the DR's northern coastal lowlands, parts of the Cordillera Central, near Haina and Santo Domingo in southern DR, and in the southwestern mountains of the Sierra de Neiba and Sierra de Bahoruco. On the Haitian mainland, it is known only from the Massif du Nord. Has also occurred on several of the major satellite islands, including Grande and Petite Cayemite, Île à Vache, Île de la Gonâve, and Islas Alto Velo and Beata. Breeding has been reported from the Cayemites and Île de la Gonâve.
Status: Rare endemic, surely Hispaniola's most critically endangered species. Recently it has become much more localized. There are no available reports from Haiti for 20 years, though it may still survive. In DR, possibly still found in Sierra de Neiba, very rare in (or extirpated from) Sierra de Bahoruco where it has not been seen with certainty since at least 1994. Most numerous in Los Haitises National Park, but it is also becoming extremely scarce there; also recorded recently in the Samaná area. Recent surveys located about 50 pairs in Los Haitises, with 31 pairs attempting to nest in 2004, and only 11 pairs successful in fledging at least 1 young. Low productivity and survival rate of nestlings appear to be problems, but more fieldwork needs to be done. Status on satellite islands is not currently known.
Comments: Often allows close approach before flying. Feeds on lizards, snakes, rats, bats, and some amphibians, insects, and small birds. Home ranges average nearly 60 ha.
Nesting: Lays 2 chalky cream eggs, heavily marked with orange-red mottling. Builds a nest of sticks and epiphytes high in a tree or palm in dense vegetation. Breeds March to June. Display flights begin in January and occur most often in late morning, with nest-building, done mostly by the male, beginning in late February. Females perform virtually all incubation, which begins by late March and takes 10–12 weeks. Males capture nearly all of the prey brought to nest.
Range: Endemic to Hispaniola, including the Cayemites and Île de la Gonâve.
Local names: DR: Gavilán de los Bosques, Gavilán de Ridgway, Guaraguaíto; **Haiti:** Ti Malfini Savann.

BROAD-WINGED HAWK—*Buteo platypterus* **Plate 11**
Vagrant

Description: 34–44 cm long; male 420 g, female 490 g. A medium-sized, chunky, soaring hawk with a relatively short tail and broad wings with slightly pointed tips. Adult is dark brown above, whitish below, with reddish brown barring; tail boldly banded with broad black and white bars. Underwings have a dark border on trailing edge. Immature has white underparts heavily streaked with dark brown, and tail with more numerous, indistinct bands than adult's; dark border of underwing is less distinct.
Similar species: Slightly larger Swainson's Hawk has a gray tail with many thin, black bands and a wide subterminal black band; primaries and secondaries are darker from below. Red-tailed Hawk is much larger and has a white breast with a band of dark streaks across belly. In flight, Broad-winged Hawk flaps more often than Red-tailed. Sharp-shinned Hawk has a similar color pattern but has a long, narrow tail rather than a fan-shaped one, and flies with a rapid spurt of wing beats alternating with short glides, rather than by soaring.
Voice: Thin, shrill whistle, *pweeeeeeeeee*.
Hispaniola: Generally prefers dense broadleaf, mixed broadleaf and pine, or plantation forests at all elevations; found less frequently in open woodlands. Few DR records, all from lowland areas, including La Estrella near Santo Domingo, south of Sabana de la Mar, Bayahibe, and Isla Saona.
Status: Vagrant.
Comments: Feeds mainly on invertebrates, amphibians, reptiles, and small birds.
Range: Breeds from c. British Columbia, c. Manitoba, s. Quebec, and Nova Scotia south to e. Texas, Gulf Coast, and n. Florida. Occurs in non-breeding season from s. Florida and s. Mexico south to Bolivia and s. Brazil. In West Indies, resident on Cuba and Puerto Rico and in Lesser Antilles from Antigua and Dominica south to Grenada.
Local names: DR: Gavilán Bobo, Guaraguao de Bosque; **Haiti:** Malfini Rak Bwa.

SWAINSON'S HAWK—*Buteo swainsoni* **Plate 11**
Vagrant

Description: 48–56 cm long; male 908 g, female 1,069 g. A long-tailed, long-winged hawk of open habitats, with an overall slender appearance. Occurs in different color phases. Adult is generally dark brown above, whitish below, with a rufous to dark brown bib. Less common darker phase birds may have bib extending through breast and belly. Immature is similar to adult but underparts variably streaked and spotted brown. In flight, soars like a Turkey Vulture on wings held in a dihedral; dark flight feathers contrast with lighter underwing coverts. Gray tail with numerous thin black bands, wide subterminal black band, and buffy tip.
Similar species: Broad-winged Hawk has broad black and white tail bands, and light flight feathers contrasting with darker underwing coverts.
Voice: Drawn-out scream similar to that of Red-tailed Hawk but higher, clearer, and weaker.
Hispaniola: One record of a light-phase bird in extreme northwestern DR near Monte Cristi in April 1996 is the only report known for the West Indies.
Status: Vagrant.
Range: Breeds from ec. Alaska and British Columbia, s. Manitoba, and w. Illinois south to s. California, Mexican highlands south to Durango, s. Texas, and w. Missouri. Occurs in non-breeding season primarily on pampas of Argentina and Uruguay, with smaller numbers north to Costa Rica and Panama. In West Indies, recorded only once in DR.
Local names: DR: Gavilán de Swainson; **Haiti:** Malfini Swenson.

RED-TAILED HAWK—*Buteo jamaicensis* **Plate 11**
Breeding Resident; Non-breeding Visitor

Description: 45–65 cm long; male 1,028 g, female 1,224 g. A large, conspicuous hawk most often seen soaring on broad, rounded wings. Adult is dark brown above and buffy to whitish below, with a distinctly contrasting dark belly band and a reddish tail, though the belly band tends to be lighter and more diffuse on Hispaniola. Wings have a dark bar and wrist patch on leading edge. Immature has a faintly barred, grayish brown tail; underparts are more heavily streaked.
Similar species: Immature Broad-winged Hawk is smaller, has broad tail bands, and flaps more in flight. Smaller Ridgway's Hawk has barred tail.
Voice: Sharp, raspy scream, *keeer-r-r-r,* slurring downward.
Hispaniola: Found from coastal lowlands to over 2,440 m in the Cordillera Central, DR, and recorded from nearly all the major satellite islands. Although occurring in nearly all habitat types, it is most common in the Sierra de Bahoruco, DR, in pine and broadleaf forest. Occasionally, birds from mainland North America occur in the non-breeding season, though these are probably indistinguishable from resident birds.
Status: Common breeding resident and occasional non-breeding visitor islandwide, perhaps slightly more numerous in foothills and mountains than in the lowlands. Formerly considered common, but its numbers have apparently declined since the 1930s, particularly in the DR lowlands because of habitat disturbance and shooting, despite legal protection.
Comments: Feeds mainly on introduced rodents, lizards, snakes, birds, and large invertebrates. Some of the densest populations in Hispaniola are in interior forests, unlike in North America where the highest concentrations are in open, drier areas.
Nesting: Lays 2–3 white eggs in a bulky stick nest placed in a tall tree or on a cliff face. Breeds January to July.
Range: Breeds from wc. Alaska east to Nova Scotia, south to Sonora, s. Texas, and Florida; also in highlands of Middle America south to w. Panama. Occurs in non-breeding season in southern portion of breeding range. In West Indies, a breeding resident in Greater Antilles, and in Lesser Antilles south to Nevis.
Local names: DR: Guaraguao, Cola Roja; **Haiti:** Gros Malfini.

FALCONS: FAMILY FALCONIDAE

Falcons constitute a cosmopolitan family of raptors that differ from hawks and their allies mainly in internal structural characters. Wings are narrow and pointed, and the tail is long and generally tapered. Falcons have a relatively large, rounded head and a strong, hooked bill. Flight is swift and powerful, and rarely includes soaring or gliding. Falcons feed on birds, mammals, reptiles, and insects such as dragonflies.

AMERICAN KESTREL—*Falco sparverius* **Plate 10**
Breeding Resident

Description: 22–31 cm long; 115 g. A small, common falcon with a reddish brown, barred back; reddish brown tail with a broad, black terminal band; and boldly patterned head with two distinctive black facial bars. Underparts vary from whitish to reddish brown and are spotted brown to black. Adult males are distinguished by blue-gray wings; adult females have reddish brown wings. Immatures are similar to respective adults but are more streaked than spotted on breast. Often perches along roadsides, bobbing tail.
Similar species: Merlin has bluish gray (male) or dark brown (female) upperparts, heavy streaking below, and a single black facial bar; flight is faster and more powerful.
Voice: High-pitched *killi-killi-killi* or *kli-kli-kli-kli-kli,* from which some of common names are derived.
Hispaniola: Found in nearly all habitat types including suburbs and municipal parks from coastal lowlands into the pine zone of interior foothills, but especially in dry forest and scrub. Less numerous at higher elevations, the highest documented locality being a nest with recently fledged young at 2,840 m at Valle Nuevo, Cordillera Central, DR. Very scarce in or absent from dense rainforest. Reported from nearly all the major satellite islands and breeds on some of them.
Status: Common to abundant resident islandwide. Considered common by all early writers; may have declined somewhat since 1900, although quantitative historical data are lacking. Although birds that breed in North America are known to visit the Bahamas, Cuba, and Cayman Islands, to date there is no documentation of non-breeding visitors on Hispaniola. The subspecies *F. s. dominicensis* is endemic to Hispaniola and associated islands.
Comments: Although it is sometimes thought that the local Dominican name *cernícalo* is a modification of "Tsar Nicholas," a formerly "high-perched" monarch, it is more likely derived from the Spanish verb *cernerse*. The kestrel is frequently seen hovering while hunting.
Nesting: Lays 2–4 tan eggs flecked with brown in a cavity in a tree, telephone pole, or cliff. Breeds at least January to August and possibly other times. Some pairs raise 2 broods per year.
Range: Breeds from c. Alaska east to Newfoundland and south to Tierra del Fuego; northern populations occur in non-breeding season as far south as Panama. In West Indies, resident in Bahamas and Greater Antilles, rare or local in s. Lesser Antilles.
Local names: DR: Cuyaya, Cernícalo, Cernícalo Americano; **Haiti:** Grigri.

MERLIN—*Falco columbarius* **Plate 10**
Passage Migrant; Non-breeding Visitor

Description: 24–30 cm long; male 163 g, female 218 g. A small, dark, and compact falcon with a pale tan supercilium. Adult male is distinguished by pale gray, blue-gray, or blackish upperparts and rufous feathering on legs. Underparts are heavily streaked, and tail is barred black with 2–5 narrow gray or whitish to buffy bands. Adult female has dark brown upperparts; tail is barred black with brown or buff. Immatures resemble females. In flight, Merlin's

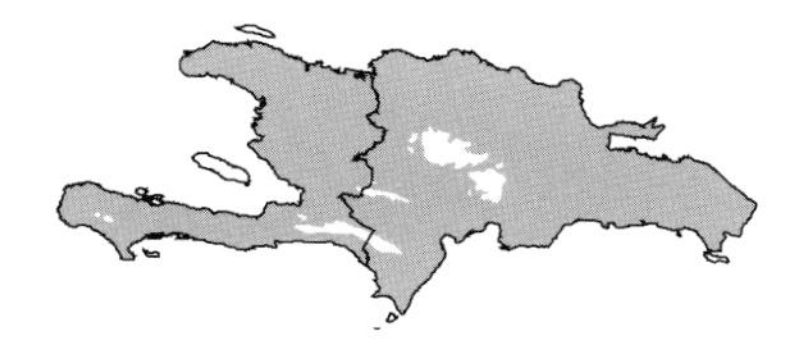

great speed and agility, long, pointed wings, and long, narrow tail are characteristic. It captures prey in midair with strong, level flight.
Similar species: Peregrine Falcon is much larger, with a wide and dark, pronounced malar mark; tail has more than 5 light bands. American Kestrel has rufous on back or tail. In flight, Merlin may be mistaken for a Rock Pigeon, thus its common name Pigeon Hawk, but Merlin's flight is more rapid with a quicker, steadier wing beat.
Voice: Rapid, accelerating series of strident, trilling calls, *ki-ki-kee*, rarely given on Hispaniola.
Hispaniola: Prefers open to semiopen areas; may occur in cities. Found from the coastal lowlands, especially near lakes and lagoons where shorebirds abound, well into the mountains, the highest elevation recorded being 1,465 m in pine forest of the Sierra de Bahoruco, DR.
Status: Regular spring and fall passage migrant and non-breeding visitor islandwide. Non-breeding-season numbers may fluctuate from year to year, though quantitative documentation of the variation is lacking.
Comments: Feeds predominantly on small birds. Also feeds on bats and large insects such as dragonflies.
Range: Breeds from nw. Alaska east to Newfoundland and south to e. Oregon, n. Minnesota, s. Quebec, and Nova Scotia. Occurs in non-breeding season to n. Peru and Venezuela. Also widespread in Eurasia. In West Indies, a non-breeding visitor and passage migrant throughout.
Local names: DR: Merlin, Halcón, Halconcito; **Haiti:** Grigri Mòn, Faucon Émerillon.

PEREGRINE FALCON—*Falco peregrinus* **Plate 10**
Passage Migrant; Non-breeding Visitor

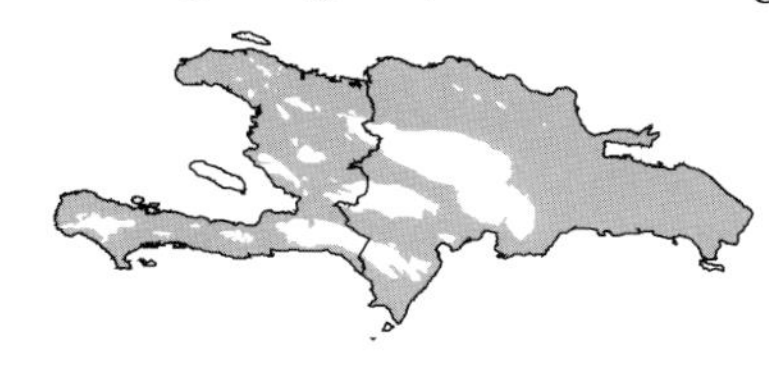

Description: 36–58 cm long; male 611 g, female 952 g. A large, stocky falcon with pointed wings; long, tapered tail; and rapid, powerful flight. When perched, its wide, dark malar mark and crown give it a distinctive "helmeted" appearance. Adult is dark brown to slaty gray above, cream colored to whitish below with dark barring. Immature is brown above; underparts are heavily streaked brown.
Similar species: Merlin is distinctly smaller, with less pronounced facial markings.
Voice: Harsh, scolding *kak-kak-kak-kak*.
Hispaniola: Most often seen near the coast, offshore cays and rock outcroppings, and other localities with abundant shorebirds, seabirds, or waterfowl prey. Recorded from only one satellite island: Navassa Island. Only rarely found in the mountains; highest elevation recorded is 1,440 m at Valle del Tetero, Armando Bermúdez National Park. Peak fall migration period is mid-October to late November; peak northbound migration period in spring is March.
Status: Regular spring and fall passage migrant and non-breeding visitor in small numbers. Always considered rare, but apparently found more often since the late 1970s, about the time the eastern North American breeding population began to recover from its lowest levels resulting from pesticide poisoning.
Comments: In 2001–2002, a young Peregrine Falcon was tracked by satellite from Toronto, through Cuba and Haiti, to the DR, where it spent much of the non-breeding season.
Range: Breeds in scattered locales from n. Alaska east to Baffin Island and Labrador, south to s. Baja California, coastal Sonora, s. Arizona, and w. Texas; since the 1980s, reestablished locally in e. U.S. Occurs in non-breeding season from s. Canada south to Tierra del Fuego. One of the most broadly distributed terrestrial vertebrates worldwide, nearly cosmopolitan

outside the Western Hemisphere. In West Indies, a non-breeding visitor and passage migrant throughout.
Local names: DR: Halcón de Patos, Halcón Peregrino; **Haiti:** Grigri Peleren.

RAILS, GALLINULES, AND COOTS: FAMILY RALLIDAE

This cosmopolitan family consists of small to medium-sized wetland birds with long legs and toes, a heavy body, and short, rounded wings. **Rails** inhabit marshes where their secretive, skulking behavior rarely brings them into view. Most are primarily nocturnal, and much more frequently heard than seen. Broadcasting tapes of their calls is an effective way to document their presence. Rails rarely flush, preferring to run for cover, and their short flights appear labored, with legs dangling conspicuously beneath them. **Gallinules** and **coots** are larger than rails and more strictly aquatic, spending much time swimming. Both superficially resemble ducks, but the distinctive bill extends onto a horny frontal shield, and they characteristically jerk their head while swimming.

BLACK RAIL—*Laterallus jamaicensis* **Plate 13**
Non-breeding Visitor? Breeding Resident?—*Threatened*

Description: 10–15 cm long; 34 g. A tiny, dark rail. Slaty gray overall, with scattered white spots on back and wings and a dark reddish brown nape. Underparts are dark gray with white barring on flanks. Bill is short and black.
Similar species: Downy young of gallinules, coots, and other rails are uniformly black; other adult rails are considerably larger.
Voice: Male's territorial call is a rich, nasal *keeic-keeic-kueerr* or *kic-kic-kic-kurr*, the first 2–3 syllables typically high-pitched whistles, followed by a distinctly lower note. Females occasionally give a low, cooing *croocroocroo*, reminiscent of Least Bittern. The defense call is an emphatic and irregularly pulsing cackle. Rarely vocalizes during the day.
Hispaniola: Occurs in wet grassy marsh edges, both saline and fresh. Recorded in DR only once: a single individual caught in a grassy field between Monte Plata and Bayaguana near the Río Yaví in 1984. A small colony was suspected, but the possibility of the birds being only non-breeding-season visitors was not ruled out.
Status: Rare and local, possible non-breeding visitor or breeding resident.
Comments: Shy and usually active only after dark, so easily overlooked. Runs rather than flies when disturbed and rarely leaves the cover of marsh grasses. Feeds on insects, small crustaceans, and seeds. Focused surveys are needed to determine the species' status in the DR and Haiti.
Nesting: Lays 5–7 buff white eggs with fine reddish brown spots. The nest is concealed in dense marsh grass.
Range: Breeds irregularly and in scattered locales from s. California east to Kansas, n. Illinois, sw. Ohio, and New York, south to Florida and Gulf Coast in e. Texas. Occurs in non-breeding season in Florida, along Gulf Coast of Texas, and in Mexico. In West Indies, breeding-season records from Cuba and Jamaica; rare in DR, Bahamas, and Antigua, where only recorded in non-breeding season.
Local names: DR: Gallito Negro, Gallinetita Negra, Polluela Negra; **Haiti:** Ti Rato Nwa.

CLAPPER RAIL—*Rallus longirostris* **Plate 13**
Breeding Resident

Description: 32–41 cm long; male 323 g, female 271 g. A large, relatively long-necked rail with a long bill, and the habit of stalking among mangrove roots. Upperparts olive-brown to

gray-brown and mottled with black, with gray auriculars. Underparts are grayish, with variable amounts of cinnamon-buff on throat and belly, and flanks faintly barred whitish.

Similar species: Common Moorhen has much shorter bill. Virginia Rail is much smaller and darker overall, with reddish bill and gray face contrasting with the rich rusty chest and rufous upperwing coverts.

Voice: Call is a long, loud, grating series of rising and falling *kek* notes, slowing at the end. The cackle of one rail often sets off a chorus of others.

Hispaniola: Inhabits salt marshes and mangroves that have been left largely undisturbed. Now known along most of the northern coast from Caracol, Haiti, to at least Sabana de la Mar, DR; in coastal La Altagracia Province near Bávaro, DR; at Lago Enriquillo, DR; and along the border of Port-au-Prince Bay and the marshes of the Artibonite River mouth, north at least to Gonaïves harbor, Haiti. Since the 1980s, reported from Laguna de Oviedo, Cabo Rojo, and Las Salinas, DR; therefore almost surely present in suitable habitat on the southwestern DR coast between those points. Also recorded from satellite islands of Grand Cayemite, Île à Vache, and Île de la Gonâve.

Status: Moderately common breeding resident. Certainly has become less numerous since the 1930s because of hunting and habitat loss.

Comments: Far more often heard than seen. It is most active at dawn and dusk. Feeds on crabs, small mollusks, aquatic insects, fish, and plants.

Nesting: Lays 5–9 creamy white eggs with spots in a nest of sticks among mangrove roots. Breeds April to July.

Range: Breeds on Atlantic and Gulf coasts of North America from Massachusetts to Florida, s. Texas to Yucatán Peninsula, Panama, and coastal South America south to se. Brazil; northernmost populations are partially migratory. In West Indies, found in Bahamas, Greater Antilles, and Barbuda; rare at St. Kitts, Guadeloupe, and Martinique, casual at Barbados.

Local names: DR: Pollo de Manglar; **Haiti:** Rato Mang, Râle Gris .

SORA—*Porzana carolina* **Plate 13**

Passage Migrant; Non-breeding Visitor

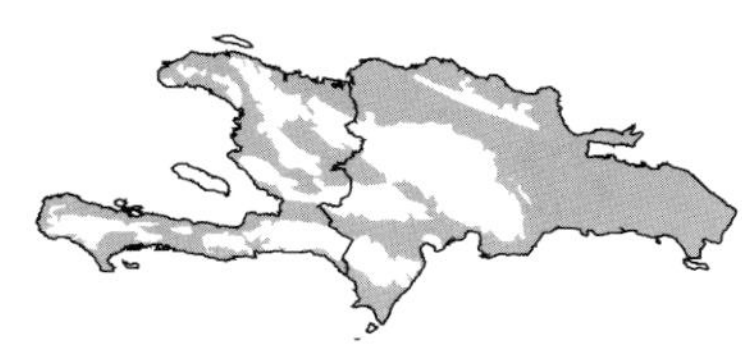

Description: 20–25 cm long; 75 g. A relatively small, brownish gray rail with a black mask and distinctive stubby, yellow bill. Breeding adult has blackish face, throat, and chest, and belly is barred black and white. In non-breeding plumage, throat and chest are mostly gray. Black is absent in immatures, which have buffy brown face, throat, and chest.

Similar species: Juvenile Yellow-breasted Crake is much smaller, with black crown and eyeline, whitish supercilium, and darker bill.

Voice: Clear, descending whinny, *ko-WEEee-ee-ee-ee-ee-ee*, and a plaintive, rising whistle, *koo-wee*.

Hispaniola: A non-breeding visitor to rice fields, mangroves, and freshwater marshes in the lowlands and middle elevations. Seldom reported but surely overlooked because of its secretive habits and because it does not normally call on the non-breeding grounds. Records from Laguna de Rincón at Cabral, Cabo Rojo, and Baní, DR, and Trou Caïman, Étang Saumâtre, and Étang de Miragoâne, Haiti. Careful fieldwork would probably show this species to be a moderately common passage migrant September–October and March–April and regularly present in modest numbers November–February.

Status: Regular passage migrant and non-breeding visitor.

Range: Breeds from s. Alaska to n. Ontario and Newfoundland, south to n. Baja California, sw. Tennessee, and Maryland. Occurs in non-breeding season from mid-Atlantic and mid-

Pacific coasts of U.S. south to Peru and Guyana. In West Indies, a regular passage migrant and non-breeding visitor virtually throughout.
Local names: DR: Gallito, Polluela Sora; **Haiti:** Ti Rato Gòj Nwa.

YELLOW-BREASTED CRAKE—*Porzana flaviventer* **Plate 13**
Breeding Resident

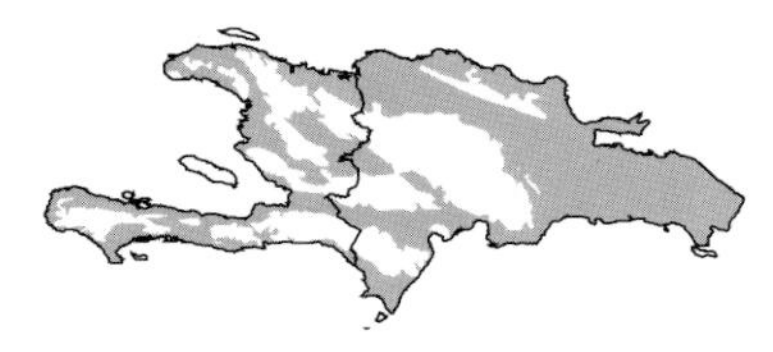

Description: 14 cm long; 25 g. Rarely seen well, this pale rail is distinguished by its tiny size, rich buffy sides of neck and breast, and bold black and white barring on flanks. At close range, note the blackish crown, white supercilium, black eye-line, and small, dark bill. When flushed, flies weakly a short distance, the feet dangling and head drooping.
Similar species: Sora is much larger with a stouter, paler bill.
Voice: Low, rough, rolled or churring *k'kuk kuh-kurr* and a high-pitched, plaintive *peep*.
Hispaniola: Resident of freshwater marshes, swamps, and canals with borders of short grass or other water plants. Sight records indicate that it formerly was common in appropriate habitat in many parts of the Río Yaque del Norte, Río Cana, and Río Yuna drainages, east to El Valle, Pimentel, and Laguna de Rincón at Cabral, DR; also reported from Trou Caïman, Fort Liberté, near Glore on Étang Saumâtre, and Trou des Roseaux, Haiti. Most records since the 1970s are from Laguna de Rincón at Cabral and Pimentel, DR.
Status: Scarce and local resident. Accurate population estimates are difficult to make because of its secretive behavior. Numbers probably substantially reduced since the early 1900s because of habitat alteration for the cultivation of sugar and rice.
Comments: Typically observed only by being flushed from the shallow water of a swamp edge. Feeds on plants and small invertebrates.
Nesting: Lays 3–5 pale cream colored eggs lightly spotted with brown or lavender. The nest is woven atop a floating plant. Breeds March to June.
Range: Resident locally in Mexico from Guerrero and Veracruz south through South America east of Andes to n. Argentina. In West Indies, resident in Cuba, Hispaniola, Jamaica, and Puerto Rico.
Local names: DR: Guineíta, Gallito Amarillo, Polluela Pechiamarilla; **Haiti:** Ti Rato Jòn.

SPOTTED RAIL—*Pardirallus maculatus* **Plate 13**
Breeding Resident?

Description: 25 cm long; 171 g. A medium-sized, strikingly plumaged rail with long, pinkish red legs and a long, greenish yellow bill, slightly decurved and red at the base. Adult is slate black, with head and upperparts heavily spotted white; underparts are boldly barred white; undertail coverts are white. Iris is red. Immature is browner with fewer and duller spots, bill and legs are duller, and iris is brownish.
Similar species: None; bold spotting is unique among rails.
Voice: Peculiar high, guttural screech, which may be preceded by a short grunt. Also an accelerating *tuk-tuk-tuk-tuk* or *wuh-wuh-wuh*.
Hispaniola: Inhabits freshwater swamps with dense emergent vegetation in which it can hide. First discovered in DR in 1978 when 1 was caught alive at a rice field in Pimentel; 2 more were found at Madre Vieja, including 1 in late June, suggestive of a breeding resident population. Since then only 2 reports: at least 1 seen and several heard at Laguna Villa Isabella, Santo Domingo, DR, in March and April 1999; 1 captured in mangroves and photographed at Caño Hondo, Los Haitises National Park, in early August 2002. Although large areas of the lower Río Yuna basin, where the first birds were found, have been converted from freshwater marshes to rice plantations, reducing potential habitat, this resourceful species likely survives in small numbers.

Status: Probable resident breeder in small numbers, but current status poorly known.
Comments: Generally solitary and very secretive. Much more often heard than seen.
Nesting: No records from Hispaniola, but elsewhere lays 3–7 white or cream colored eggs with small reddish brown or bluish gray spots. Nest is a platform of grasses built low in marsh vegetation. Breeding season in West Indies is unknown.
Range: Resident in Mexico from Nayarit and Veracruz south locally through Middle America to nw. Peru and n. Argentina. In West Indies, resident in Cuba, Jamaica, and DR.
Local names: DR: Gallito Manchado, Pollo Manchado; **Haiti:** Rato Tache.

PURPLE GALLINULE—*Porphyrula martinica* — Plate 13
Breeding Resident

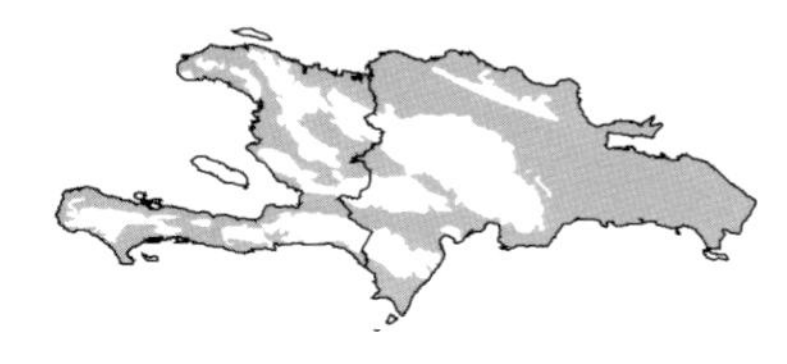

Description: 27–36 cm long; 235 g. A colorful, marsh-dwelling rail of blue, green, and purple plumage. Adult is bluish purple overall, with an iridescent olive green back and wings, and white undertail coverts. Red bill is tipped yellow and is topped by a pale blue frontal shield. Legs are bright yellow. Immature is pale buffy brown, with an olive green sheen on wings and back; bill is duller with a smaller, grayish frontal shield.
Similar species: Immature is similar to Common Moorhen but is browner, lacks the flank stripe, and has greenish wings. Immature American Coots are also grayer, have dark undertail coverts, darker legs and feet, and lighter bill.
Voice: Varied vocalizations include a high-pitched, melodious *klee-klee* and many cackling and guttural notes.
Hispaniola: Resident of freshwater marshes and rivers islandwide. Found in relatively undisturbed localities along major interior river drainages (e.g., Artibonite, Yaque del Norte, Yuna) and major lakes (e.g., Trou Caïman, Étang de Mirogoâne, Laguna de Rincón at Cabral, Laguna Saladilla). A few migrants probably arrive occasionally from the North American mainland in the non-breeding season.
Status: Moderately common but local breeding resident. Numbers probably somewhat diminished by alteration of marsh habitat for rice and sugar agriculture.
Nesting: Lays 3–12 pinkish buff and spotted eggs. A low, bulky nest of vegetation is constructed among cattails or rice grass. Breeds July to September.
Range: Breeds from Delaware and Nayarit south along coasts of Middle and South America to Argentina and Chile; also an isolated population in interior U.S. In West Indies, resident in Greater Antilles east to Puerto Rico, casual elsewhere.
Local names: DR: Gallareta Pico Azul, Gallareta Azul, Gallareta Purpura; **Haiti:** Poule Sultane, Poul Dlo Tèt Ble, Poule d'eau à Cachet Bleu.

COMMON MOORHEN—*Gallinula chloropus* — Plate 13
Breeding Resident; Non-breeding Visitor

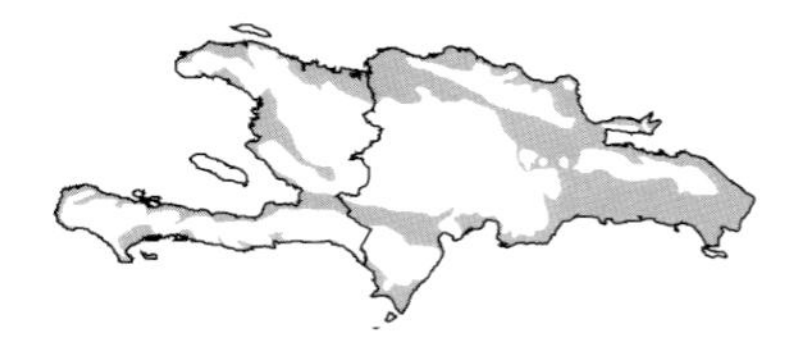

Description: 32–35 cm long; 305 g. Adult is slaty gray overall but brownish gray on upper back, flanks, and sides, with a prominent white stripe along upper flanks. Undertail coverts are white with a dark central stripe. Red bill is tipped yellow, with a high, red frontal shield. Immature is gray and brown and displays the white flank stripe; bill is dusky.
Similar species: Clapper Rail has much longer bill. Immature Purple Gallinule is buffy brown, lacks a flank stripe, and has bluish green wings. Immature American and Caribbean coots are entirely dark gray with whitish bill.
Voice: Variety of clucks, cackles, and grunts, the most common being a piercing series of clucks slowing and ending with long, whining notes, *ki-ki-ki-ki-ka, kaa, kaaa*.

Hispaniola: Occurs in lowland freshwater marshes islandwide, less often in brackish or more saline water, but not infrequently in coastal lagoons bordered by mangroves. Birds may move around seasonally within Hispaniola to access changing food sources, particularly in January and February. Also reported from Île à Vache.
Status: Generally a common year-round resident. At least some birds come from the North American mainland in the non-breeding period, though they are visually indistinguishable from local birds.
Comments: A fairly conspicuous waterbird, and not particularly shy; it often strays from dense vegetation of the pond edge to swim in calm, open water. Swims with a characteristic bobbing motion of the head. Feeds on aquatic plants, mollusks, worms, and fruit.
Nesting: Lays 3–9 grayish white to light brown eggs with reddish spots. The nest is usually a bulky platform of plant material suspended over water. Breeds all year but with peak from May to September.
Range: Breeds from n. California east to Minnesota, sw. Quebec, New Brunswick, and Nova Scotia, south to n. Chile and n. Argentina. Widespread in Eastern Hemispere. Found throughout West Indies.
Local names: DR: Gallareta Pico Rojo, Gallareta Común; **Haiti:** Poul Dlo Tèt Wouj.

AMERICAN COOT—*Fulica americana* **Plate 13**
Non-breeding Visitor; Breeding Resident

Description: 38–40 cm long; male 724 g, female 560 g. A stocky, gray, chickenlike marsh bird. Adult head and neck are black, contrasting sharply with the white bill; the rest of body is slaty gray, except for white undertail coverts. Frontal shield of bill is white with a maroon base but may be all white in some birds. White tips on secondaries are visible in flight. Immature is paler than adult, gray-brown above, dull grayish below. Coots appear very buoyant on water but must run over the surface to take flight.
Similar species: Distinguishable from Caribbean Coot only by the lack of a full white frontal shield extending well onto crown. Immatures of the two coot species are virtually identical. Immature Purple Gallinule is buffy brown with a greenish sheen on the back and wings. Immature Common Moorhen has a white flank stripe and dusky bill.
Voice: Variety of short clucks, croaks, and cackles, usually lower pitched than in Common Moorhen.
Hispaniola: Occurs islandwide at freshwater ponds and lakes and along river courses in areas with much submerged vegetation.
Status: Primarily a non-breeding visitor from the North American mainland in variable numbers from one year to the next, but also a breeding resident in small numbers.
Comments: Often flocks. Swims with a bobbing motion of the head. Dives proficiently. Varied diet includes aquatic plants and algae, mollusks, crustaceans, insects, tadpoles, and small fish. Interbreeding with Caribbean Coots has been observed on several islands, including among birds that frequent the ponds surrounding the golf course in the Bavaro Barceló Resorts, north of Punta Cana.
Nesting: Lays 6–12 beige eggs heavily spotted with dark brown and black. A nest of dried grass and aquatic plants is placed on the ground near water, or is floated in emergent vegetation. Breeds primarily March to June.
Range: Breeds from c. Alaska, nw. Saskatchewan, w. Ontario, and Nova Scotia south to s. Baja California, Nicaragua, nw. Costa Rica, Gulf Coast, and s. Florida. Occurs in non-breeding season from extreme s. Alaska and s. New England south to e. Panama and n. Colombia. In West Indies, occurs in Bahamas and Greater Antilles east to St. John, casual in non-breeding period south through Lesser Antilles to Grenada.
Local names: DR: Gallareta Pico Blanco Americana; **Haiti:** Poul Dlo Jidèl, Judelle, Poule d'eau à Cachet Blanc.

CARIBBEAN COOT—*Fulica caribaea* **Plate 13**
Breeding Resident—*Threatened*

Description: 33–36 cm long; mass unavailable but approximately 650 g. Grayish black overall, with white undertail coverts and a white frontal shield (sometimes tinged yellow) that extends well up onto crown. Immature is paler than adult. Coot run over water to take off.
Similar species: American Coot has a white bill but lacks the white frontal shield extending onto crown. Immatures of the two species are virtually identical.
Voice: Variety of croaking and cackling sounds, not known to be distinguishable from those of American Coot.
Hispaniola: Resident of primarily open freshwater lakes, ponds, streams, and rivers, occasionally in slightly brackish coastal sloughs, usually near river mouths. There are few reports of large numbers, but may be found at Laguna Limón, DR; Savane Desolée near Gonaïves and on the eastern shore of Étang Saumâtre, Haiti; and at Laguna de Rincón at Cabral, DR, where 50–100 are present year-round.
Status: Uncommon and local year-round resident on Hispaniola. Formerly considered locally common, but the population has been sharply reduced by hunting, drainage and alteration of habitat, and egg-robbing. It is considered threatened.
Comments: Typically occurs in flocks. Swims with a bobbing head motion and dives proficiently. Some individuals may wander among islands in response to changing habitat conditions. Interbreeding with American Coots has been observed on several islands, including among birds that frequent the ponds surrounding the golf course in the Bavaro Barceló Resorts, north of Punta Cana.
Nesting: Lays 4–7 bright pale olive-buff eggs spotted finely with plumbeous or slaty black dots and spots that are distributed uniformly and rather closely across the surface. The nest floats in emergent vegetation. Breeds year-round with peaks from April to June and September to November.
Range: Found in Caribbean area only. In West Indies, resident in most of Greater Antilles, uncommon to casual in Lesser Antilles south to Barbados and Grenada.
Local names: DR: Gallareta Pico Blanco Caribeña; **Haiti:** Poul Dlo Tèt Blan, Poule d'eau à Cachet Blanc.

LIMPKIN: FAMILY ARAMIDAE

The Limpkin is the sole member of this Western Hemisphere family. It is a large, but relatively inconspicuous, ibislike bird that dwells primarily in swamps and forested riparian zones along streams, wading among vegetation in search of snails on which it specializes. The Limpkin is solitary, largely nocturnal, and often roosts in trees. Its distinctive call is loud and carrying and is frequently heard at dusk.

LIMPKIN—*Aramus guarauna* **Plate 9**
Breeding Resident—*Threatened*

Description: 69 cm long; 1,080 g. A large, long-legged, and long-necked bird, entirely brown with white streaking and triangular spotting in its plumage, especially on the neck, mantle, and wings. The long, slightly down-curved bill is distinctive.
Similar species: Limpkin could be confused with young night-herons but can be distin-

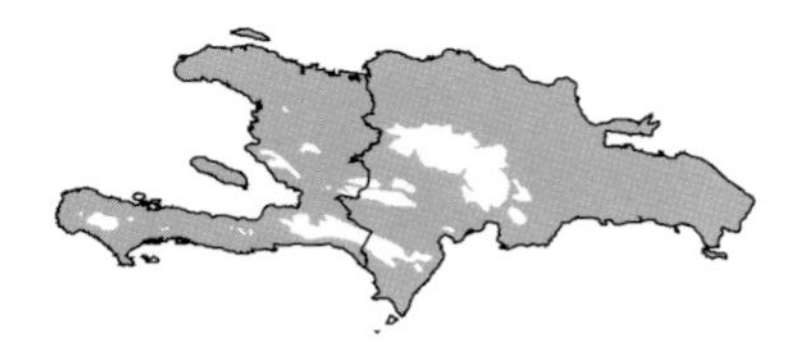

guished by its larger size and longer, paler, downcurved bill. Immature ibises have more slender and strongly decurved bills.
Voice: Loud, piercing scream, *carrao*, which gives rise to the species' local name.
Hispaniola: Occurs in a variety of habitats, including grassy freshwater swamps, marshes, wooded floodplains of rivers, upland dry and humid forest, cacao plantations, and savannas. Appears now to be more abundant at moderate elevations in the mountains than at lowland sites, especially in dry forests of the Sierra de Bahoruco. Reported at nearly 1,600 m in the Sierra de Neiba. Rarely reported from Haiti since the 1980s, although 1 was found near Pignon in 2000. Also has occurred at Île de la Gonâve, Île de la Tortue, and Isla Saona.
Status: Increasingly local and scarce resident. Formerly considered locally common but now very reduced in numbers. Considered rare and threatened in both La Visite National Park and Macaya Biosphere Reserve, Haiti; numbers in Los Haitises National Park, DR, declined substantially between 1974–1976 and 1996. The Hispaniolan population is considered threatened because of habitat loss, hunting, and possibly the impact of introduced predators.
Comments: A secretive bird, most active at dawn and dusk, often located by its call. Limpkins on Hispaniola are noted for their reluctance to fly; this has led to the suggestion that they may be evolving toward flightlessness.
Nesting: Lays 4–8 pale creamy buff or olive buff eggs spotted with brown. The nest is a loose platform, usually built in a thick tangle of vegetation in undergrowth or in a tree not far above ground. May breed April to October, but most nests in DR reported April to June.
Range: Resident in se. Georgia and Florida and from c. Mexico south through Middle and South America to Ecuador, s. Bolivia, and n. Argentina. In West Indies, occurs in Bahamas, Cuba, Jamaica, and Hispaniola.
Local names: DR: Carrao; **Haiti:** Rele, Gran Kola, Poule-à-jolie.

THICK-KNEES: FAMILY BURHINIDAE

The thick-knees are large, long-legged, upland shorebirds that resemble robust plovers with a heavy bill and long, stout legs. They inhabit open country and pastures. Thick-knees are typically shy and swiftly run away when disturbed, sometimes making short yet strong flights. Their large, yellow eyes are an adaptation to their crepuscular and nocturnal habits. During the day they often crouch on the ground and are difficult to see.

DOUBLE-STRIPED THICK-KNEE—*Burhinus bistriatus* **Plate 19**
Breeding Resident—*Threatened*

Description: 38–43 cm long; 787 g. A large, ploverlike shorebird, mostly brown streaked tawny and white. Face, neck, and chest are grayish buff with dark streaking, throat and belly are white, and breast is striped. Note the long, sturdy legs; heavy bill; large, yellow iris; and whitish supercilium. Camouflaged plumage makes the species difficult to detect. In flight, it shows conspicuous white wing patches. Immature is duller and grayer.
Similar species: Much larger than all the plovers.
Voice: Loud rattling or barking *ca-ca-ca-ca-ca-ca-ca-ca!*, rising in volume, then descending in pitch and fading away. Usually calls at dusk, dawn, and during the night.

Hispaniola: A bird of grasslands, savannas, arid lowland scrub, and agricultural areas. Formerly reported as locally fairly common in the northern part of the DR and extreme northeastern Haiti, as well as near Hinche, Haiti, and San Juan, DR. But its DR range has been greatly reduced, and its current status in Haiti is unknown. The distribution of this species is now restricted in the DR to small savannas of Santiago, Moca, and in the east around Monte Plata, Sierra de Agua, Guerra, and the savannas near Bayaguana. Also reported from Isla Beata, but its presence there needs to be confirmed.
Status: Formerly a common year-round resident on Hispaniola, but now uncommon and local because of hunting and habitat destruction. The thick-knee inhabits cultivated country but does not occur in sugarcane. Most of the savannas have been destroyed to plant sugarcane, consequently numbers of this species are greatly reduced. Also, the thick-knee's habit of nesting on the ground makes it easy prey for the introduced mongoose which preys on nestlings and eggs as well as adults. The species is now considered threatened. The subspecies *B. b. dominicensis* is endemic to Hispaniola.
Comments: Secretive. Primarily nocturnal and ground dwelling. Feeds on insects, worms, mollusks, and sometimes small lizards and small rodents. Generally found in pairs or family groups. Occasionally kept by people to eat cockroaches and other household vermin, and to serve as a "watchdog," as its call can carry over large distances.
Nesting: Lays 2 eggs in a shallow scrape on the ground. Eggs are light brown to pale olive buff, heavily marked with brown or maroon spots, and are well camouflaged. Breeds in April and May.
Range: Resident in Middle America from c. Mexico south to nw. Costa Rica, and in South America from n. Colombia east to nw. Brazil. In West Indies, thought to be resident only on Hispaniola, but 1 pair found breeding on Great Inagua, Bahamas, in 2003.
Local names: DR: Búcaro; **Haiti:** Kòk Savann, Poul Savann, Courlis de Terre.

PLOVERS: FAMILY CHARADRIIDAE

Plovers are small to medium-sized chunky shorebirds with a relatively large head, short neck, large eyes, and a short, stout bill which is slightly swollen toward the tip. Plumages are characterized by browns, grays, black, and white; many species have striking patterns of neck or breast markings. Plovers typically inhabit open, often fairly dry, areas, where they feed with a distinctive stop-start motion, often running several steps, pausing motionless, pecking at a food item, then continuing. Diets include a wide variety of invertebrates, which are picked from the substrate surface. Calls are mostly plaintive whistles and sharp chips.

BLACK-BELLIED PLOVER—*Pluvialis squatarola* **Plate 14**

Passage Migrant; Non-breeding Visitor

Description: 26–34 cm long; 220 g. A large plover with a heavy body and thick bill. Non-breeding birds have lightly mottled gray upperparts; underparts are whitish with light streaking on breast, and belly is white. There is indistinct contrast between the gray crown and whitish supercilium. Breeding birds are mottled white and black above, with solid black underparts from the chin through the belly. In flight, white uppertail coverts, white tail with dark bars, and distinct white wing stripe are good field marks, as is the conspicuous black patch on axillaries.
Similar species: Very similar American Golden-Plover occurs more typically in fields than along water edges. Black-bellied Plover is larger, has a heavier bill, and in non-breeding

plumage, is grayer. Golden-plovers tend to be more golden yellow. In breeding plumage, note Black-bellied Plover's light crown and nape. It has white, rather than dark, undertail coverts, and lighter flecks above.
Voice: Single, plaintive *klee* and a sweet, slurred whistle, *klee-a-lee* or *pee-u-wee*.
Hispaniola: Found on tidal mudflats around the entire DR coast and along other water edges. Possible in every month of the year, but peak fall migration period is mid-August to mid-November, peak spring migration period March to mid-April. Recorded in large numbers from such coastal sites as Port-au-Prince Bay, Haiti, and Las Salinas and Puerto Viejo, DR. Also routinely found inland at saltwater lakes such as Étang Saumâtre, Haiti, and Lago Enriquillo, DR, and occasionally at inland freshwater lakes such as Laguna de Rincón at Cabral. Among satellite islands, reported from Île de la Gonâve, Cayos Siete Hermanos, and Islas Beata and Saona.
Status: Regular spring and fall passage migrant and non-breeding visitor in variable but moderate numbers, a few usually lingering through June and July.
Comments: Typically in loose flocks. Feeds on marine invertebrates, especially worms and small clams.
Range: Breeds from n. Alaska east to Devon Island and south to Yukon River and Southampton Island. Occurs in non-breeding season from s. coastal British Columbia and Massachusetts south to Argentina and Chile. Widespread in Eurasia. In West Indies, a regular passage migrant and non-breeding visitor throughout, with non-breeding birds occasionally lingering through breeding season.
Local names: DR: Chorlo Gris, Playero Gris; **Haiti:** Plivye Vant Nwa, Pluvier Argenté.

AMERICAN GOLDEN-PLOVER—*Pluvialis dominica* **Plate 14**
Passage Migrant

Description: 24–28 cm long; 145 g. A fairly large and stocky plover with a short bill. In non-breeding plumage, mottled gray throughout, with distinct contrast between the dark crown and whitish supercilium. Breeding birds have mottled black and golden brown upperparts separated from entirely black underparts by a bold white stripe running from the forecrown down the sides of the neck to the upper breast. Underwings are gray.
Similar species: Very similar to more common Black-bellied Plover, but American Golden-Plover frequents fields rather than water edges. In non-breeding plumage, American Golden-Plover is darker above, particularly on the crown, which highlights the white supercilium. In breeding plumage, which is rarely seen on Hispaniola, it has black, rather than white, undertail coverts, and golden flecks on upperparts. In flight, distinguished from Black-bellied Plover by dark tail and uppertail coverts, and absence of a white wing stripe; from below, by lack of black axillaries.
Voice: Variety of calls, including a high, sharp *quit* and a soft, warbled whistle, *queedle*, sometimes given more loudly, and often in flight.
Hispaniola: A rare passage migrant, usually found in grasslands, agricultural lands, and savannas, often near freshwater. Known records include individuals at Étang Saumâtre and Trou Caïman, Haiti, and Sabana San Thome near San Juan, Las Salinas, Lago Enriquillo, and Cabo Rojo, DR.
Status: Generally a rare southbound migrant, primarily August to November, and very rare during its northward migration in March and April.
Comments: Typically flocks. Usually occurs in uplands, unlike other regularly occurring plovers.
Range: Breeds from n. Alaska east to Baffin Island, south to nw. British Columbia, and n. Ontario. Occurs in non-breeding season from n. South America south to Chile and Argentina. In West Indies, a regular passage migrant principally on eastern islands in fall.
Local names: DR: Chorlo Americano, Playero Americano; **Haiti:** Plivye Savann, Pluvier Bronzé.

SNOWY PLOVER—*Charadrius alexandrinus* **Plate 14**
Breeding Resident—*Threatened*

Description: 15–17 cm long; 41 g. A very small, pale plover with a slender black bill, dark neck marks which do not meet to form a collar, and black or dark grayish legs. Upperparts are uniformly pale sandy gray, underparts are white. In breeding plumage, note the black auricular patch and black forehead bordered below by a white forecrown and supercilium. Immatures lack the black markings of adults. In flight, the tail and uppertail coverts are edged white.
Similar species: Piping Plover is similarly pale but larger and stockier, with a relatively shorter, stubbier bill, orangish legs, and completely white uppertail coverts. Semipalmated and Wilson's plovers are larger, browner above, and normally have complete breast bands.
Voice: Weak, husky whistle, *ca-WEE*, and a low, slightly rough *quip* or *krut*. Breeding display song, given on the ground, is a repeated whistle, *tuEEoo*.
Hispaniola: Found on beaches and lagoon borders with extensive saltflats around most of the coast. It has been found routinely at Las Salinas, DR, and at interior sites such as Lago Enriquillo and Laguna de Rincón at Cabral, DR. Also reported from Isla Catalina.
Status: Moderately common breeding resident, but considered threatened rangewide because of habitat destruction and depredation by introduced predators.
Comments: This plover may move among islands, particularly smaller ones, after breeding. Feeds on invertebrates, mainly worms, insects, and mollusks.
Nesting: Lays 3–4 sand-colored eggs with scrawled markings in a depression in the sand, sometimes lined with shell fragments. Breeds January to August, though at Las Salinas, DR, nesting is most common April to July.
Range: Breeds on Pacific coast from s. Washington south to s. Baja California; very local in interior U.S. east to se. Colorado, c. Kansas, nc. Texas east to Florida; also ne. Tamaulipas to coastal Oaxaca, Mexico. Widespread in Eurasia and Australasia. In non-breeding period, found locally through much of breeding range south to Guatemala. In West Indies, found locally in s. Bahamas, Cuba, Hispaniola, Virgin Islands, St. Martin, and St. Barthélemy; casual at Jamaica, Cayman Islands, Antigua, Guadeloupe, and Barbados.
Local names: DR: Chorlito Niveo, Playero Corredor; **Haiti:** Ti Plivye Blanch, Pluvier Blanc.

WILSON'S PLOVER—*Charadrius wilsonia* **Plate 14**
Breeding Resident

Description: 16–20 cm long; 55 g. A common plover, slightly larger than other banded species, with a distinctive heavy, black bill. Breeding plumage is grayish brown above, white below, with a single broad breast band and a white forecrown and supercilium. Legs are dull pinkish. The single, wide breast band and forehead patch are black in males but brown in females, non-breeders, and immatures.
Similar species: Semipalmated Plover is smaller with a narrower breast band, much shorter and stubbier bill, and orange legs. Piping and Snowy plovers are smaller and pale sandy gray above.
Voice: Call is an emphatic, raspy whistled *peet*, *quit*, or *wheet*. When flushed, a high-pitched, hard *dik* or *kid* is uttered. Gives grating or rattling notes, *jrrrrid-jrrrrid*, when agitated or during flight displays.
Hispaniola: Occurs along beaches, mudflats, and other coastal sites, and especially along the borders of salt ponds. Only rarely reported from interior saline lakes. Has occurred at most satellite islands, including the Cayemites, Île de la Gonâve, Île de la Tortue, and Islas Beata, Catalina, and Saona.

Status: Common breeding resident.
Comments: Feeds on insects and larvae, crustaceans, worms, and small snails.
Nesting: Lays 2–4 light buff-colored eggs with splotches in a depression in the sand, sometimes lined with bits of shell. Breeds primarily March to July.
Range: Breeds from c. Baja California south along Pacific coast to Peru, and on Atlantic coast from Virginia south through Florida and Gulf Coast to Belize. Occurs in non-breeding season from Baja California south to Peru on Pacific coast, and along Atlantic coast from n. Florida and coastal Texas south to n. South America. In West Indies, found in Bahamas, Greater Antilles, and Virgin Islands south to Barbuda in Lesser Antilles; casual at Cayman Islands; local population in Grenadines.
Local names: DR: Chorlito de Wilson, Corredor, Cabezón; **Haiti:** Kolye, Pluvier de Wilson.

SEMIPALMATED PLOVER—*Charadrius semipalmatus* **Plate 14**
Passage Migrant; Non-breeding Visitor

Description: 17–19 cm long; 47 g. A small, common, distinctive plover with a very short bill. Upperparts are brown, underparts white. Non-breeding birds have a grayish brown crown and line through eyes, a white forecrown and supercilium, and a single grayish brown band across breast. Bill is stubby and dark, legs are yellowish orange. Breast band is sometimes incomplete and then shows only as bars on either side of breast. Breeding birds have black replacing the grayish brown on breast, and a black line through eyes extends onto crown. Bill is dark and often shows orange at base. Immature resembles winter adult, with a slightly scaled appearance on upperparts.
Similar species: Piping and Snowy plovers are much paler above, with incomplete breast band. Wilson's Plover is larger, with a thicker breast band, much heavier bill, and pinkish legs.
Voice: Flight call, often given when flushed, is a plaintive *weet* or a husky, plaintive whistled *chu-WEE*.
Hispaniola: Most often found on coastal mudflats but also recorded regularly inland at saline Lago Enriquillo, and occasionally at inland freshwater lakes such as Laguna de Rincón at Cabral, DR. Also recorded from Île de la Gonâve and Islas Beata and Saona. Peak spring migration period is March to mid-April, peak fall migration late August to early October. Regular in modest but variable numbers November to February.
Status: Regular spring and fall passage migrant and non-breeding visitor.
Comments: Typically flocks, often with other shorebirds.
Range: Breeds from n. Alaska east to n. Labrador, south to British Columbia and Nova Scotia. Occurs in non-breeding season coastally from n. California and s. Virginia south to Argentina and Chile. In West Indies, a regular passage migrant throughout, but most common in fall.
Local names: DR: Chorlito Semipalmado, Playerito Semipalmado; **Haiti:** Kolye Janm Jòn, Collier à Pattes Jaunes.

PIPING PLOVER—*Charadrius melodus* **Plate 14**
Non-breeding Visitor—*Threatened*

Description: 17–18 cm long; 55 g. A small plover, distinguished by its pale sandy gray upperparts, white underparts, short stubby bill, and yellow-orange legs. Black breast band on non-breeding birds may be partial or absent, and is usually replaced by brownish gray patches on sides of upper breast; bill is black. Breeding birds have a single, narrow black breast

band that is usually broken but may be complete; a narrow black band across head bordered below by white forecrown; and an orange base of bill. In flight, shows a white band across uppertail coverts and a black spot near tip of tail.
Similar species: Snowy Plover is smaller and has a thinner, longer bill and grayish legs. Semipalmated Plover is decidedly browner above, with darker facial markings. Wilson's Plover is darker above and has a much heavier bill and gray to pinkish gray legs.
Voice: Clear, mellow whistled *peep* or *peep-lo*.
Hispaniola: Occurs along sandy water edges, both fresh and saline. All known reports are of 1–4 birds at Bucán de Base, Punta Arena Gorda, Cabo Engaño, and Las Salinas, DR, as well as Isla Tercero, Cayos Siete Hermanos, and Islas Beata and Saona.
Status: Rare non-breeding visitor from North America. This plover is threatened primarily by loss and degradation of its breeding habitat in North America, but few research and conservation efforts have been directed toward non-breeding areas.
Range: Breeds locally in interior North America from sc. Alberta and n. Minnesota south to ne. Montana, e. Colorado, n. Iowa, and Great Lakes region; and from Newfoundland south along Atlantic coast to South Carolina. Occurs in non-breeding season on Atlantic and Gulf coasts from North Carolina to Yucatán Peninsula. In West Indies, occurs in Bahamas and Greater Antilles east to Virgin Islands; casual in Lesser Antilles south to Barbados.
Local names: DR: Chorlito Silbador; **Haiti:** Ti Plivye Siflè, Pluvier Siffleur.

KILLDEER—*Charadrius vociferus* **Plate 14**
Breeding Resident; Non-breeding Visitor

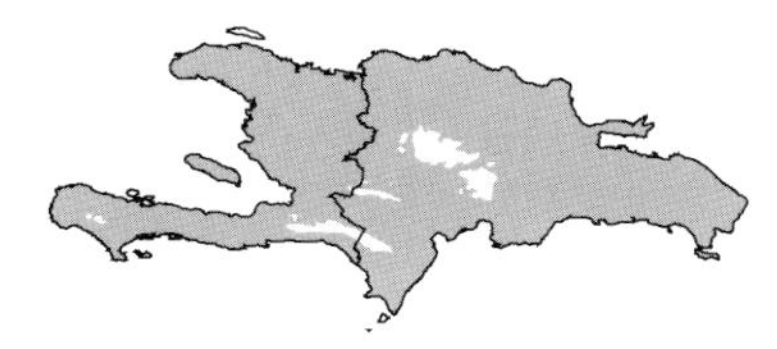

Description: 20–28 cm long; 95 g. A familiar, slender plover, with relatively long wings and tail. Upperparts are grayish brown, underparts are whitish, with two conspicuous black bands across breast. In flight, bright reddish brown rump and uppertail coverts are distinctive. Wings show a broad white stripe. Bill is relatively long; legs are dusky pink.
Similar species: Distinguished from all other plovers by the two black breast bands.
Voice: A noisy, conspicuous bird, giving a strident, high-pitched *kee* and *kill-deer* or *ti-íto*, reminiscent of its English and Spanish names, respectively. When agitated, repeats a loud, clear *teeee di di*.
Hispaniola: Inhabits relatively undisturbed open meadows, fields, savannas, beaches, and edges of freshwater ponds islandwide. Most often reported from the coastal lowlands but also at many interior sites, up to elevations of 1,400 m, often but not always near bodies of water. Reported from the satellite islands of Île à Vache, Île de la Gonâve, and Isla Saona. The resident population is likely increased significantly by migrants in the non-breeding season.
Status: Locally common breeding resident and non-breeding visitor from North America.
Comments: Often found near human habitations. This species is well known for its broken-wing displays to distract potential predators from the nest or chicks. Feeds on terrestrial invertebrates, especially earthworms, grasshoppers, beetles, and snails; sometimes small vertebrates and seeds.
Nesting: Lays 3–4 olive-buff eggs spotted irregularly with black, with the larger, heavier spots on the large end of the egg. The nest is a shallow depression in the ground, sometimes lined with pebbles, plant fragments, or wood chips. Breeds March to October.
Range: Breeds from c. Alaska east to Nova Scotia, south to Guerrero and Florida. Occurs in non-breeding season from s. British Columbia and s. New England south to Venezuela and Chile. In West Indies, a regular breeder and passage migrant in Bahamas and Greater Antilles, uncommon in Cayman Islands and n. Lesser Antilles, casual in s. Lesser Antilles.
Local names: DR: Ti-íto, Chorlito Tildio; **Haiti:** Kolye Doub, Collier Double, Chevalier de Terre.

OYSTERCATCHERS: FAMILY HAEMATOPODIDAE

The oystercatchers form a small but cosmopolitan family of large, stocky, short-necked shorebirds. The bill is red or orange, long and stout, and laterally compressed as an adaptation for opening bivalve mollusks, the chief food. Oystercatchers are exclusively coastal, inhabiting sandy or rocky shorelines.

AMERICAN OYSTERCATCHER—*Haematopus palliatus* **Plate 18**

Breeding Resident; Non-breeding Visitor

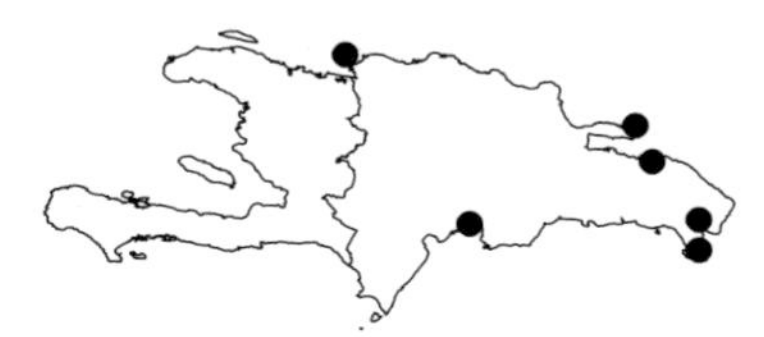

Description: 40–44 cm long; 632 g. A large, pied shorebird with black hood and back, white breast and flanks, and long, heavy bill. Adult has orange-red bill and pinkish legs. Immature has dark-tipped, dull pinkish bill and gray legs. In flight, white uppertail coverts and broad white wing stripe are distinctive.

Similar species: Unmistakable.

Voice: Loud, emphatic, coarsely whistled *wheep* or *queet*, often running into a piping chatter.

Hispaniola: Typically inhabits stony beaches and rocky headlands of offshore islands and cays. Definite records of 1–3 birds seen at various coastal sites, including Jovero, Playa Tortuguero near Azua, Punta Algibe, Las Cañas in Del Este National Park, Monte Cristi, and at Las Galeras, Samaná Peninsula, DR; also reported from Isla Tercero, Cayos Siete Hermanos, and Isla Saona.

Status: Rare breeding resident and non-breeding visitor, perhaps slightly more numerous than recorded.

Comments: Restricted exclusively to a relatively scarce habitat type seldom occupied by other bird species except perhaps Ruddy Turnstone.

Nesting: The nest is a scrape in sand, pebbles, or coral rubble on a remote coastal beach or islet. Lays 1–3 buff-colored eggs spotted with dark brown. Breeds April to July.

Range: Breeds on Atlantic and Gulf of Mexico coasts from Massachusetts to Yucatán Peninsula, and on Pacific coast from Gulf of California to Chile. Eastern population winters to northern coast of South America. In West Indies, a local resident in Bahamas, Greater Antilles, and in Lesser Antilles at St. Barthélemy, Guadeloupe to St. Lucia, and Grenadines; casual at Jamaica.

Local names: DR: Caracolero, Ostrero Americano; **Haiti:** Gwo Bekasin Bèk Jòn.

STILTS: FAMILY RECURVIROSTRIDAE

This small family of medium-sized, long-legged wading birds is characterized by its slender, graceful appearance and noisy, gregarious habits. The head is relatively small and rounded. Most species are boldly patterned, often in black and white, as is the sole member of the family occurring on Hispaniola. These birds frequent open, shallow wetlands where they pick food from the surface of the water or probe in the mud.

BLACK-NECKED STILT—*Himantopus mexicanus* **Plate 19**

Breeding Resident

Description: 35–39 cm long; 166 g. A slender, strikingly patterned shorebird, black above and white below, with long red or pink legs. Bill is very fine and straight. In flight, underwings are black; underparts and tail are white, extending as a white V onto lower back. Female has

a browner back than male; immature resembles female with pale edgings to back feathers, and paler legs.
Similar species: Unmistakable.
Voice: Loud, barking series of short notes, *wit wit wit wit* or *kek kek kek kek*.
Hispaniola: Resident of coastal lagoons and marshes, less often but regularly found at interior freshwater and saline lakes in the lowlands. Also reported from almost all of the major satellite islands.
Status: Local breeding resident. Generally common and widespread.
Comments: Often seen foraging in large flocks, wading with an almost exaggerated grace. When disturbed, conspicuous and noisy, calling loudly and frequently. Feeds on aquatic invertebrates such as crayfish, brine shrimp, adult and larval beetles, and snails; also tadpoles, fish, and occasionally seeds.
Nesting: Lays 2–4 eggs which are sand colored with many dark scrawls. The nest, built near water, is a scrape or shallow platform lined with grass, other plant materials, pebbles, or shells. Breeds primarily late April to August, occasionally in September and October.
Range: Breeds on Atlantic, Gulf of Mexico, and Caribbean coasts from Delaware to Argentina; also an isolated population in central and interior se. U.S. Occurs in non-breeding season from n. California, Sonora, Gulf Coast, and c. Florida south through Middle America to limit of its South American range. In West Indies, breeds in Bahamas, Greater Antilles, and Lesser Antilles south to Antigua and Montserrat.
Local names: DR: Viuda, Zancudo; **Haiti:** Pèt-pèt, Echasse.

JACANAS: FAMILY JACANIDAE

Jacanas represent a small family that superficially resembles gallinules and inhabits tropical swamps, ponds, and lake edges, particularly where there are water lilies. With their long legs and extremely long toes, jacanas are able to walk on floating vegetation without sinking. However, they can swim and will dive to escape danger. Their flight is slow and labored. Unusual among birds, males are smaller than females, build the nest, incubate the eggs, and care for the precocial young. In some species the female may be mated simultaneously to several males, which defend their territories against one another.

NORTHERN JACANA—*Jacana spinosa* — Plate 13
Breeding Resident

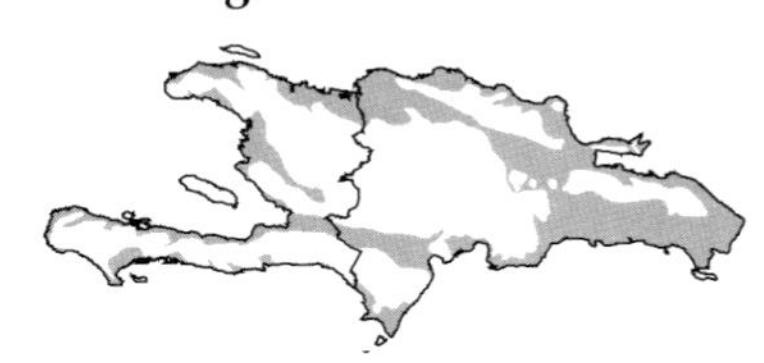

Description: 22–24 cm long; male 82 g, female 135 g. A distinctive wading bird with large yellow wing patches and extremely long, slender, greenish toes. Adult's body is deep reddish brown, contrasting with blackish head and neck. Bill and forehead shield are yellow. Immature is whitish below, olive-brown above, with whitish supercilium and face and brown eye-line, crown, and nape. Sexes are similar in plumage, but females are conspicuously larger than males. Flight is characteristically low over the water with shallow wing beats and dangling legs. Jacanas often raise their wings after landing, exposing yellow flight feathers.
Similar species: Unmistakable.
Voice: Loud, sharp cackle or squawking *scraa scraa scraa*, especially in flight. Also rasping, chattering, and clacking notes.
Hispaniola: Occurs in lowlands, principally at both freshwater and saline ponds in the inte-

rior, such as Laguna de Rincón at Cabral, north of Punta Cana at the Bavaro Barceló Resorts, and sparingly in marshes associated with rivers and streams islandwide.
Status: Locally common breeding resident.
Comments: Very active and noisy. Food consists of mollusks, small fish, crustaceans, insects, and aquatic seeds, which jacanas seek while walking slowly over floating plants.
Nesting: Lays 3–4 light brown eggs, sometimes spotted with black or green. Builds a small platform nest of plant material among dense, floating aquatic vegetation. Breeds April to September.
Range: Resident from Sinaloa to s. Texas south on both slopes of Middle America to w. Panama. In West Indies, resident on Cuba, Jamaica, and Hispaniola; casual on Puerto Rico.
Local names: DR: Gallito de Agua; **Haiti:** Doktè, Chevalye Dore, Poul Dlo Dore, Poule d'eau Dorée.

SANDPIPERS, PHALAROPES, AND ALLIES: FAMILY SCOLOPACIDAE

The sandpipers and allies are a nearly cosmopolitan, fairly diverse family of shorebirds that present an identification challenge to many birders. Of the 25 species that occur on Hispaniola, all but the Willet are passage migrants or visitors from their breeding grounds, primarily in the far north. Sandpipers are characterized by their fairly long legs and neck and long, thin, straight to slightly down-curved or upturned bill. Most wade in shallow water or on wet, exposed flats where they probe in the mud or sand for invertebrates. Most species are highly gregarious and often occur in mixed feeding assemblages. This sociality aids in their identification because it often allows close comparison of several similar species. The most useful characters to note are relative size, structure, bill shape, and call. Plumages are individually and seasonally variable, and several species remain late enough in spring that they can be seen in their breeding plumage. Among all species that occur on Hispaniola, both sexes have similar plumages.

GREATER YELLOWLEGS—*Tringa melanoleuca* **Plate 19**
Passage Migrant; Non-breeding Visitor

Description: 29–33 cm long; 171 g. A fairly large shorebird with a long, straight bill and long, orangish yellow legs. In non-breeding plumage, has gray to grayish brown upperparts with pale spots, and pale underparts with faint streaking on throat and chest. Breeding plumage is darker above and more conspicuously spotted; underparts are more heavily streaked, with dark bars on flanks. In flight, dark above and lacking wing bars, but with white uppertail coverts.
Similar species: Often occurs with more common Lesser Yellowlegs, allowing size comparison. Greater Yellowlegs has a relatively longer, thicker bill that often appears slightly upturned and two-toned, being paler at the base. Willet has a thicker, straight bill, gray legs, and prominent white wing stripe in flight.
Voice: Loud, ringing whistle of 3–4 notes, *tew tew tew* or *klee-klee-cu*. This species' sharp 3- to 4-note call is louder than the more mellow 1- to 2-note call of Lesser Yellowlegs.
Hispaniola: Regularly occurs in coastal marshes, mudflats, and lagoons, and may be found in smaller numbers at inland saline and freshwater lakes, ponds, and along river courses. Among satellite islands, reported from Île à Vache, Île de la Gonâve, and Islas Beata, Catalina, and Saona. Peak spring migration period is March and April, the last migrants usually gone by

late May. Non-breeding stragglers are found occasionally in June. The first presumed southbound migrants are known from early July, and peak fall migration period is mid-August to early October.
Status: Common passage migrant and regular non-breeding visitor in variable numbers.
Comments: Forages actively in shallow water, often running after prey. Often bobs head and body emphatically when alarmed. Feeds on small aquatic and terrestrial invertebrates, small fish, frogs, and tadpoles.
Range: Breeds from s. Alaska east to c. Labrador, south to c. British Columbia, s. Quebec, and Nova Scotia. Occurs in non-breeding season from s. Oregon and se. New York south to Tierra del Fuego. In West Indies, a regular passage migrant and non-breeding visitor; non-breeders occasionally summer.
Local names: DR: Patas Amarillas Mayor, Patas Amarillas Grande, Patiamarillo Mayor; **Haiti:** Gwo Bekasin Janm Jòn, Bécasse à Pattes Jaunes, Grand Chevalier.

LESSER YELLOWLEGS—*Tringa flavipes* **Plate 19**
Passage Migrant; Non-breeding Visitor

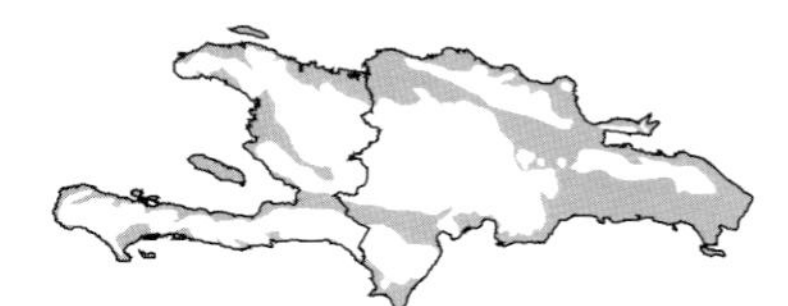

Description: 23–25 cm long; 81 g. A medium-sized, slender, long-necked and long-legged shorebird. Distinctive orangish yellow legs and thin, straight bill help identify this species. Winter plumage is mottled gray-brown above, white below, with faint gray streaking on neck and breast. Breeding plumage is darker and more heavily mottled above; underparts have heavy brownish streaking on throat and breast, and flanks are barred irregularly blackish. In flight, dark above with white uppertail coverts.
Similar species: Most easily distinguished from Greater Yellowlegs by its thinner, shorter bill (only slightly longer than head), and its 1- or 2-note call. See also Solitary and Stilt sandpipers.
Voice: A 1- (usually) or 2-note call, *tu* or *tu-tu*, softer and flatter than that of Greater Yellowlegs.
Hispaniola: Occurs at coastal lagoons and marshes and at inland lakes and ponds, principally at low elevations. Also reported from almost all of the major satellite islands. Peak spring migration period is March to mid-April, peak fall migration period mid-August to September.
Status: Regular spring and fall passage migrant and non-breeding winter visitor in variable numbers.
Comments: Typically flocks; often seen together with Greater Yellowlegs. Feeds actively, picking and jabbing at invertebrate prey, and often running through shallow water.
Range: Breeds from c. Alaska south to c. British Columbia, se. Manitoba, and c. Quebec. Winters from coastal c. California and New Jersey south to Tierra del Fuego. In West Indies, a regular passage migrant and winter resident; non-breeders occasionally summer.
Local names: DR: Patas Amarillas Menor, Patiamarillo Menor; **Haiti:** Bekasin Janm Jòn, Bécassine à Pattes Jaunes, Petit Chevalier.

SOLITARY SANDPIPER—*Tringa solitaria* **Plate 15**
Passage Migrant; Non-breeding Visitor

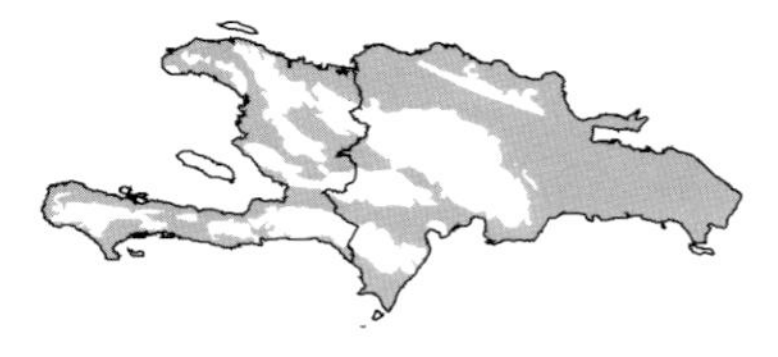

Description: 19–23 cm long; 48 g. A slender, fairly small, dark sandpiper with a conspicuous narrow white eye-ring. Non-breeding plumage is olive-brown above, finely spotted with white. Head, neck, and chest are grayish and faintly streaked white; rest of underparts are white. Breeding plumage is darker above with more pronounced white spotting, and head, throat, and breast are more heavily streaked dark. Legs are olive to olive green in all plumages; bill is thin, straight, and of medium length. In flight, entirely dark above, including upper- and underwings, and rump and central tail feathers are dark with heavy bars on a white-edged tail. The flicking, erratic wing beats are distinctive. When landing, frequently holds wings straight up, then slowly closes them.

Similar species: Smaller than similar Lesser Yellowlegs; white eye-ring and spotting on back are distinctive, and legs are greenish instead of yellow. Spotted Sandpiper is smaller, lacks white speckling above, and has distinctive white wing stripe in flight.

Voice: Clear, high, rising series of whistles when alarmed, *peet-weet-weet*. Also a soft *pip* or *weet* when undisturbed.

Hispaniola: Occurs primarily at freshwater lakes, ponds, and riverbanks in the lowlands and at middle elevations islandwide. Also reported from Isla Catalina.

Status: Regular spring and fall passage migrant and non-breeding visitor in variable numbers.

Comments: Generally occurs singly, even during migration. Like yellowlegs, frequently bobs and teeters its body.

Range: Breeds from coastal Alaska east to s. Labrador, south to c. British Columbia and n. Minnesota. Winters from se. Texas and w. Mexico south to Peru and Argentina. In West Indies, a fairly common passage migrant throughout and rare non-breeding visitor.

Local names: DR: Playero Solitario, Andarios Solitario; **Haiti:** Bekasin Dlo Dous, Chevalier à Aile Noire, Chevalier Solitaire.

WILLET—*Catoptrophorus semipalmatus* **Plate 18**

Non-breeding Visitor; Breeding Resident

Description: 33–41 cm long; 215 g. A large, stocky, light gray shorebird with gray legs and a long, thick bill. This species is often inconspicuous until it opens its wings to display a bold black and white wing pattern. Non-breeding birds are uniformly pale gray above and on throat and breast, but white on belly and undertail coverts. Breeding birds are mottled grayish brown above, streaked and barred black on head, neck, and breast. In flight, prominent white stripe on black flight feathers, and white rump, are distinctive, as are loud calls.

Similar species: Other large shorebirds, including Whimbrel and both yellowlegs, may be mostly grayish brown in the non-breeding season, but none have Willet's distinctive wing pattern.

Voice: Often recognized by its clear, loud, piercing whistle, *kyaah yah* or *kyeh yeh-yeh*, and other sharp calls, including a persistent, clipped *keh-keh-keh* or *kleep-kleep*. Territorial song is a clear, rolling *pill WILL WILLET*.

Hispaniola: In the non-breeding season Willets may be distributed around the entire coastline, including Île de la Gonâve, Île à Vache, and Islas Catalina and Saona, but the species is a very local breeder or summer resident with nest records only from Isla Beata and Las Salinas, DR. Peak fall migration period is September to October, and peak northbound migration period is late March to early May.

Status: Regular but local breeding resident and moderately common non-breeding visitor at the coast. Field surveys suggest its numbers may have increased since the 1930s.

Comments: This is the only species of sandpiper known to nest on Hispaniola. Very vocal and territorial during the breeding season. Feeds mainly on small crustaceans, mollusks, aquatic insects, worms, and occasionally small fish.

Nesting: Lays 3–4 buff-colored eggs which are heavily splotched. The nest is a slightly lined depression in the sand. Breeds April to July.

Range: Breeds on Atlantic and Gulf Coasts of U.S. from Nova Scotia to ne. Mexico; also an isolated inland population in w. U.S. Occurs in non-breeding season from n. California and c. Virginia along coasts throughout Middle and South America to Brazil and Chile. In West Indies, breeds in Bahamas, most of Greater Antilles, and n. Lesser Antilles south to St. Martin; uncommon to rare elsewhere.

Local names: DR: Chorlo, Vadeador Aliblanco; **Haiti:** Bekasin Zèl Blanch, Bécasse à Aile Blanche, Salomon, Chevalier Semipalmé.

SPOTTED SANDPIPER—*Actitis macularius* **Plate 16**
Passage Migrant; Non-breeding Visitor

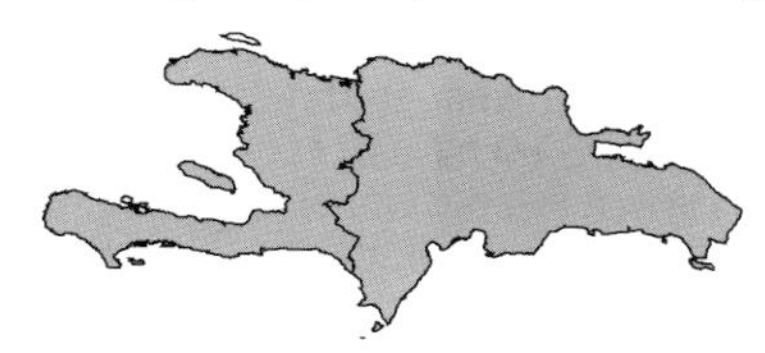

Description: 18–20 cm long; 40 g. A fairly small, short-necked sandpiper with a near-constant, exaggerated rocking motion. Non-breeding birds are plain grayish brown above, with gray extending from neck onto breast; white below. Legs are greenish; bill is dusky to fleshy. Breeding birds have upperparts with indistinct black spotting and barring, white underparts with bold black spotting, and a black-tipped orange bill. In flight, shows a short, white wing stripe. Flies low over the water with shallow, stiff, rapid wing beats.
Similar species: Rocking motion is unmistakable.
Voice: Flight call is a high, whistled *weet-weet* or *peet-weet-weet-weet,* lower and slower than Solitary Sandpiper. Invariably calls when flushed.
Hispaniola: Most common in the coastal lowlands but also regularly found at the edges of ponds, lakes, rivers, and stream courses in the interior, and reported from Île à Vache, Île de la Gonâve, and Islas Beata and Saona. Peak fall migration period is early August to mid-September. Peak spring migration period is late March and April.
Status: Common spring and fall passage migrant and non-breeding visitor, rarely lingering from late May to mid-July, and though observed courting, never known to have nested. General status and abundance appear to be little changed since 1900.
Comments: Usually seen singly or in small groups. Not generally found on mudflats.
Range: Breeds from c. Alaska east to ne. Manitoba and Newfoundland, south to s. California, n. Texas, and North Carolina. Occurs in non-breeding season from sw. British Columbia and South Carolina south to Argentina and Chile. In West Indies, a common passage migrant and non-breeding visitor; non-breeders are occasionally found in summer.
Local names: DR: Playerito Manchado, Andarios Maculado; **Haiti:** Bekasin Zèl Tramble, Chevalier Branle-queue, Chevalier Grivelé.

WHIMBREL—*Numenius phaeopus* **Plate 18**
Passage Migrant; Non-breeding Visitor

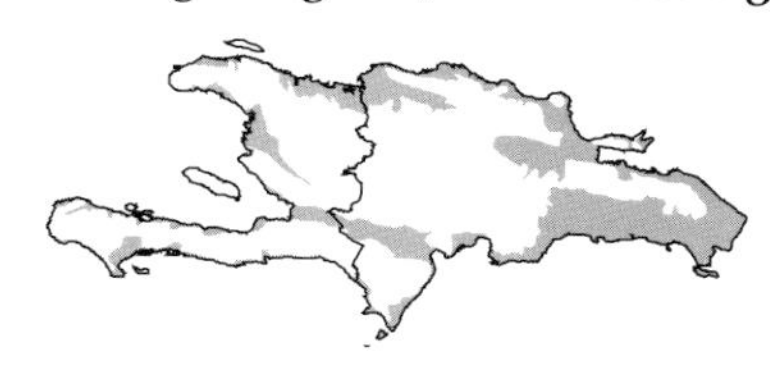

Description: 38–46 cm long; male 355 g, female 404 g. A large, sturdy shorebird with a long, down-curved bill. Upperparts are dark brown marked with pale buff; underparts are pale buff streaked dark brown on neck and breast; lower belly is pale and mostly unmarked. Distinctly dark crown has a well-defined buff median stripe. In flight, shows grayish brown and barred underwings.
Similar species: No other species on Hispaniola has a similarly long, down-curved bill.
Voice: Flight call is a hard, rapid, tittering whistle, *pip-pip-pip-pip* or *kee-kee-kee-kee.*
Hispaniola: Occurs in the lowlands at ponds, swamps, and marshes, and along the coastline islandwide. Has occurred annually on migration from late July to October, although most often in August and September and again in April to May. May rarely linger in summer, but has not bred. Also recorded in December, suggesting that this species occurs as a non-breeding-season resident in small numbers. Favorite locales include Las Salinas, Bahía de las Calderas, and Puerto Alejandro, DR; also reported from Islas Beata and Saona.
Status: Uncommon spring and fall passage migrant and rare non-breeding visitor.
Range: Breeds from n. Alaska east to Southhampton Island, south to sw. Alaska and James Bay. Occurs in non-breeding season from coastal c. California and South Carolina south to Chile and s. Brazil. Also widespread in Eurasia. In West Indies, a regular passage migrant and rare non-breeding-season and summer visitor.
Local names: DR: Zarapito Trinador, Playero Picocorvo; **Haiti:** Kouli, Kalanderik, Courlis.

HUDSONIAN GODWIT—*Limosa haemastica* **Plate 18**
Vagrant

Description: 36–42 cm long; male 222 g, female 289 g. A large, dark-legged shorebird with a long, slightly upturned bill, pinkish at base but dark at tip, and black underwing coverts. Non-breeding birds are plain gray overall; paler below, especially on belly, with a distinct, short white supercilium. Breeding birds have dark brown upperparts mottled with light spots, and a gray neck; underparts are dark reddish brown with scattered dark barring. Female is paler below than male. Immatures resemble non-breeding adults, but upperparts browner with scaly feather edgings. In flight, shows narrow white stripe on upperwing, white underwing stripe bordered by distinctive blackish underwing coverts, and black tail with a broad, white base.
Similar species: Marbled Godwit is buffy brown overall, with plain rump and tail, and lacks a white wing stripe. Willet has a shorter, heavier, straight bill, blue-gray legs, and broader, bolder white wing stripe.
Voice: Usually silent during migration, but may give a high *god-wit* or *quik-quik*, with each note rising, and strongly accented on the second note. Also a single *whit* or *week* note when alarmed.
Hispaniola: Inhabits grassy freshwater pond edges and mudflats. Only 3 records: 2 collected in 1930 at Hato del Yaque, DR; another collected in Haiti in 1930 at an unknown locale; and 1 seen at Puerto Alejandro, DR, in September 1998.
Status: Vagrant.
Comments: Breeds very locally in northern Canada and Alaska and then undertakes an extraordinary migration to its wintering grounds in southern South American. This flight includes nonstop legs of several thousand kilometers over the Atlantic Ocean. It is believed that large numbers pass over the West Indies at this time, only rarely stopping. The return migration northward is more westerly, apparently along the Central American coast. The worldwide population of this species is estimated to number only 50,000 birds.
Range: Breeds in scattered locales from w. and s. coastal Alaska to nw. British Columbia and Hudson Bay. Migrates in fall from maritime provinces and New England by transoceanic flights to s. Brazil and Argentina. In West Indies, recorded in fall rarely from Bahamas, Cuba, Hispaniola, Puerto Rico, Virgin Islands, St. Martin, St. Barthélemy, Antigua, Guadeloupe, Dominica, Martinique, and St. Lucia; regular at Barbados.
Local names: DR: Barga Aliblanca; **Haiti:** Kouli Vant Blanch, Barge Hudsonienne.

MARBLED GODWIT—*Limosa fedoa* **Plate 18**
Vagrant

Description: 42–48 cm long; male 320 g, female 421 g. A large shorebird with a long, bicolored, slightly upturned bill and long, dark legs. Non-breeding plumage is tawny brown above, speckled and barred dark brown and black; underparts are buffy and unmarked. Breeding plumage is similar, except for fine black streaking on breast, and barring elsewhere on underparts. In flight, from above shows blackish primary wing coverts and a plain rump; from below, shows cinnamon-colored underwing coverts contrasting with paler flight feathers.
Similar species: Hudsonian Godwit is grayish overall with a black and white tail and white wing stripe. Whimbrel has a long, decurved bill and lacks cinnamon-colored underwing coverts.
Voice: Non-breeding birds give a nasal, laughing chatter, *ah-ha* or *ah-ahk*.
Hispaniola: Occurs at mudflats and marshy areas. Only 2 records, both from the harbor mudflats at Port-au-Prince, Haiti, in November 1961 and March 1963.
Status: Vagrant.
Range: Breeds from Alaska Peninsula east to James Bay, south to c. Montana, North Dakota, and nw. Minnesota, but principally in northern prairies. Occurs in non-breeding season coastally from c. California and North Carolina south to Colombia, Venezuela, and rarely

n. Chile. In West Indies, rare in Cuba, Jamaica, Haiti, Virgin and Cayman islands, St. Kitts, Guadeloupe, Martinique, Barbados, Carriacou, and Grenada.
Local names: DR: Barga Jaspeada; **Haiti:** Kouli Takte, Barge Marbrée.

RUDDY TURNSTONE—*Arenaria interpres* **Plate 16**
Passage Migrant; Non-breeding Visitor

Description: 21–26 cm long; 115 g. A medium-sized, stocky shorebird with short, orange legs and a small, wedge-shaped, slightly upturned, black bill. Non-breeding adults are mottled grayish brown and blackish above; head is brownish with smudgy black and white markings; breast has a broad, brownish black band; and belly is white. Breeding birds have reddish brown upperparts with blackish patches, and striking black and white facial markings extending to neck and throat. In flight, shows a distinctive pattern of white markings on upperwings, back, and tail.
Similar species: Orange legs and striking pattern on upperside are distinctive.
Voice: Flight call is a low, guttural rattle, *ki-ti-tuk* or *kut-a-kut,* increasing in volume; also a single, sharp *kek.*
Hispaniola: Inhabits coastal lagoons, mudflats, and beaches; only rarely found at interior saline lakes such as Étang Saumâtre, Haiti, and Lago Enriquillo, DR. The first fall migrants may arrive by mid-July, but peak fall migration period is late August to late September. Peak spring migration period is early March and April, and the last northbound migrants are usually gone by mid- to late May. Recorded from several major satellite islands, including Île de la Gonâve, Cayos Siete Hermanos, and Islas Beata, Catalina, and Saona.
Status: Common spring and fall passage migrant and non-breeding visitor. A few non-breeding birds occasionally linger through the summer.
Comments: Typically forms looser flocks than most other sandpipers. Its common name derives from its distinctive foraging behavior of overturning small stones and shells in search of invertebrate prey. On Caribbean beaches, turnstones use their bill and forehead to push over piles of seaweeds in search of prey.
Range: Breeds from n. Alaska east to Baffin Island, south to Yukon River delta and Mansel Island. Occurs in non-breeding season coastally from n. Califiornia and Massachusetts south to Tierra del Fuego. Also widespread in Eurasia. In West Indies, a regular passage migrant and non-breeding visitor; non-breeders rarely linger in summer.
Local names: DR: Playero Turco, Vuelvepiedras Rojizo; **Haiti:** Eryis, Tournepierre.

RED KNOT—*Calidris canutus* **Plate 15**
Passage Migrant

Description: 23–25 cm long; 135 g. A stocky, medium-sized shorebird with short legs, neck, and bill. Non-breeding birds are plain gray above, dingy white below, with indistinct dusky streaking on breast and flanks. Breeding birds are dark gray above with buffy and rusty feather edgings and spots and an orangish red face and underparts. Immature resembles winter adult but has scaly pattern on back and a faint buffy wash on breast. In all plumages, legs are dull greenish. In flight, shows pale gray underwing coverts; narrow, white wing stripe; and a unique pale gray rump with fine barring.
Similar species: Non-breeding Dunlin are similarly pale and nondescript, but the bill is longer and drooping. Dowitchers have much longer bills and a white wedge extending from tail to lower back.

Voice: Soft, low, hoarse 2-note *chuh chuh* or *kuh kuh*.
Hispaniola: Occurs at sandy tidal flats islandwide during migration. Most migrants typically overfly the entire West Indies. Spring records include birds from early March to mid-April at mudflats at Port-au-Prince harbor, Las Salinas, and Monte Cristi, DR, and Isla Saona. The bulk of southbound migration in fall is well to the east over the Atlantic Ocean. May rarely remain throughout the non-breeding season.
Status: Rare passage migrant in fall, but more common during northbound migration in spring.
Comments: Noted for its extraordinary long-distance migrations, which include nonstop flights that may exceed 2,500 km between staging areas. May congregate in huge migratory flocks, especially at spring staging areas on the mid-Atlantic Coast, where tens of thousands of northbound migrants gather to feed on eggs of horseshoe crabs.
Range: Breeds from nw. Alaska east to Ellesmere Island, south to Victoria and Southampton islands. Occurs in non-breeding season at coastal sites from n. California and Massachusetts south to Tierra del Fuego. Also widespread in Eurasia. In West Indies, regular but scarce migrant in Bahamas and Greater Antilles, casual in Lesser Antilles except Barbados where regular.
Local names: DR: Playero Pechirrojo; **Haiti:** Mòbèch, Bécasseau Maubèche.

SANDERLING—*Calidris alba* **Plate 16**

Passage Migrant; Non-breeding Visitor

Description: 18–22 cm long; 57 g. A small, plump, active shorebird with a short, stout bill. Non-breeding plumage is the palest among sandpipers, with white underparts, white face, and light gray upperparts. Often shows a black mark on bend of folded wing. Bill, legs, and feet are black. In breeding plumage, head, neck, and breast are reddish brown; back is brown mottled black and rufous; belly and undertail coverts are white. Immature resembles winter adult, but upperparts are blackish spangled with white and buff. In flight, shows a black leading edge of wing bordered by a conspicuous white wing stripe above; underwing is bright white.
Similar species: Non-breeding Red Knot is larger, darker, and lacks black shoulder patch. Sanderling is distinctly paler than all other small *Calidris* sandpipers.
Voice: Distinctive short, sharp *kwit* or *klip*.
Hispaniola: Found at coastal lagoons, mudflats, and beaches, rarely at inland saline lakes such as Étang Saumâtre, Haiti, and Lago Enriquillo, DR. Regularly reported in small flocks of 5–10 at favored localities in the non-breeding season. First fall migrants may appear by early August, but peak migration period is early September to mid-October. Peak spring migration period is March to mid-April, and all northbound migrants have usually left by late May. Recorded from several major satellite islands, including Île de la Tortue, Cayos Siete Hermanos, and Isla Beata.
Status: Moderately common spring and fall passage migrant and non-breeding visitor, with non-breeding birds rarely remaining through June and July.
Comments: Often seen along sandy beaches, where flocks typically run smoothly and rapidly, advancing and retreating with each wave. Forages by picking and probing for small crustaceans, bivalve mollusks, and marine worms.
Range: Breeds from n. Alaska east to Ellesmere Island, south to Northwest Territories and Baffin Island. Occurs in non-breeding season coastally from British Columbia and Maine south to Tierra del Fuego. Widespread in Eurasia. In West Indies, a regular passage migrant and non-breeding visitor, occasionally lingering through breeding season.
Local names: DR: Playerito Blanquito, Playero Arenero; **Haiti:** Bécassine Blanche, Bekasin Blanch, Bécasseau Sanderling.

SEMIPALMATED SANDPIPER—*Calidris pusilla* **Plate 17**

Passage Migrant; Non-breeding Visitor

Description: 13–15 cm long; 31 g. In all plumages, identified by its small size, black legs, and medium-length, straight black bill. This is the most common small sandpiper on Hispaniola and a standard for identification against which all others should be compared. In non-breeding plumage, grayish brown above with an indistinct whitish supercilium; whitish below with diffuse streaking on chest. Breeding birds have upperparts with reddish brown and buffy feather edgings, brightest on the crown and auriculars; breast is more heavily steaked dark brown. Immature resembles breeding adult and is more uniformly scaly and less rusty above, with a dingy buff-colored wash on chest. In flight, shows a narrow white wing stripe and black center of rump and tail.
Similar species: Western Sandpiper is very similar and difficult to distinguish but has a slightly longer bill, often down-curved at the tip. These two species are best identified by calls. White-rumped Sandpiper has longer wings projecting beyond the tip of tail at rest, and a white rump. Least Sandpiper is smaller, browner above, with greenish to yellowish legs.
Voice: Typical flight call is a short, sharp, husky *chrit* or *chert*. Feeding flocks often are very vocal, twittering or murmuring.
Hispaniola: Found at coastal lagoons, mudflats, and beaches but also regularly though uncommonly on the shores of interior ponds, lakes, and rivers. Earliest southbound migrants can appear in late July, with peak fall migration period mid-August to early October. Peak spring migration period is March to late April, and all northbound birds are usually gone by early May. Not recorded from late May to early July. In fall migration, counts of 20–50 are not unusual, though the primary southbound migration route for this species' eastern breeding population lies well to the east over the Atlantic Ocean. Recorded from several major satellite islands, including Île de la Gonâve, Cayos Siete Hermanos, and Islas Beata and Saona.
Status: Common spring and fall migrant and non-breeding visitor. During the peak of its southward migration from August to October, it is probably the most abundant shorebird in the West Indies.
Comments: Typically occurs in large flocks. Feeds on small aquatic and marine invertebrates captured by pecking and probing; generally pecks faster than other *Calidris* sandpipers. Surprisingly little is known of this species' non-breeding-season ecology, although aerial surveys have estimated more than 2 million birds in South America.
Range: Breeds from n. Alaska east to n. Labrador, south to Yukon River and s. Labrador. Occurs in non-breeding season from Honduras and Hispaniola south along coasts of South America, primarily Suriname and French Guiana. In West Indies, a common fall and regular spring migrant throughout; casual at other seasons.
Local names: DR: Playerito Semipalmado; **Haiti:** Ti Bekasin Janm Nwa, Bécassine à Pattes Noires, Maringouin, Bécasseau Semipalmé.

WESTERN SANDPIPER—*Calidris mauri* **Plate 17**

Passage Migrant; Non-breeding Visitor

Description: 14–17 cm; 23 g. A small sandpiper distinguishable only with care from Semipalmated Sandpiper. Bill is black, relatively long and heavy at base, and narrower and drooping at tip. Best identified by its voice. Non-breeding plumage is uniform brownish gray above, whitish below, with a finely streaked breast band. In breeding plumage, upperparts are brownish mottled black, with a reddish brown crown, auriculars, and upper scapu-

lars; upper breast is more heavily streaked. Immature is similar to breeding adult, but chest is finely streaked and washed buffy. In flight, shows a narrow white wing stripe; rump and center of tail are black, with white outer tail feathers.
Similar species: Relatively long bill, heavy at the base and down-curved at the tip, and relatively pale face help distinguish Western Sandpiper from the very similar Semipalmated Sandpiper. However, both species overlap substantially with regard to bill shape and size. They are most reliably separated by flight calls. See also White-rumped and Baird's sandpipers.
Voice: Typical flight call is a high, thin *kreep* or *cheet,* coarser and more querulous than that of Semipalmated Sandpiper.
Hispaniola: Found on coastal mudflats and beaches, occasionally at inland saline lakes. The first southbound migrants are to be expected by mid-August, with the peak fall migration period in September and October. Scattered non-breeding-season records exist, but more common again during spring migration from March to mid-April. Recorded from two major satellite islands: Cayos Siete Hermanos and Isla Saona.
Status: Uncommon but regular spring and fall migrant, some birds remaining throughout the non-breeding season. The species may be more common than recorded, but it is often overlooked because of its similarity to the more abundant Semipalmated Sandpiper with which it often occurs.
Comments: Western Sandpiper generally feeds in deeper water than Semipalmated Sandpiper and probes more methodically.
Range: Breeds on St. Lawrence and Nunivak islands and along Alaskan coast from Bristol Bay to Pt. Barrow. Occurs in non-breeding season from Washington and s. New Jersey south to Suriname and Peru. In West Indies, a regular passage migrant and non-breeding visitor; non-breeders are rare in summer.
Local names: DR: Playero Occidental; **Haiti:** Ti Bekasin Bèk Long, Maubèche.

LEAST SANDPIPER—*Calidris minutilla* **Plate 17**
Passage Migrant; Non-breeding Visitor

Description: 13–15 cm long; 23 g. The world's smallest shorebird, with yellowish green legs and a short, thin bill that droops slightly at the tip. Also distinguished by its overall brown coloration and finely, but heavily, streaked breast. Non-breeding birds are mottled grayish brown above and on breast, with a white belly. Breeding plumage is darker brown and more heavily mottled above, with reddish brown tints. In flight, Least Sandpipers are dark above with a faint wing stripe.
Similar species: Yellowish green legs and darker brown upperparts distinguish Least Sandpiper from all other small sandpipers.
Voice: Typical flight call is a high, reedy trill, *brreep* or *krreet;* also a thin, mellow twittering in flocks.
Hispaniola: Found principally at coastal mudflats and lagoons and less frequently inland at the edges of ponds and lakes. First southbound migrants are to be expected by late July, sometimes in flocks of hundreds. Peak fall migration period is from early August to early October. Peak spring migration period is March to mid-April. All northbound birds are usually gone by mid-May, though the species may occur rarely from late May to mid-July. Recorded from several major satellite islands, including the Cayemites, Île de la Gonâve, and Islas Beata and Saona.
Status: Common spring and fall passage migrant and regular non-breeding visitor in variable numbers.
Comments: Freely intermingles with other *Calidris* sandpipers, but tends to prefer drier margins and upper edges of mudflats.
Range: Breeds from w. Alaska east to n. Quebec and n. Labrador, south to nw. British Columbia, ne. Manitoba, and Newfoundland. Occurs in non-breeding season from Oregon and

New Jersey south to Argentina and Chile. In West Indies, a regular passage migrant and non-breeding visitor.
Local names: DR: Playerito Menudo; **Haiti:** Ti Bekasin Piti, Bécasseau Minuscule.

WHITE-RUMPED SANDPIPER—*Calidris fuscicollis* **Plate 17**
Passage Migrant

Description: 15–18 cm long; 35 g. A medium-sized sandpiper with long wings and a distinctive white rump. Bill is short and straight, and lower mandible has a pale, reddish base that is sometimes visible. Wings are longer than in any other *Calidris* sandpiper, with primary tips extending well beyond the end of tail at rest. In non-breeding plumage, gray to brownish gray above and on upper breast, giving a hooded appearance; rest of underparts are whitish. Breeding birds are browner, with reddish brown tints on crown, upper back, and auriculars; streaking heavier and more extensive on breast and sides. Immature resembles breeding adult, with breast and sides less distinctly streaked and washed buffy gray. In all plumages, shows a distinct whitish supercilium, thickest behind eye. In flight, white rump distinguishes this from other small sandpipers.
Similar species: White-rumped Sandpiper is distinctly larger than Semipalmated and Western sandpipers, and is darker gray in non-breeding plumage and darker brown in breeding plumage. Breast is more heavily streaked, and when standing, wings extend beyond tail. Best distinguished from Baird's Sandpiper by entirely white rump.
Voice: Distinctive high, thin, mouselike squeak, *peet* or *jeet*. Also gives a thin, high-pitched twitter in flocks.
Hispaniola: Found at mudflats and borders of still water, principally along the coast. All known reports from Hispaniola are from DR. Spring records from mid-March to May at Isla Beata, Barahona, Las Salinas, and Lago Enriquillo. Fall records from late August to late September at Hato Nuevo, Puerto Plata, Las Salinas, and Cabo Rojo. One at Las Salinas, DR, in 1996 may have remained through the non-breeding season. This sandpiper normally follows a migratory path that takes it well to the east of Hispaniola.
Status: Uncommon passage migrant. Because this species undertakes a series of long-distance, nonstop flights from its arctic breeding grounds, it is particularly reliant on high-quality stopover sites where food is abundant.
Comments: Associates freely with other sandpipers. Regularly occurs on tidal flats, but the species is easily overlooked and this may account for the paucity of records. Consumes invertebrate animal prey found in muddy substrates away from emergent vegetation by making 2–3 quick probes, then running a short distance to repeat.
Range: Breeds from n. Alaska east to Bathurst Island, south to Yukon River and s. Baffin Island. Occurs in non-breeding season in South America east of Andes from s. Brazil south to Tierra del Fuego. In West Indies, a regular fall and rare spring passage migrant.
Local names: DR: Playero de Rabadilla Blanca; **Haiti:** Bekasin Ke Blanch, Bécassine Queue Blanche, Bécasseau à Croupion Blanc.

BAIRD'S SANDPIPER—*Calidris bairdii* **Plate 17**
Vagrant

Description: 14–18 cm long; 40 g. A relatively broad-breasted and short-legged sandpiper with a short, straight, black bill and very long wings which extend well beyond the tail at rest. In non-breeding plumage, gray-brown above with faint reddish brown tints and pale edges to feathers, resulting in a decidedly scaled appearance. Breast is buffy and finely streaked brown; belly is white. Breeding birds are similar to non-breeding individuals but brighter buff on face and breast, with scaling effect on upperparts more pronounced. Immatures resemble

breeding adults but are often brighter buff on face and chest, and less distinctly streaked below. In flight, note the long wings and indistinct white wing stripe.
Similar species: Similarity of Baird's Sandpiper to other *Calidris* species requires extreme care in its identification. Distinguished from Semipalmated and Western sandpipers by its larger size, and wings that extend noticeably beyond tail. White-rumped and Baird's sandpipers are similarly long winged and best separated in flight. White rump of Baird's Sandpiper is divided into lateral patches by a dark central stripe, whereas in White-rumped Sandpiper this white patch is continuous. Baird's Sandpiper has a buffier breast than most similar species; it also picks for its food rather than probes.
Voice: In non-breeding period, flight call is a rough, somewhat trilled *krreep*.
Hispaniola: Two DR reports of single birds identified in direct comparison with other sandpipers near Cabo Rojo in late September 1999 and early April 2000. To be looked for primarily around upper edges of inland wetland habitats, including dry edges among vegetation. Often occurs some distance from water.
Status: Vagrant.
Comments: Typically occurs singly or in small groups, unlike other sandpipers which are more gregarious. The status of this species is poorly known since it is easily overlooked because of its similarity to more common sandpipers. It should be expected primarily during its southbound migration in September and October and to a lesser extent on its northward return in March and April.
Range: Breeds from Wrangel Island, ne. Siberia, and nw. Alaska east to Ellesmere Island, south to c. Alaska, and east to n. Mackenzie, Baffin Island, and nw. Greenland. Occurs in non-breeding season in South America from Ecuador and Chile south to Tierra del Fuego. In West Indies, a vagrant in Cayman Islands, DR, Virgin Islands, Martinique, Dominica, St. Lucia, Barbados, and St. Vincent.
Local names: DR: Playero de Baird; **Haiti:** Bekasin Zèl Long.

PECTORAL SANDPIPER—*Calidris melanotos* **Plate 17**
Passage Migrant

Description: 20–24 cm long; male 70 g, female 52 g. A medium-sized sandpiper with yellowish green legs and a sharp demarcation between heavily streaked, brownish breast and white belly. Non-breeding birds have gray-brown upperparts, head, and breast. Breeding plumage is more mottled brown above, with the breast more heavily streaked with black. In flight, the sharp breast demarcation is evident; also note a fine white wing stripe, and white rump divided by a black bar.
Similar species: Only Pectoral Sandpiper shows such a sharp breast demarcation.
Voice: Low, reedy, harsh *trrip* or *brrrp*.
Hispaniola: Typically found on mudflats of fresh or brackish bodies of water, in agricultural areas and savannas, usually in the coastal lowlands but occasionally in the interior. Peak fall migration period is mid-August to October. Numbers seen at this time are typically 1–5, but counts up to 100 have been recorded. The only known non-breeding-season records are for Las Salinas. There are only 2 known spring records: from Lago Enriquillo and Isla Beata, DR.
Status: Uncommon passage migrant, much more numerous in fall than spring; occasional in the non-breeding season; once recorded in early summer.
Comments: Typically flocks, but not in association with other sandpipers. After nesting in the Arctic, most Pectoral Sandpipers migrate southward over the western Atlantic Ocean, many birds probably passing through the West Indies en route to their non-breeding grounds in southern South America. The northward migration generally bypasses most of the West Indies.
Range: Breeds from n. Alaska east to Baffin Island, south to w. Alaska and n. Ontario. Occurs in non-breeding season in South America from Peru and s. Brazil south to Chile and Ar-

gentina. In West Indies, a passage migrant, regular in fall, rare in spring, casual in non-breeding period.
Local names: DR: Playero Pectoral; **Haiti:** Bekasin Fal Nwa, Bécassine à Poitrine Noire.

DUNLIN—*Calidris alpina* **Plate 16**
Vagrant

Description: 16–22 cm long; 60 g. A medium-sized sandpiper identified in all plumages by its relatively long and heavy bill with distinctively drooping tip; short-necked, hunched appearance; and black legs. Non-breeding birds have a light brownish gray back with a diffuse brownish wash on breast and head; rest of underparts are white. Breeding birds have a reddish brown back and cap, whitish underparts with fine streaking on breast, and a large, distinct, black belly patch. In flight, shows a white wing stripe, and a white rump divided by a black bar.
Similar species: Red Knot is similarly pale in non-breeding plumage but larger, with a short, straight bill. Non-breeding birds are also similar to Western and Semipalmated sandpipers, but both are smaller and have distinctly shorter bills. White-rumped Sandpiper has a distinctly streaked breast and sides, all-white rump, and longer wings. Breeding plumage is unmistakable.
Voice: Flight call is a distinctive, raspy, rolled *tzeep* or *jeeep*.
Hispaniola: One spring record of 1 bird in bright breeding plumage on the coast near Barahona, DR, in late April 1996; 1 fall record of 2 birds at Las Salinas, DR, in late August 1999; and 1 record in the non-breeding period of 2 individuals at Las Salinas in December 1998. To be looked for at borders of still water, particularly mudflats along the coast.
Status: Vagrant.
Range: Breeds from n. Alaska east to n. Mackenzie and Baffin Island, south to ne. Manitoba and s. Hudson Bay; also in n. Palearctic. Occurs in non-breeding season on Atlantic coast from Massachusetts south to Florida, west to Texas, and south to Yucatán Peninsula. Widespread in Eurasia. In West Indies, reported from Bahamas, Greater Antilles, St. Kitts, Dominica, and Barbados.
Local names: DR: Dunlin; **Haiti:** Bekasin Vant Nwa, Bécasseau Variable.

STILT SANDPIPER—*Calidris himantopus* **Plate 15**
Passage Migrant; Non-breeding Visitor

Description: 20–23 cm long; male 54 g, female 61 g. A medium-sized sandpiper with relatively long, greenish legs and a long bill that droops slightly at tip. In all plumages, shows a well-defined whitish supercilium. Non-breeding birds are unmarked, pale brownish gray above, whitish below, with light streaking on neck and upper breast. Breeding birds are dark brown above, edged whitish to pale rusty, with a reddish brown auricular; underparts are whitish and heavily barred. Immature resembles breeding adult above but is unbarred below and lacks rusty auriculars. In flight, shows a white rump and pale gray to whitish tail; wings are unmarked.
Similar species: Both dowitcher species are larger with longer, straighter bills. Yellowlegs are also larger, with straight bills and bright yellow legs. Solitary Sandpiper has a shorter, straight bill, distinct eye-ring, and boldly barred tail.
Voice: Low, soft, muffled *jeew* or *toof*.
Hispaniola: Occurs at coastal lagoons and mudflats but also in the interior at ponds, lakes, and rice-growing areas in the valleys of the Ríos Yuna and Yaque del Norte, DR. Peak fall migration period is mid-August to mid-September. Non-breeding-season numbers are variable, but flocks as large as 100 birds have been recorded. Four spring records of presumed migrants in March. Recorded from two major satellite islands: Islas Beata and Saona.

Status: Moderately common fall migrant and non-breeding visitor; less frequent as a spring migrant.
Comments: Typically occurs in large flocks where it often forages in belly-deep water with rapidly repeated probing while submerging the head.
Range: Breeds from n. Alaska east to Victoria Island, south to Yukon River and n. Ontario. Occurs in non-breeding season primarily in interior South America from Bolivia and Brazil south to Argentina and Chile; also locally in extreme s. U.S., and from Mexico south to Costa Rica. In West Indies, a fairly common passage migrant in fall, rare in spring, casual in non-breeding period.
Local names: DR: Playerito Pata Largas, Playero Zancudo; **Haiti:** Bekasin Janm Long.

BUFF-BREASTED SANDPIPER—*Tryngites subruficollis* **Plate 16**
Vagrant

Description: 18–20 cm long; male 71 g, female 53 g. A medium-sized, ploverlike shorebird with a slender build and strong buff wash on the face and breast in all plumages. Upperparts of adults have a scaled appearance; underparts are pale buffy with the sides of neck and breast lightly spotted blackish. Large, dark eye stands out on the clean buffy face; bill is thin and black; legs and feet are yellow; and long wings extend beyond short tail when at rest. In flight, shows bright white underwing coverts. Immatures are similar to adults.
Similar species: Baird's and Pectoral sandpipers both have buffy breasts, but the buff is divided sharply from the white belly. Legs are black in Baird's Sandpiper.
Voice: Flight call is a low, quiet, trilled *preeet* or *greeet*; also short, soft *tick* and *chup* notes.
Hispaniola: One record of an individual collected at Sabana San Thome near San Juan, DR, in October 1928; 1 individual seen at Trou Caïman, Haiti, in mid-October 2003. To be looked for in fields, pastures, and areas with short grass.
Status: Vagrant.
Comment: Sometimes migrates with American Golden-Plovers. Both are most likely to be found after easterly storms. This sandpiper is an extreme long-distance migrant, passing between arctic North America and southern South America.
Range: Breeds from n. Alaska east to Devon Island, south to Victoria and King William islands. Occurs in non-breeding season in Argentina, Paraguay, and Uruguay. In West Indies, a rare passage migrant in fall, vagrant in spring.
Local names: DR: Playero de Pecha Crema; **Haiti:** Bekasin Savann, Bécasseau Roussâtre.

SHORT-BILLED DOWITCHER—*Limnodromus griseus* **Plate 15**
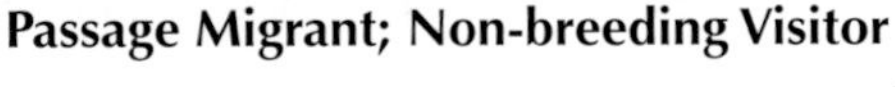
Passage Migrant; Non-breeding Visitor

Description: 25–29 cm long; 112 g. A medium-sized, stocky shorebird with a long, straight bill. Non-breeding birds are dull brownish gray above, paler below, with a dark tail, white upper rump, and whitish supercilium. In breeding plumage, upperparts are reddish brown mottled with black and buff, with a pale reddish brown head and breast, blending to white on belly. Breast is finely barred, and flanks are heavily barred. Immatures are similar to breeding adults but are more brownish and indistinctly marked below. In flight, all plumages show a conspicuous white upper rump patch extending as a sharp V well up the back, and pale secondaries.
Similar species: Very similar to and easily confused with Long-billed Dowitcher, which averages larger and is shorter winged and longer billed. Best identified by flight calls.
Voice: In flight, a soft, rapid whistled *tu-tu-tu*, harsher when the bird is alarmed.
Hispaniola: Occurs at coastal lagoons and mudflats, occasionally inland at lake shores. The first fall transients appear by late July, and peak fall migration period is mid-August to mid-

September. Peak spring migration is March to mid-April, with the last northbound birds usually gone by late April. Recorded from two major satellite islands: Islas Beata and Saona.
Status: Known as a regular fall migrant in small numbers, with a few remaining throughout the non-breeding season, and as a regular but uncommon spring migrant.
Comments: Typically flocks. Feeds by probing sediments with vertical thrusts of the bill in a "sewing machine" motion. Dowitchers on Hispaniola are most likely this species unless very carefully identified as Long-billed Dowitchers.
Range: Breeds from s. Alaska east to s. Mackenzie and Ungava Peninsula, south to nw. British Columbia, ne. Manitoba, sc. Quebec, and c. Labrador. Occurs in non-breeding season along coasts of n. California and New Jersey south to c. Peru and c. Brazil. In West Indies, a regular passage migrant, rare non-breeding visitor, and summer vagrant.
Local names: DR: Agujeta Piquicorta, Costurero, Playero Pico Largo; **Haiti:** Bekasin Mawon Bèk Long.

LONG-BILLED DOWITCHER—*Limnodromus scolopaceus* **Plate 15**
Vagrant

Description: 29 cm long; 105 g. A medium-sized, chunky shorebird with a long, straight bill. Non-breeding individuals are dull gray above, paler below, with a dark tail, white upper rump, and whitish supercilium. In breeding plumage, birds have reddish brown upperparts mottled with black and buff; underparts are uniformly reddish brown with fine barring and spotting on breast, and heavier barring on flanks. Immature resembles non-breeding adult but is darker on the back. In flight, a prominent white rump patch extends as a sharp V well up the back.
Similar species: Very difficult to distinguish from Short-billed Dowitcher; safely identified in the field only by voice. In non-breeding Long-billed Dowitcher, face is less streaked, and gray of breast tends to be darker, more even, and less streaked, and extends lower onto belly than in Short-billed Dowitcher. In breeding plumage, Long-billed Dowitcher has reddish underparts to lower belly, whereas Short-billed Dowitcher typically is white on belly.
Voice: Thin, high-pitched *keek* or *pweek*, singly or in a series.
Hispaniola: Five DR records: 1 injured bird caught and examined in the hand and then released at Las Salinas in November 1980; 11 seen and the diagnostic call note heard at Monte Cristi in March 1985; 1 in a small flock of Short-billed Dowitchers at Cabo Rojo in December 1996; 1 near Barahona in April 1996; and 1 at Cabo Rojo in early June 1998. To be looked for primarily near shallow fresh and brackish water, but also tidal mudflats.
Status: Vagrant; much scarcer than Short-billed Dowitcher.
Comments: Feeds with rapid, vertical bill thrusts. Dowitchers observed on Hispaniola should be presumed to be Short-billed Dowitchers unless identified with certainty as this species. Dowitchers in freshwater habitats should be examined especially carefully, since Long-billed Dowitcher favors that habitat.
Range: Breeds in coastal w. and n. Alaska, n. Yukon, and nw. Mackenzie; also Siberia. Occurs in non-breeding season from s. coastal British Columbia, s. New Mexico, c. Texas, and s. Florida south to Guatemala, rarely Panama. In West Indies, recorded from Cuba, Cayman Islands, Jamaica, Hispaniola, Virgin Islands, St. Kitts, Guadeloupe, and Barbados.
Local names: DR: Agujeta Piquilarga, Costurero, Playero Pico Largo; **Haiti:** Bekasin Bèk Long.

WILSON'S SNIPE—*Gallinago delicata* **Plate 15**
Passage Migrant; Non-breeding Visitor

Description: 27–29 cm long; 100 g. A medium-sized, stocky sandpiper with a long, straight bill and fairly short legs. Upperparts are black and dark brown with four bold, buffy stripes running down the back. Head is also striped black and buff. Underparts are mostly white, with neck and breast streaked brown and flanks heavily barred. Tail is reddish brown. The snipe is well camouflaged and usually seen bursting from cover in an erratic, zigzag flight, uttering its call note.

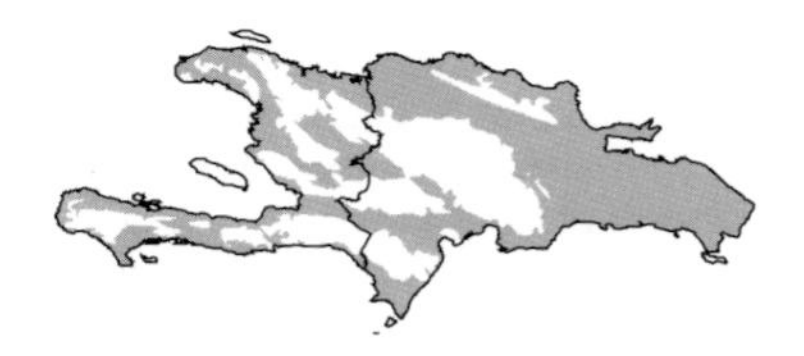

Similar species: Dowitchers are somewhat slimmer, lack striping on the back, and have a distinct white patch from the rump to upper back.
Voice: Dry, rasping, squawk, *scaap* or *rrahk,* when flushed.
Hispaniola: Occurs in wet meadows and savannas, edges of rivers, lakes, and ponds in the interior, as well as the coastal lowlands. Because of its secretive habits and camouflaged appearance, probably more numerous than records indicate. Now known as an uncommon fall transient, with earliest arrivals to be expected in September and peak migration period in October and November. Regular in the non-breeding season in small numbers, with peak spring migration period in March.
Status: Regular passage migrant and non-breeding visitor. One of the most widespread and abundant shorebirds in the Western Hemisphere.
Range: Breeds from n. Alaska east to c. Labrador, south to s. California, w. Nebraska, n. West Virginia, and n. New Jersey. Occurs in non-breeding season from se. Alaska, sw. British Columbia, s. Ontario, and se. Massachusetts south to Tierra del Fuego. In West Indies, a regular passage migrant and non-breeding visitor.
Local names: DR: Becasina, Guineita Grande, Guineíto; **Haiti:** Bekasin Janm Kout.

WILSON'S PHALAROPE—*Phalaropus tricolor* **Plate 16**
Vagrant

Description: 22–24 cm long; male 52 g, female 68 g. A small, slender sandpiper with a thin, needlelike bill. Adult non-breeding plumage is pale gray above with a white rump; thin, dark gray mark through eye; pale yellow legs; and white underparts. Breeding plumage is blue-gray above with two chestnut Vs on back and brownish gray wings. A dark reddish brown band runs from bill, through eye, and along sides of neck; supercilium is white. Throat and nape are white, with throat washed in buffy chestnut. Underparts are white, and legs are black. Breeding male is similar to female but less contrasting. Immature resembles non-breeding adult but is browner above, with a slight buffy wash on breast. In all plumages, birds in flight show dark upperparts with a contrasting white rump, and plain gray wings.
Similar species: Less distinct facial pattern, white rump, plain gray wings, and longer bill separate this from other phalaropes. Stilt Sandpiper has a heavier bill. Lesser Yellowlegs is longer legged and has fine streakings on head and upper breast.
Voice: Mostly silent away from breeding grounds. Flight call is a soft, nasal, high-pitched *creep* or *vimp*.
Hispaniola: All known records from Hispaniola are from Las Salinas, DR, and include up to 3 birds in the August-to-April period with some individuals apparently remaining throughout the non-breeding period. To be looked for primarily in shallow saline ponds and lagoons.
Status: Vagrant.
Comments: Like other phalaropes, feeds actively and rapidly, often spinning in the water to stir up food, which is plucked off the surface.
Range: Breeds from coastal British Columbia east to n. Alberta, w. and s. Ontario, and New Brunswick south to c. California, n. New Mexico, c. Kansas, and n. Ohio. Occurs in non-breeding season from Peru and Bolivia south to Argentina and Chile. In West Indies, recorded from Bahamas, Cayman Islands, Jamaica, DR, Puerto Rico, Virgin Islands, Antigua, Guadeloupe, Martinique, Barbados, and Grenada.
Local names: DR: Falaropo de Wilson; **Haiti:** Falawòp.

RED-NECKED PHALAROPE—*Phalaropus lobatus* **Plate 16**
Vagrant

Description: 18–20 cm long; 34 g. A small, delicate sandpiper with a very thin, straight black bill. Non-breeding adults have gray upperparts streaked white on the mantle, a blackish cap above a large white forecrown, and a striking black patch through the eye. Breeding female

is more brightly colored than male, with a black cap, a dark back with two prominent, buffy white Vs, reddish brown sides of the neck and lower throat, and a white throat. Rest of underparts are whitish, with a grayish breast. Male is duller than female. Immature resembles breeding male but is duller. In flight, shows a white wing stripe and white stripes on back.
Similar species: Non-breeding Wilson's Phalarope has a thin gray eye-line rather than a broad black bar, a longer bill, and no white stripes on the back.
Voice: Flight call is a short, hard *kett*.
Hispaniola: All known reports are of single birds from Las Salinas, DR, from November to March. This phalarope usually remains far out at sea, but when it comes to land it may be expected at ponds and lagoons.
Status: Vagrant.
Comments: Like other phalaropes, often seen spinning in the water to stir up food.
Range: Breeds from n. Alaska east to s. Victoria Island and s. Baffin Island, south to s. Alaska, n. Alberta, n. Quebec, and locally on Labrador coast. Found in non-breeding season primarily at sea in Atlantic Ocean off s. South America. Also breeds in n. Eurasia. In West Indies, a vagrant in Bahamas, Cuba, Jamaica, DR, and Puerto Rico.
Local names: DR: Falaropo de Cuello Rojo, Falaropo Cuellinegro; **Haiti:** Falawòp Bèk Fen.

JAEGERS, GULLS, TERNS, AND SKIMMERS: FAMILY LARIDAE

This family consists of three distinctive subgroups. **Jaegers** are found over the open ocean and rarely seen from land. They are large, heavy bodied, predatory birds with a hooked bill, often seen harassing gulls and terns to force them to drop their catches. The name "jaeger" is derived from the German word *jager*, meaning "hunter." The flight of jaegers is very swift and direct, like that of a falcon, and the wings are sharply bent at the wrist. The bases of the primaries form a distinctive white patch. All species have dark color phases, but these are very rare in the region. The long, central tail feathers in adult jaegers are diagnostic in each species. **Gulls** and **terns** form a cosmopolitan subfamily that primarily frequents coastal waters, rivers, and large lakes. Gulls are generally larger and more robust than terns, with broader wings and a fan-shaped tail. Adults are usually a combination of white, gray, and black, whereas immatures, which may take several years to acquire adult plumage, are principally mottled pale brown. Terns are slimmer birds with a graceful flight, and often have a long, notched tail. They are black about the head with a thin, pointed bill. Unlike gulls, which pick food off the water's surface or feed on shore, many terns hover and dive into the water after fish. Both terns and gulls are quite gregarious and generally nest in colonies. **Skimmers** are characterized by their stout, laterally compressed bill with a projecting lower mandible. Both the upper and lower mandibles form two blades, with the sharp edges facing one another. The birds feed by plowing the surface of calm waters with the lower bill and snapping up fish and other organisms.

POMARINE JAEGER—*Stercorarius pomarinus* **Plate 20**
Passage Migrant; Non-breeding Visitor

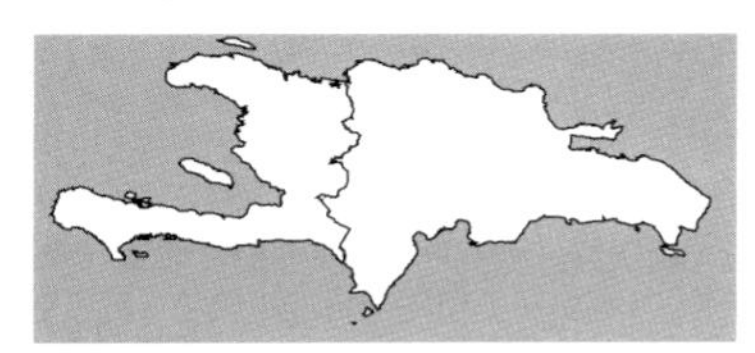

Description: 65–78 cm long; male 648 g, female 740 g. The largest jaeger, powerful and heavy bodied, with a conspicuous white base to the primaries which flashes in flight. Adult birds have elongated central tail feathers which are blunt and twisted 90 degrees, resulting in a spoonlike appearance. There are two color phases with much intermediate variation. Light-phase individuals have a blackish cap,

yellowish wash on nape and sides of the neck, and broad, dark band across the breast. Dark-phase individuals are less common and entirely dark, ranging from brown to black, slightly paler below. Subadult and juvenal plumages are usually heavily barred below, especially along the sides and under the wings. Most birds develop a pale belly as they approach adult plumage. The rounded central tail feathers typically do not extend beyond the rest of the tail in young birds.
Similar species: Parasitic Jaeger is smaller, has a more buoyant flight with a slightly faster wing beat, and lacks heavy barring on sides. Long-tailed Jaeger is smaller and more slender than either species; adults have very long, pointed tail feathers, no breast band, and a graceful ternlike flight.
Voice: Generally silent away from breeding grounds.
Hispaniola: Despite relatively few records, now known to be a regular and occasionally common southbound fall transient in November and December, particularly through the Mona Passage off the eastern DR coast. A regular non-breeding visitor in offshore waters, especially off the southern coast of the island, and a regular northbound transient in April and May.
Status: Passage migrant and regular non-breeding visitor.
Comments: Generally far offshore, which accounts for the scarcity of records. Occasionally seen close to land harrassing gulls and terns in dramatic twisting, acrobatic pursuit.
Range: Breeds from n. Alaska east to Baffin Island, south to n. Northwest Territories and nw. Quebec. Occurs in non-breeding season at sea from California and Florida south to coasts of Peru and Brazil. Also widespread in Eastern Hemisphere. In West Indies, found at sea both north and south of Greater Antilles, and both east and west of Lesser Antilles.
Local names: DR: Pagalo Pomarino; **Haiti:** Lab Pirat, Pijon-dlo.

PARASITIC JAEGER—*Stercorarius parasiticus* **Plate 20**
Passage Migrant; Non-breeding Visitor

Description: 46–67 cm long; male 421 g, female 508 g. A medium-sized jaeger with strong, direct flight and elongated, sharply pointed central tail feathers. All color phases show white patches at the base of primaries in flight. Light-phase adults are dark brownish gray above with a distinct grayish brown cap, white forecrown, and yellowish wash on nape and sides of the neck. Underparts are whitish with a partial to complete pale brown band across upper breast, and gray lower belly. Dark-phase adults are dark brown overall, perhaps slightly paler below. Subadults and juveniles are finely barred dusky below and often have a reddish brown cast to overall plumage; pointed tips of central tail feathers may protrude slightly.
Similar species: Pomarine Jaeger is decidedly larger, has a more labored flight, heavily barred sides, and twisted, blunt-tipped central tail feathers in adults. Long-tailed Jaeger is smaller and has a more graceful flight, and adults have greatly elongated central tail feathers. Juvenile and subadult Long-tailed Jaegers without long central tail feathers cannot be distinguished reliably by any single trait.
Voice: Generally silent away from breeding grounds.
Hispaniola: Few records exist, though believed to be a regular but uncommon southbound transient in November and December in small numbers, with a few remaining throughout the non-breeding season in offshore waters. A regular but uncommon northbound transient in April and May.
Status: Passage migrant and regular non-breeding visitor.
Comments: Generally far offshore, which accounts for the scarcity of records. Occasionally seen close to land harrassing gulls, terns, and other small seabirds, often in prolonged, twisting pursuit.
Range: Breeds from w. and n. Alaska east to nw. Mackenzie and Baffin Island, south to Kodiak Island, ne. Manitoba, n. Quebec, and n. Labrador. Occurs in non-breeding season in

w. Atlantic Ocean from Maine south to e. Argentina. Also widespread in Eastern Hemisphere. In West Indies, recorded from Bahamas, Cuba, Jamaica, Hispaniola, Virgin Islands, and Lesser Antilles from Guadeloupe south to Grenada.
Local names: DR: Pagalo Parásito; **Haiti:** Lab Parazit, Pijon-dlo.

LONG-TAILED JAEGER—*Stercorarius longicaudus* **Plate 20**
Vagrant

Description: 50–58 cm long, including 15- to 25-cm tail; male 280 g, female 310 g. The smallest jaeger, relatively delicate in build. Adults have no dark phase, but juveniles occur in light, intermediate, and dark phases. All adults have very elongated and sharply pointed central tail feathers, but only the first two or three primaries have white bases, so the white wing flash in flight is less noticeable than in other jaegers. Adults have a distinct blackish cap with a white collar, rest of upperparts and secondaries pale grayish brown contrasting with darker primaries; underparts white and unmarked fading to gray on lower belly; underwings uniformly dark. Light-phase subadults and juveniles are finely barred below and have fine white barring on back. Some have a pale head and nape and blue legs. Dark-phase birds are much less common; nearly uniform grayish brown with a darker cap, slightly paler below. Central tail feathers are blunt-tipped and may extend beyond rest of tail. Flight is graceful and buoyant as in a tern.
Similar species: See Parasitic and Pomarine jaegers.
Voice: Generally silent away from the breeding grounds.
Hispaniola: Two spring records: 1 at sea off Cabo Beata, DR, in March 1970, and 1 off the southern coast of Haiti in April 1983.
Status: Vagrant.
Comments: Likely more frequent than records indicate. The scarcity of records results from the species occurring far at sea. May be seen following boats.
Range: Breeds from w. Alaska east to n. Mackenzie and the n. Canadian Arctic Islands, south to c. Alaska, Southampton Island, and n. Quebec. Range in non-breeding season not well known but appears to be at sea, mainly off se. South America. Also widespread in Eastern Hemisphere. In West Indies, reported from Bahamas, Cuba, Jamaica, Hispaniola, Cayman Islands, Guadeloupe, Dominica, Martinique, and Barbados.
Local names: DR: Pagalo Rabero; **Haiti:** Lap Ke Long, Pijon-dlo.

LAUGHING GULL—*Larus atricilla* **Plate 24**
Breeding Resident; Non-breeding Visitor

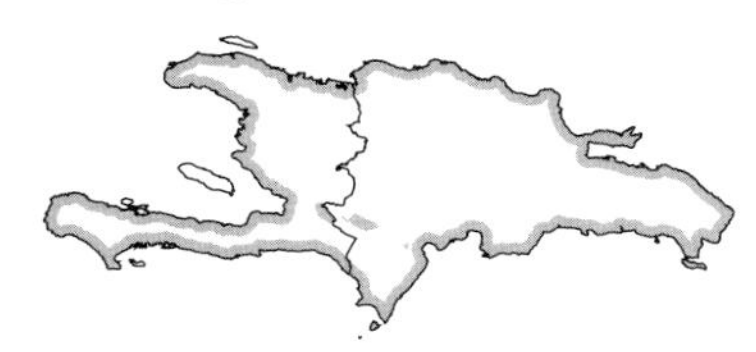

Description: 39–46 cm long; male 327 g, female 289 g. A medium-sized, hooded gull; the most common gull on Hispaniola. In breeding plumage, the black head with a narrow white eye-ring, separated from the dark gray mantle by a white collar, and black wing tips are distinctive. Underparts and tail are white. Bill and legs are reddish. Non-breeding adults have a mostly white head with a diffuse gray mark around and behind eye; bill and legs are black. First-year birds have a gray mantle and upperwings mottled gray and brown. Sides are gray and rump is white, contrasting with broad, black tail band. A partial hood and slaty mantle may show in some individuals. Juvenile has dull brown upperparts and white rump.
Similar species: Adult Bonaparte's Gull is smaller and has a pale gray mantle and white in primaries, with immature possessing a narrow black tail band and pale underwing coverts. Franklin's Gull is very similar to Laughing Gull but smaller and more slightly built, with a shorter bill, less black on primaries, and broader white crescent around back of eye.
Voice: Nasal, laughing *ka-ka-ka-ka-ka-ka-ka-kaa-kaa-kaaa-kaaaa*.

Hispaniola: Occurs at calm bays around the entire coastline of the island, and regularly at interior lakes, especially saline ones, in the coastal lowlands. Birds from the North American mainland typically arrive from September to November and leave in February or March, augmenting population numbers. The pattern of a larger non-breeding-season population at selected localities is the reverse of that found in most of the West Indies where this species is most numerous in summer. Recorded from several major satellite islands, including Île à Vache and Islas Alto Velo, Beata, and Saona.
Status: Moderately common year-round resident and non-breeding visitor from the north. Overall its numbers are increasing in the West Indies.
Nesting: Lays 2–4 grayish brown eggs with large brownish or blackish splotches. A well-woven nest is made on the ground or in a rock crevice on an offshore cay. Breeds May to July. Nesting has been documented at Isla Saona and Laguna de Oviedo, DR.
Range: Breeds on Atlantic coast from Nova Scotia south to Texas, Yucatán Peninsula, and Venezuela; also on w. Mexico coast. Eastern population ranges in non-breeding season from North Carolina south to Colombia and Brazil, as well as throughout West Indies.
Local names: DR: Gaviota Cabecinegra, Gaviota Reidora; **Haiti:** Mòv Tèt Nwa, Mauve à Tête Noire, Mouette Atricille.

FRANKLIN'S GULL—*Larus pipixcan* **Plate 24**
Vagrant

Description: 32–36 cm long; 280 g. A small gull with a dark gray mantle and black hood in breeding plumage; wing tips with a black bar bordered by white on both sides; bill and legs reddish. Non-breeding plumage is similar, but black hood is only partial and forecrown is whitish. First-year individuals show a narrow black tail band which does not extend to outermost tail feathers, white breast and underparts, gray back (washed brown and extending onto nape in juveniles), partial blackish hood with white nape, and white forecrown.
Similar species: Similar to Laughing Gull in all plumages, but Franklin's Gull is smaller with a shorter, stout bill. First-year and non-breeding adults have a more extensive partial black hood, white forecrown, and clean white underwing. First-year birds are also differentiated by their uniformly white underparts. Breeding adult Franklin's Gulls have white in wing tips, whereas Laughing Gull's are entirely black. White crescents above and below eye are broader in Franklin's Gull.
Voice: Nasal, laughing *kuk-kuk-kuk-kuk* or *kah* followed by long *keeaaahh*; also commonly gives short, hollow *keeah* or *kowii* notes.
Hispaniola: Two DR records: 1 first-year bird at Isla Saona in August 1978, and 1 first-year bird at Las Salinas in February 2000. Postbreeding wanderers should be looked for at bays and estuaries.
Status: Vagrant.
Range: Breeds from nw. British Columbia east to c. Manitoba and nw. Minnesota, south to e. Oregon, nw. Utah, and ne. South Dakota. Occurs in non-breeding season primarily on Pacific coast from Mexico to Chile. In West Indies, reported from Cuba, DR, Puerto Rico, St. Barthélemy, and Guadeloupe.
Local names: DR: Gaviota de Franklin; **Haiti:** Mòv Tèt Nwa Ti Bèk.

BONAPARTE'S GULL—*Larus philadelphia* **Plate 24**
Vagrant

Description: 28–30 cm long; 212 g. A small, graceful, ternlike gull with a slender black bill and pink legs. First-year birds have a mostly pale gray mantle and nape, black terminal tail band, and white head and underparts. Note the black ear spot, black primaries and coverts showing a narrow M across upperwings in flight, and whitish undersides to primaries. Non-breeding adults are similar to immatures, but mantle is entirely pale gray; tail and outer pri-

maries are white. Breeding adult is similar to non-breeding adult, but head is entirely black with very narrow white crescents above and below eye; legs orange-red.
Similar species: Adult Laughing Gull has a dark mantle and black primaries, whereas first-year birds have a broader tail band and darker underwing coverts. See also Black-legged Kittiwake.
Voice: Mostly buzzy or twangy, grating *keh* or *kaa-aa* notes, singly or in a descending series.
Hispaniola: One record: 1 at Las Salinas, DR, in November 1980. To be looked for in coastal areas, harbors, lagoons, and at sea.
Status: Vagrant.
Comments: Bonaparte's Gulls occurring in the West Indies tend to be subadults. Adult plumage is reached at about 2 years of age.
Range: Breeds from w. Alaska east to c. Yukon and n. Manitoba, south to se. Alaska, c. Alberta, s. Manitoba, and sc. Quebec. Occurs in non-breeding season on Atlantic coast from Massachusetts to Florida, and along Gulf Coast. In West Indies, uncommon in non-breeding season in Bahamas, Cuba, and Barbuda; vagrant in DR, Puerto Rico, Antigua, and Martinique.
Local names: DR: Gaviota de Bonaparte, Gaviota Pipizcan; **Haiti:** Mòv Janm Wòz.

RING-BILLED GULL—*Larus delawarensis* **Plate 23**
Non-breeding Visitor

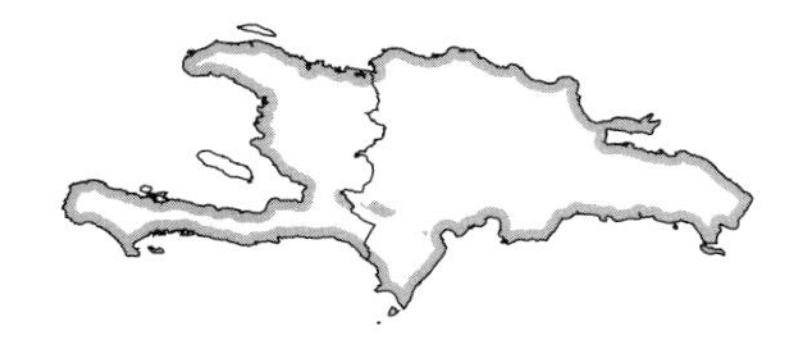

Description: 46–51 cm long; male 566 g, female 471 g. A medium-sized, white-headed gull. First-year birds have mottled grayish brown wings and gray back; head is streaked brownish. Tail has a broad black band, and bill is pinkish with a black tip. Second-year birds have mantle and upperwing coverts mostly gray in sharp contrast to black primaries, which may have only a single white spot at wing tip. Tail is densely mottled with pale gray subterminal marks which form a partial band with white at the base. Head and nape are flecked with brown, and bill is greenish with a black band at tip. Non-breeding adults are similar but have a yellowish bill with a black band, yellowish green legs, and more extensive white spots on black wing tips. Breeding adults are similar to non-breeding adults, but white head and nape are unspotted. Juveniles are brownish gray overall with a black bill and pinkish legs.
Similar species: Distinguished from Herring Gull by its smaller size, proportionately more delicate head and bill, and yellowish green or grayish green (not pink) legs even in subadult plumages. Herring Gull may have a black bill tip but almost never has a complete ring on bill.
Voice: Numerous calls, high pitched and hoarse, with a wheezy quality, including a series of *kreeeee, kow,* or *kah* notes, and a longer but variable *kah, keeeeeaaaah, keeeeeaaaah kah kah kah kah kah.*
Hispaniola: An annual non-breeding visitor in small numbers from September to late April at coastal sites such as Monte Cristi, Cabrera, Sánchez, Punta Arena Gorda, and Barahona, DR. Reported slightly more often on the northern coast. To be looked for in harbors, lagoons, and occasionally grassy fields. Often frequents urban areas. Seldom occurs far at sea.
Status: Uncommon but regular non-breeding visitor. First reported from Port-au-Prince harbor, Haiti, in 1961, and has since been found with increasing frequency.
Comments: Ring-billed Gulls in the West Indies are primarily subadults. Adult plumage is reached at about 3 years of age.
Range: Breeds from nc. Ontario east to Newfoundland, south to Wisconsin and New Hampshire; also an isolated population in nw. U.S. Eastern population occurs in non-breeding season from s. New England south to Gulf of Mexico coast and Yucatán Peninsula. In West Indies, regular in non-breeding period in Bahamas and Greater Antilles; casual but occurring with increasing frequency elsewhere.
Local names: DR: Gaviota Piquianillada; **Haiti:** Mòv Bèk Jòn, Goéland.

HERRING GULL—*Larus argentatus* **Plate 23**
Non-breeding Visitor

Description: 56–66 cm long; male 1,226 g, female 1,044 g. A large gull, variably plumaged, with a heavy bill. First-year individual's back and wings are mottled grayish brown; underparts are lightly streaked pale brown; bill is pinkish at the base and black at the tip; and tail has a broad dark band. Legs are pinkish. Second-year individuals are similar to first-year birds but with variable amounts of gray on the back and wings, outer primaries are black, and bill is pinkish with a pale gray band beyond nostril. Third-year birds have a white tail with a broad black band and a yellowish bill with a dark band. Non-breeding adult's head and underparts are white, head and nape are flecked with pale brown, and mantle is gray. Legs are pink, bill is heavy and yellow with a red spot near tip of lower mandible. Breeding adult is similar to non-breeding adult, but head and nape are clean white.
Similar species: Lesser Black-backed Gull is slightly smaller, and adult has a dark grayish black mantle and yellowish legs. Great Black-backed Gull is larger and has a massive bill, and adult has jet black back and wings. See also Ring-billed Gull.
Voice: Clear, flat, bugling call, often with paired syllables, but also yelping, single *kyow* or *klaaw* notes.
Hispaniola: An annual non-breeding visitor in small numbers, usually appearing only at localities where other gulls congregate, from October to April. Reported about equally on the northern and southern coasts. To be expected in coastal areas, harbors, and lagoons.
Status: Uncommon non-breeding visitor. Since the 1930s, the eastern North American population of this species has increased greatly and begun to occur in the West Indies more often. Herring Gull numbers appear to continue to increase on Hispaniola and elsewhere.
Comments: Herring Gulls wandering to the West Indies are primarily subadults. Adult plumage is reached at about 4 years of age.
Range: Breeds from n. Alaska east to n. Labrador, south to British Columbia and South Carolina. Occurs in non-breeding season from southern part of breeding range south to Florida and s. Mexico coast. Widespread in Eastern Hemispere. In West Indies, an uncommon non-breeding visitor nearly throughout, but most frequently reported from Bahamas and Greater Antilles.
Local names: DR: Gaviota Arenquera, Gaviota del Norte, Gaviota Argentea; **Haiti:** Mòv Gri, Goéland Argenté.

LESSER BLACK-BACKED GULL—*Larus fuscus* **Plate 22**
Vagrant

Description: 53–63 cm long; 800 g. A large but slender gull with relatively long and narrow wings. First-year birds are mottled brownish gray overall, with paler head and streaked underparts. Bill is black. In flight, shows dark primaries, broad tail band, and pale rump. Second-year birds have a dark gray back; broad black tail band; white rump; and brownish gray wings with no white spots at tips. Bill is pinkish with a large black band near tip. Third-year birds have dark gray mantle and upperwing coverts, yellowish bill with black band around tip. Non-breeding adults have dark grayish black mantle, pale yellow legs, mottled brown head and neck, and yellow bill with a red spot near tip. Breeding adult is similar to non-breeding adult, but head and neck are mostly white.
Similar species: Adult Great Black-backed Gull is larger and has jet black mantle and pink legs. First- and second-year birds are distinguished by their larger size and massive bill. Adult Herring Gull has paler mantle and pink legs, with first- and second-year birds having a less pronounced white rump patch.
Voice: Guttural calls, intermediate between the less nasal Herring Gull and the deeper Great Black-backed Gull.

Hispaniola: Three DR reports: 1 in first-year plumage at Las Salinas remained for several months in the non-breeding season of 1997–1998. What may have been the same individual returned for the next 2 years. Another bird was seen at Isla Saona, DR, in November 2000. A third bird was reported at Laguna de Oviedo in January 2002.
Status: Vagrant.
Comments: Wanderers to the West Indies tend to be subadults. Adult plumage is reached at about 4 years of age. Often congregates with gulls of other species.
Range: Widespread in Eastern Hemispere. In North America, occurs in non-breeding season from Labrador and Great Lakes south to Gulf Coast, Florida, and west to Missouri. In West Indies, recorded from Bahamas, Cuba, DR, Puerto Rico, Virgin Islands, St. Martin, St. Barthélemy, Antigua, and Barbados.
Local names: DR: Gaviota Sombria; **Haiti:** Ti Mòv Do Nwa, Goéland Brun.

GREAT BLACK-BACKED GULL—*Larus marinus* **Plate 22**

Non-breeding Visitor

Description: 71–79 cm long; male 1,829 g, female 1,488 g. The largest and heaviest gull in the Western Hemisphere. Identified in all plumages by its large size and massive head and bill. First-year birds are mottled grayish brown above and below; head is white with pale flecks on the rear and nape; bill is black; tail has a broad dark band; rump is white. Second-year birds have a pinkish bill with a large black band near the tip, white rump patch, and grayish mantle with black blotches. Third-year birds have a slaty mantle, unmarked and slightly paler upperwing coverts, and yellowish bill with a narrow black band at tip. Non-breeding adults have a black mantle, pale flecks on head, bright yellow bill with small red spot near the tip of lower mandible, and pink legs. Breeding adult is similar to non-breeding adult, but head is entirely white.
Similar species: See Herring and Lesser Black-backed gulls.
Voice: Deep, hoarse trumpeting, similar to Herring Gull but much lower in pitch.
Hispaniola: All records from Hispaniola are from the DR, and most are of second- or third-year birds. Records exist from Sabana de la Mar, Laguna Limón, Gulf of Samaná, Laguna de Oviedo, Las Salinas, and Cabo Rojo.
Status: Uncommon non-breeding visitor. This species has begun to appear much more often in the West Indies since the mid-1990s.
Comments: Wanderers to the West Indies tend to be subadults. Adult plumage is reached at about 4 years of age.
Range: Breeds in North America from n. Quebec and n. Labrador south along Atlantic coast to North Carolina and at Lakes Huron, Ontario, Oneida, and Champlain. Occurs in non-breeding season in North America on coast from Newfoundland to Florida, Great Lakes, and rarely on Gulf Coast. Widespread in Eurasia. In West Indies, reported from Bahamas, Cuba, DR, St. Barthélemy, and Barbados.
Local names: DR: Gavión; **Haiti:** Gwo Mòv Do Nwa, Goéland Marin.

BLACK-LEGGED KITTIWAKE—*Rissa tridactyla* **Plate 24**

Vagrant

Description: 38–41 cm long; 400 g. A medium-sized gull with a pearly gray mantle and wings in all plumages. First-year birds have a white head with a black ear spot and horizontal bar at the base of nape, a black bill, and a narrow, black terminal tail band. In flight, gray wings and mantle are boldly marked with a contrasting M from wing tip to wing tip. Non-breeding adults have a yellow bill and white head with a dull black mark behind eye; nape is

washed gray. Mantle is gray, and wings are tipped black with no white. Breeding adult is similar to non-breeding adult, but head is entirely white.
Similar species: First-year bird is distinguished from immature Bonaparte's Gull by black half-collar on nape and, in flight, by white trailing edge of secondaries.
Voice: Generally silent away from the breeding grounds.
Hispaniola: One DR record of a weakened individual at Las Salinas in February 1997.
Status: Vagrant. May occur more frequently than records indicate, but may go unobserved, as non-breeding birds generally occur far out at sea.
Comments: Wanderers to the West Indies tend to be subadults. Adult plumage is reached at about 3 years of age.
Range: Breeds on Bering Sea coast east through Arctic Ocean islands to Newfoundland; regularly occurs in non-breeding season at sea south to Nayarit in Pacific and North Carolina in Atlantic. Widespread in Eastern Hemispere. In West Indies, vagrant to Bahamas, Cuba, Jamaica, DR, Guadeloupe, and St. Lucia.
Local names: DR: Gaviota Tridactila; **Haiti:** Mòv Pye Nwa, Mouette Tridactyle.

GULL-BILLED TERN—*Sterna nilotica* **Plate 21**

Passage Migrant; Non-breeding Visitor

Description: 33–38 cm long; 170 g. A chunky, gull-like tern with a proportionately heavy black bill, broader and longer wings than other terns, and a very shallow fork to tail. Legs are relatively long and black. Breeding adult is pale gray above with a clean white body and a black crown and nape. Non-breeding adult identified by its whitish crown with pale gray flecks, and distinct gray smudge behind eye. Immature birds have back and scapulars mottled with grayish brown, and bill is brownish.
Similar species: Stout black bill separates this tern from others except Sandwich Tern, which is more lightly built and has a longer, more deeply forked tail.
Voice: Raspy 3-syllabled *za-za-za* or 2-syllabled, nasal *kay-wek, kay-wek.*
Hispaniola: Occurs at coastal lagoons and beaches, also at both fresh and saline lakes in the lowlands. Uncommon spring and fall transient, presumably with birds passing to and from nesting colonies in the Bahamas or along the Gulf Coast. More common as a spring migrant, especially from April to early May, but regular as a fall migrant from early August to October. Occasional in both the summer and non-breeding season. Also reported from Île de la Gonâve and Isla Beata.
Status: Uncommon passage migrant and non-breeding visitor.
Range: Breeds on Atlantic and Gulf of Mexico coasts from New York to Tamaulipas, Mexico; populations also occur on western coast of Mexico. Occurs in non-breeding season from sw. Florida and s. Mexico south along coasts to Argentina and Peru. Widespread in Eastern Hemisphere. In West Indies, a local breeder in Bahamas and Virgin Islands; uncommon to rare elsewhere.
Local names: DR: Gaviota Pico Corto, Charrán Piquigrueso; **Haiti:** Mòv Bèk Nwa.

CASPIAN TERN—*Sterna caspia* **Plate 21**

Passage Migrant; Non-breeding Visitor

Description: 47–54 cm long; 655 g. The largest tern, heavy bodied and gull-like. Mostly white with pale gray upperparts; black crest; massive, dark red bill with dusky tip; and dark gray undersides to outer primaries. Tail is only slightly forked, and legs are black. Breeding adult has entirely black crest; non-breeding adult has black crest flecked with white. Wing beats are stiff and shallow.
Similar species: Royal Tern is distinguished by its smaller size, bright orange-red bill, pale underside to primaries, and in non-breeding plumage, white forecrown.

Voice: Call is a hoarse, low croak, *rau* or *kraau*, and other similar rasping notes.
Hispaniola: Regular but uncommon transient in spring and fall, a few present irregularly in both summer and the non-breeding season at coastal lagoons and both fresh and saline lakes in the lowlands. Never suspected of breeding; those seen are thought to be non-breeders or postbreeding birds that have dispersed from North American colonies.
Status: Uncommon passage migrant and non-breeding visitor. Regularly occurring, but usual numbers are only 1–4 in any season.
Range: Breeds along Atlantic and Gulf of Mexico coasts at scattered localities from Newfoundland to Florida and Texas; also in coastal areas and interior of w. U.S. and Canada, and on Mexican coast south to Sinaloa. Eastern population occurs in non-breeding season from North Carolina south to Columbia and Venezuela. Widespread in Eastern Hemisphere. In West Indies, a rare visitor to Bahamas, Greater Antilles, St. Kitts, Antigua, Dominica, Martinique, St. Lucia, and Barbados.
Local names: DR: Gaviota Real, Gaviota Picorojo, Charrán Piquiroja; **Haiti:** Gwo Mòv Bèk Jòn, Sterne Caspienne.

ROYAL TERN—*Sterna maxima* **Plate 21**
Breeding Resident; Non-breeding Visitor

Description: 45–50 cm long; 470 g. A large tern, pale gray above, white below, with a bright orange-red bill and a cap across the top and back of head with elongated black feathers forming a shaggy crest. Legs are black; tail is moderately forked. Breeding adults have entirely black crown. Non-breeding adults and immatures have black crown streaked with white, white forecrown and lores, and paler bill.
Similar species: Caspian Tern is slightly larger, has a dark red bill with a dusky tip, and has more extensive blackish on underside of primaries.
Voice: Harsh, high-pitched *krit-krit* and throaty, rolling *keer-reet*.
Hispaniola: Occurs in coastal lagoons, beaches, and harbors, and regular inland at both fresh and saline lakes in the lowlands. Numbers present from September to April are augmented by birds from North America. Also reported from nearly all major satellite islands.
Status: Uncommon breeding resident and fairly common non-breeding visitor. Numbers appear to have increased since the 1930s.
Nesting: Lays a single creamy white egg heavily spotted with blackish. Nest is a shallow depression in the sand or seaweed along tideline. Breeds May to July. Approximately 10 pairs have been breeding on Cayo Puerto Rico in Laguna Oviedo since at least 2003.
Range: Breeds from New Jersey south at scattered localities along Atlantic and Gulf of Mexico coasts to Yucatán Peninsula, Venezuela, and n. Argentina; also along Pacific Coast and coastal West Africa. Non-breeders occur from s. California and North Carolina coasts south to Panama and Guianas. Occurs throughout West Indies.
Local names: DR: Gaviota Real, Charrán Real; **Haiti:** Mòv Wayal, Sterne Royale.

SANDWICH TERN—*Sterna sandvicensis* **Plate 21**
Breeding Resident; Non-breeding Visitor

Description: 35–45 cm long; 208 g. A relatively large tern, and the only crested tern with a black bill. Breeding adult has pale gray upperparts, appearing white at a distance, with a shaggy black crest, and long, slender black bill with conspicuous yellow tip. Tail is moderately forked and does not extend beyond folded wings at rest. Legs are black. Non-breeding

plumage is similar to breeding plumage, but forecrown is extensively white to behind eye, and cap is often flecked with white. Immature is like adult, but back, rump, scapulars, and uppertail coverts are pale gray mottled with brown, gray, or blackish.
Similar species: Gull-billed Tern has a short, thick, entirely black bill, and a black line along trailing edge of primaries.
Voice: Metallic, grating *kirrik* or *kerr-ick*.
Hispaniola: Occurs at coastal lagoons, beaches, and harbors, less often but frequently at saline lakes. Birds from North America augment the local population from late August to October. Some birds typically remain throughout the non-breeding period. Peak spring migration period is late March and April. A few usually linger May to August. Recorded from several major satellite islands, including Islas Alto Velo, Beata, and Saona.
Status: Uncommon breeding resident and regular non-breeding visitor year-round. Appears to have become more numerous since the 1930s, and counts of 5–10 are not unusual.
Comments: The South American form of Sandwich Tern, the Cayenne Tern (*S. s. eurygnatha*), considered by some authorities to be a separate species, has identical plumage characters, but the entire bill is dull yellow. Cayenne Terns interbreed with Sandwich Terns at large colonies off Venezuela and in smaller colonies off Puerto Rico and the Virgin Islands. Cayenne Tern is occasionally recorded on Hispaniola at coastal locations such as Monte Cristi and Laguna de Oviedo.
Nesting: Lays 1–2 pale buff to creamy white eggs variably marked with dark brownish spots, blotches, and scrawls. Breeds May to July. Nest is a shallow depression in sand or coral rubble, sometimes in areas of sparsely vegetated soil. A breeding colony of 25–35 pairs exists on Cayo Puerto Rico in Laguna de Oviedo.
Range: Breeds locally on Atlantic and Gulf of Mexico coasts from Virginia to Texas and Yucatán Peninsula; also along coasts of Venezuela, French Guiana, Trinidad, and Argentina. Non-breeding birds occur from Florida, locally along Gulf Coast, and along both coasts of Central America to Colombia. Widespread in Eastern Hemisphere. In West Indies, a local breeder in Bahamas and Greater Antilles, casual elsewhere; uncommon transient nearly throughout.
Local names: DR: Gaviota Pico Agudo, Charrán de Sandwich; **Haiti:** Mòv Kojèk.

ROSEATE TERN—*Sterna dougallii* **Plate 21**

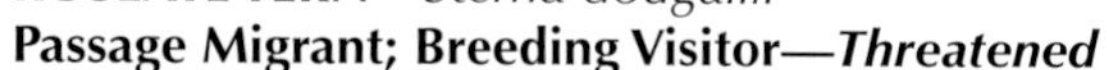

Passage Migrant; Breeding Visitor—*Threatened*

Description: 33–41 cm long; 110 g. A medium-sized, pale tern with a long, deeply forked tail that projects well beyond wing tips at rest. Mantle is pale gray; cap entirely black; primaries pale gray with underwings appearing pure white and translucent. Breeding adult has a thin, black bill with some red at the base, and dark red legs. Non-breeding adult has a blackish bill and a white forehead to above the eye. Immature birds have a dark forehead and crown, blackish bill and legs, mottled back, and shoulders with indistinct markings.
Similar species: Adult Common Terns have a darker gray mantle, and primaries with noticeable blackish on underside. Immature Common Tern has a distinct black shoulder mark. Roseate Tern flies with faster, deeper wing beats than Common Tern. Roseate Terns in the Caribbean possess much more red in the bill than individuals from North America. This has led to their frequent misidentification as Common Terns. Non-breeding Forster's Tern has a bold, black patch behind eye.
Voice: Raspy, high-pitched *krek* and soft, distinctive 2-syllabled *tu-ick* or *chi-vik*. Call notes differ noticeably from the harsh *kee-arr-r* of Common Tern.
Hispaniola: Expected at harbors, lagoons, and other sites, only along the coast. Few sight

records exist, in part because of the difficulty of distinguishing this species from Common and Forster's terns at any distance. Spring migrants occur over a short period, primarily in late April and early May. Presumed breeding birds are present through the summer, while some non-breeding birds that originated in mainland colonies may also occur. Migrants from North America pass through primarily in September and October, when local breeding birds also depart; only stragglers are to be expected after that time.

Status: Uncommon but regular spring and fall passage migrant and presumed local breeding visitor in small numbers. The subspecies of Roseate Tern native primarily to the Atlantic Ocean has declined dramatically in the north temperate zone. Similar evidence of the bird's decline in the West Indies is not available, but there have been known pressures of habitat disturbance, egg harvesting, and rat predation; there is no documentation of breeding on Hispaniola since the 1930s. All subspecies, including the West Indian race, are considered threatened.

Comments: Some Roseate Terns from the West Indies are known to spend the non-breeding season along the coast of Brazil.

Nesting: Lays 1–3 greenish gray to buff eggs which are heavily spotted with dark brownish purple. Breeds May to July. The nest is in a sand or coral scrape or in a rock depression, usually in colonies on an offshore cay. The birds choose different sites in different years, though fidelity is strong for a general nesting area. Breeding was documented in the 1920s and 1930s from Cayos de Los Pájaros, Bahía de San Lorenzo, DR, and from Isla Beata, but there are no records since then. Reported to breed at least occasionally on Cayos Siete Hermanos, but no documentation is available.

Range: Declining and threatened, though still a regular local breeder along Atlantic Coast from Nova Scotia to Florida; also off Honduras and Venezuela. Occurs in non-breeding season in e. Caribbean south to Guianas and casually to e. Brazil. Widely distributed in Eastern Hemisphere. In West Indies, a local breeder throughout Greater and Lesser Antilles.

Local names: DR: Gaviota Palometa, Charrán Rosado; **Haiti:** Mòv Blanch, Sterne de Dougall.

COMMON TERN—*Sterna hirundo* **Plate 21**

Passage Migrant; Non-breeding Visitor

Description: 31–35 cm long; 120 g. A medium-sized tern, light gray above, paler gray below. Breeding adult is distinguished by black cap, orange-red bill with black tip, and red legs. Both leading and trailing edges of outer primaries are black, showing as a dark wedge in flight. Tail is moderately forked and does not extend beyond tips of folded wings at rest. Non-breeding adult shows a blackish bill, shoulders with distinct dark bars, black crown and nape, and white forecrown. Immature birds have a similar head pattern and bill; mantle, rump, and uppertail coverts are gray mottled with light brownish; primaries and shoulders are dark.

Similar species: Immature Roseate Tern has a paler mantle, less black in primaries, darker bill, and less distinct black shoulder mark. Common Tern's flight is slower with shallower wing beats than in Roseate Tern. Non-breeding Forster's Tern has a black patch around eye.

Voice: Strong, harsh *kee-arr-r*, dropping in pitch, particularly heard in defense of the nest. Also a single, sharp *kip*.

Hispaniola: Occurs along coastal areas and in harbors and lagoons. Migrants from North America arrive by mid-August and pass southward, occasionally in large numbers and especially through the Mona Passage in mid- to late September. A few remain from December to early February, but most birds leave by early March. Comparatively few transients occur during spring migration, and a very few non-breeders are present from May to July. Recorded from several major satellite islands, including Islas Alto Velo, Beata, and Saona.

Status: Regular spring and fall passage migrant and non-breeding visitor year-round in small numbers, but never suspected of breeding.

Range: Breeds from n. Alberta east to Nova Scotia, south to Washington, nw. Indiana, and Vermont, and locally south along Atlantic and Gulf of Mexico coasts. Occurs in non-breeding season south to Argentina and Peru. Widespread in Eurasia. In West Indies, a local breeder in Greater Antilles, Netherlands Antilles, Dominica, and Guadeloupe.
Local names: DR: Gaviota Común, Charrán Común; **Haiti:** Mòv Bèk Wouj, Sterne Pierregarin.

FORSTER'S TERN—*Sterna forsteri* **Plate 21**
Vagrant

Description: 33–36 cm long; 158 g. A mostly immaculate white tern with a deeply forked tail and with broader wings and a thicker bill than similar species. Non-breeding adult shows silvery white primaries, a large black "mask" through the eye and across the auriculars, and a forked tail projecting beyond the folded wings at rest. Bill is black. Breeding adult is white with pale gray wings and mantle and a black cap. White rump does not contrast sharply with gray of the back. Bill is orange with a black tip. Immature is similar to non-breeding adult but has a shorter tail and generally darker primaries.
Similar species: Slightly larger and bulkier than Common and Roseate terns. Noticeably whiter than Common Tern, which also has broader dark wing tips. Roseate Tern has pure white underwings, more deeply forked tail, and stiffer and faster wing beats.
Voice: Descending *kerrr,* similar to that of Common Tern but lower and more raspy.
Hispaniola: Five DR records from Punta Arena Gorda, Puerto Alejandro, and Barahona. Surely occurs more often than noted but is difficult to distinguish in breeding plumage, especially from Roseate Tern, except at close range. To be looked for at coastal areas, harbors, and lagoons.
Status: Vagrant.
Range: Breeds in interior North America from se. British Columbia east to c. Saskatchewan and s. Ontario, south to n. Baja California, w. Nevada, ne. Colorado, n. Iowa, and e. Michigan, and on Atlantic coast from Massachusetts south to South Carolina, and on Gulf Coast from Alabama to Tamaulipas. Occurs in non-breeding season from n. California and s. New Jersey south to Costa Rica and rarely Panama. In West Indies, reported from Bahamas, Greater Antilles, Antigua, Montserrat, and St. Vincent.
Local names: DR: Gaviota de Forster, Charrán de Forster; **Haiti:** Mòv Ke Fann, Sterne de Forster.

LEAST TERN—*Sterna antillarum* **Plate 21**
Passage Migrant; Breeding Visitor

Description: 21–23 cm long; 43 g. The smallest tern in the West Indies. Besides its small size, breeding adult is distinguished by a generally light gray mantle with white underparts, black crown contrasting with V-shaped white forecrown, pale yellow bill with a black tip, and yellow legs. In flight, white wings contrast with black outer two primaries; tail is short and moderately forked. Non-breeding adult shows a less sharply defined head pattern, and bill is blackish. Immatures have a black bill and eye stripe; back and hindcrown are whitish flecked with grayish brown; and dark outer primaries and a bar on shoulder form a distinctive M on upper wing.
Similar species: Small size is distinctive.
Voice: Shrill, chattering *kip* or *ki-dik* and a high-pitched, rising *zreeep* alarm call are characteristic.
Hispaniola: Occurs at coastal areas, harbors, and lagoons. Arrives mid- to late April. Regularly present through breeding season. First southbound migrants arrive by early August, and most migrants are gone by end of September. A few exceptional non-breeding-season reports

from December and January exist. Recorded from several major satellite islands, including Île de la Gonâve and Islas Alto Velo, Beata, and Saona.
Status: Regular spring and fall passage migrant and local breeding visitor in favored localities along the coast; occasional inland in lowlands. The race of Least Tern inhabiting the West Indies also breeds on both coasts of the U.S., where some local populations are considered endangered. Whereas human disturbance and introduced predators may have affected the Hispaniolan and West Indian populations, the limited information available on the species' status in the Caribbean does not currently warrant this tern being classified as threatened.
Comments: Varied diet includes small fish, shrimp, other crustaceans, and some insects.
Nesting: Lays 1–3 buffy white eggs spotted with chocolate brown. Nest is a scrape in a sand bar, coral rubble spit, or dried mudflat. Nests in single pairs or loose colonies from April to July. Since the 1970s, nest records from Isla Beata, Playa Tortuguero near Azua, Las Salinas, and Laguna de Oviedo, DR. From 1997 to 1999 at Las Salinas, about 40 pairs nested annually. Courtship began during the second week of March, the first eggs were found in late March or early April, and peak laying occurred from the second week of April to the third week of May. Most nests contained 2 eggs, with a few containing 3. Most chicks were fully feathered and flying by the first week of July.
Range: Breeds along Atlantic and Gulf of Mexico coasts from Maine to Texas, and a widespread but local breeder in interior U.S. Occurs in non-breeding season on South American coast from Colombia to e. Brazil. In West Indies, a local breeder in Bahamas, and from Greater Antilles south and east to Barbuda; otherwise only a widespread transient and summer resident.
Local names: DR: Gaviotica, Gaviota Chiquita, Charrán Menor; **Haiti:** Ti Mòv Piti.

BRIDLED TERN—*Sterna anaethetus* **Plate 20**
Breeding Visitor

Description: 30–32 cm long; 100 g. A medium-sized, strictly marine tern. Breeding adults are grayish brown above and white below, with a black cap and white V-shaped patch across the forecrown. A narrow white collar on the nape separates the black cap from the dark back. Tail is deeply forked with a narrow gray median stripe and extensive white on outer feathers. Bill and legs are black. Non-breeding adults are similar but paler with black cap streaked with white. Immature is similar to non-breeding adult, but upperparts are scaled with pale gray, and head markings are paler and more diffuse.
Similar species: Sooty Tern is blacker above, lacks a white nape, and white on forecrown does not extend behind eye.
Voice: Puppy-like *yep* or whining *yerk*. Also a continuous *ah-ah-ah*.
Hispaniola: Regular resident in coastal areas and bays during breeding season; inhabits offshore waters at other times.
Status: Regularly recorded during breeding season, but numbers are poorly known because of infrequent visits to breeding areas and because species lives far at sea most of the year.
Comments: Seen easily on its breeding islets, otherwise typically observed only from a boat at sea. Feeds on small fish, crustaceans, and aquatic insects which it typically picks by hovering above the ocean surface. Makes vertical plunge-dives less often than other tern species.
Nesting: Lays 1–3 pale buffy gray eggs which are marked with numerous fine, dark brown spots. The nest is a concealed scrape under vegetation or an overhanging rock, or in a crevice on an islet or cay. Nests in small groups or loose colonies April to July. Confirmed nesting on Cayos Siete Hermanos in May 1997, though the colony was heavily depredated by human egg robbers. Also nests in colonies on rocks near Isla Beata, Navassa Island, and on cliffs of Cayo Limón and Cayo Canas off Samaná Peninsula. Probably breeds at Isla Alto Velo, DR, but not yet confirmed.

Range: Breeds in Western Hemisphere off western coast of Costa Rica and Panama, and widely in Caribbean region. Non-breeding range poorly known, but occurs only in offshore waters. Cosmopolitan in warm oceans. In West Indies, a widespread but local breeder.
Local names: DR: Gaviota Oscura, Gaviota Monja, Charrán Embriado; **Haiti:** Mòv Gri Ke Blanch, Sterne Bridée.

SOOTY TERN—*Sterna fuscata* **Plate 20**
Breeding Visitor

Description: 36–45 cm long; 200 g. A medium-sized, dark tern. Adult is blackish above and white below, with a white forecrown extending to, but not behind, eye. In flight underwing shows dark primaries contrasting with white coverts; tail is deeply forked and mostly black with narrow, white outer margins. Immature is dark brown overall with whitish spots on mantle and wings; the tail is less deeply forked than adult's, and undertail coverts and underwing coverts are whitish.
Similar species: Bridled Tern is grayer above, with white on forecrown extending to behind eye, a white nape, and more extensive white in outer tail.
Voice: Call note is a distinctive, plaintive *wide-a-wake* or *wacky-wack*, often given at night.
Hispaniola: Common in inshore waters all around the island during breeding season, especially near nesting colonies, including those on Mona and Desecheo islands off the eastern coast of DR. The birds are usually far out at sea the rest of the year. Has occured rarely at interior saline lakes such as Lago Enriquillo, DR. Recorded from major satellite islands, including Navassa Island, Cayos Siete Hermanos, and Islas Alto Velo, Beata, and Saona.
Status: Commonly seen only during the breeding season. On Hispaniola, a substantial decline in numbers has occurred since the mid-1970s because of large-scale egg robbing, which continues today in areas such as Cayos Siete Hermanos. A large breeding colony at Isla Alto Velo, estimated at 350,000 birds in 1951, contained only an estimated 40,000 to 50,000 pairs in 1980. Surveys are needed to provide updated population estimates.
Comments: One of the most abundant of the world's tropical seabirds, with some breeding colonies numbering more than a million birds. Very aerial, rarely alighting on water. Captures small fish and squid by dipping, and occasionally by shallow surface dives. Away from its breeding islets, usually seen only from a boat at sea. A common English vernacular name, "Wideawake," refers both to the species' common call and to its habit of being active at all hours of the day.
Nesting: Lays 1–2 buff-colored eggs marked with dark brown spots. Nest is a scrape on a coral rubble beach or under overhanging vegetation, mostly in large, gregarious colonies. Generally breeds April to August, but timing of breeding may vary considerably from year to year, even at the same location. Known colonies exist at Cayos Siete Hermanos and Isla Alto Velo.
Range: Breeds on islands off w. Mexico and off coasts of Yucatán Peninsula, Texas, Louisiana, Florida, and North and South Carolina, as well as in Caribbean region. Non-breeding range of North American birds is not precisely known, but mainly in mid-Atlantic Ocean. Cosmopolitan in tropical oceans. A local breeder virtually throughout West Indies.
Local names: DR: Gaviota Oscura, Bubí, Charrán Tiznado; **Haiti:** Mòv Nwa Vant Blanch, Sterne Noire, Sterne Fuligineuse.

BLACK TERN—*Chlidonias niger* **Plate 20**
Passage Migrant

Description: 23–26 cm long; 65 g. An unmistakable, small, dark tern, often likened to a nighthawk or swallow. Non-breeding adult is gray above. Forecrown, nape, and underparts are white except for smudgy dark patches at sides of breast, and a prominent dark patch be-

hind eye and on hindcrown. Breeding adult has a black head, breast, and belly and white undertail coverts. Underwings are gray, and bill is black in all plumages; tail is only slightly forked. Immature is similar to adult, but upperparts are washed with brownish and sides with grayish. Flight is buoyant and slightly erratic as the bird darts and flits after insects, and often hovers to pluck food from the water surface. Frequently forages in flocks.
Similar species: Unmistakable.
Voice: Sharp, complaining *kip* or *kek*, higher than similar voice of Black-necked Stilt.
Hispaniola: Occurs in near-shore waters or at lagoons, and at both fresh and saline lakes in the lowlands. Since the mid-1990s, reports from Las Salinas, Laguna Salinas near Monte Cristi, and Lago Enriquillo, DR. Most Caribbean migrants are believed to pass at sea far from sight of land. Earliest southbound migrants arrive in late July, but most often recorded in August in flocks of up to 150 birds. Rarely recorded after September. Only 3 records from spring. Recorded from three major satellite islands: Islas Alto Velo, Beata, and Saona.
Status: Uncommon fall migrant, usually in small numbers but occasionally in flocks.
Comments: Associated with marshes during breeding season, but primarily frequents marine habitats in the non-breeding season. Often feeds on insects, also small fish and aquatic invertebrates.
Range: Breeds from c. and ne. British Columbia east to nw. Saskatchewan, s. Quebec, and c. Nova Scotia, south to sc. California, n. Utah, Nebraska, n. Ohio, w. New York, and Maine. Occurs in non-breeding season on Pacific Coast from Jalisco, Mexico, to Peru and on northern coast of South America from Colombia to Suriname. Widespread in Eastern Hemisphere. In West Indies, a common migrant in Puerto Rico and Jamaica, uncommon to rare in Bahamas and rest of Greater Antilles; casual in Lesser Antilles.
Local names: DR: Gaviota Negra, Fumarel Negro; **Haiti:** Mòv Nwa, Sterne Noire.

BROWN NODDY—*Anous stolidus* **Plate 20**
Breeding Visitor

Description: 40–45 cm long; 198 g. A medium-sized, ternlike bird of marine habitats. Adult is entirely dark brown except for silvery white forecrown and crown fading to gray-brown on nape. Tail is also dark brown and wedge shaped, with shorter interior tail feathers giving a unique "double-rounded" appearance. Bill and legs are black. Immature is similar to adult, but only forecrown is white, and upperwing coverts are often paler.
Similar species: Similar in flight to some dark jaegers and shearwaters, but differentiated by unique tail shape and lack of light patches on underparts or underwings.
Voice: Low, harsh, grating *karrk* or *brraak*.
Hispaniola: Common around most of the coastline of the entire island during the breeding season, especially near breeding colonies, including those on Mona and Desecheo islands off the eastern coast of DR. Usually remains in offshore waters far from land at other times of year. Recorded from most major satellite islands.
Status: Commonly seen only during the breeding season. On Hispaniola, known to suffer from egg robbers, who continue to harvest eggs at Cayos Siete Hermanos and likely other locales as well.
Nesting: Lays a single pale buffy white egg that is sparsely marked with reddish brown at the large end. The nest can be in a bare rock depression or crevice sparingly decorated with pebbles, shells, or feathers; on flotsam situated on steep cliff walls; or on sparsely vegetated flat ground. Nests may also be elaborately constructed of seagrass and twigs in a tree or low bush. Breeds April to August. Known to nest on rocks off Isla Beata, and in the Cayos Siete Her-

manos where it nests in mixed colonies with Sooty Terns. Nesting is also suspected, but not yet confirmed, at Navassa Island, Isla Alto Velo, and nearby Piedra Negra.
Range: Breeds in Florida Keys, Bahamas, Greater Antilles, islands off Yucatán Peninsula and Belize, Isla San Andrés, islands off Venezuela and French Guiana, and in South Atlantic; also off Pacific Coast. Cosmopolitan during non-breeding season in pelagic waters of warm oceans.
Local names: DR: Severo, Bubí, Tiñosa Parda; **Haiti:** Mòv Tèt Blanch, Sterne Grise.

BLACK SKIMMER—*Rynchops niger* **Plate 24**
Vagrant

Description: 40–50 cm long; male 365 g, female 265 g. A distinctive, coastal waterbird. Adult is solid blackish above, white below, with a unique scissorlike black and orange bill with lower mandible extending 2–3 cm beyond upper. Immatures are similar but have upperparts mottled blackish brown. A graceful, buoyant flier on long wings, with head always held below the level of tail.
Similar species: Unmistakable.
Voice: Unique, nasal, doglike barking, *yep, yep,* given singly or in a series.
Hispaniola: One DR record of 3 birds at Puerto Alejandro in October 1997; 1 record of 3 birds in Haiti in fall 2003. To be looked for at calm coastal waters, harbors, and lagoons.
Status: Vagrant.
Comments: A tactile feeder, often foraging at night. Skimmers have the unique habit of skimming shallow water with their longer lower mandible slicing through the surface. When a small fish is encountered, the bill automatically snaps closed.
Range: Breeds on Atlantic coast from Massachusetts south to Florida, Gulf of Mexico coast from Florida to Yucatán Peninsula; also on Pacific Coast and in South America. In West Indies, a rare migrant or vagrant reported from Bahamas, Cuba, DR, Virgin Islands, Jamaica, Cayman Islands, and Grenada.
Local names: DR: Pico de Tijera; **Haiti:** Bèk Sizo, Bec-en-ciseaux Noir.

PIGEONS AND DOVES: FAMILY COLUMBIDAE

The pigeons and doves are a cosmopolitan family of plump, heavy-bodied birds with a small, rounded head. Wings and legs are relatively short, and the bill is slender, often with a distinct fleshy base. Most family members are gregarious in habit, though some, such as the quail-doves, are solitary. Flight is strong and direct, and wings often make a conspicuous clapping or whirring sound during takeoff. There is no formal distinction between pigeons and doves, but doves are generally smaller and longer-tailed than pigeons. All give low, cooing sounds, and these are important in species recognition. Crops are well developed for food storage and production of "pigeon's milk," a rich fluid fed by regurgitation to nestlings. Nests of all species are crude, flimsy platforms of twigs placed in bushes or trees. Many species raise several clutches of young per year. Many columbids on Hispaniola are threatened by illegal hunting and clearing of forest habitats.

ROCK PIGEON—*Columba livia* **Plate 57**
Introduced Breeding Resident

Description: 29–36 cm long; 355 g. A medium-sized, stocky, short-necked pigeon with highly variable plumage. Most birds feature a combination of black, gray, and white, including a black tail band, white underwings with dark border, two dark wing bars, and conspicuous white rump.

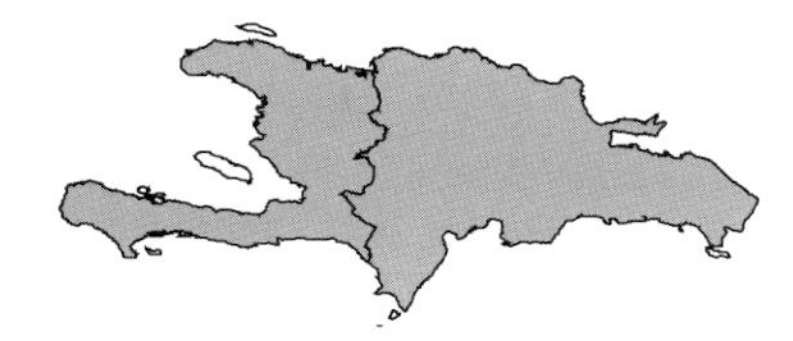

Similar species: Generally unmistakable, but in flight the large body size and sharply pointed wings appear almost falconlike. Distinguished by its stockier build, short tail, and white rump. White-crowned Pigeon is dark gray with a white cap; Scaly-naped Pigeon is also dark gray but with sides of neck suffused with iridescence.
Voice: Varied assortment of low, gentle cooing notes.
Hispaniola: Date of introduction to Hispaniola is not known, but feral populations are now present in and near most major towns and cities. Rarely occurs in forests or more remote habitats occupied by other large pigeons.
Status: Abundant year-round resident.
Comments: Commonly known as Rock Dove, Domestic Pigeon, or Pigeon.
Nesting: Lays 2 white eggs. Nests on a ledge of a building, bridge, or cliff. May breed year-round.
Range: Introduced from Eastern Hemisphere, resident virtually throughout. Present on nearly all inhabited islands in West Indies.
Local names: DR: Paloma Doméstica; **Haiti:** Pijon, Pigeon Domestique.

SCALY-NAPED PIGEON—*Patagioenas squamosa* Plate 25
Breeding Resident—*Threatened*

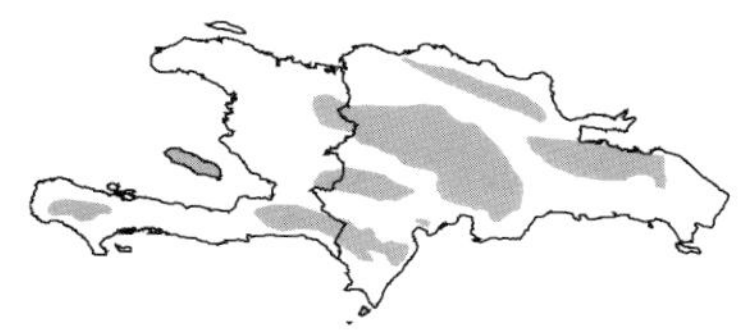

Description: 36–40 cm long; 250 g. A large pigeon that appears entirely slate gray at a distance. At close range head, neck, and breast have a purplish red tint which may be glossy in good light. Bare skin around eye is yellow to reddish-orange. Immature has a more reddish brown head and neck than adult.
Similar species: Scaly-naped Pigeon lacks the white on head of White-crowned Pigeon and is much larger and darker than doves. Plain Pigeon has a much lighter purplish red color on wings and breast, a white leading edge on closed wing, and a conspicuous white band across upperwing in flight.
Voice: Distinctive call heard frequently in the early morning is an emphatic, mournful *cruu, cruu-cru-CRUUU* with the heaviest accent on the fourth syllable. The very soft first syllable is separated by a pause. The last 3 syllables sound like *Who are YOU.*
Hispaniola: Essentially a species of moist broadleaf forest, no longer common in the lowlands and now most numerous in the more remote forested hills and mountains of the interior. Although it may occur in appropriate habitat at sites up to 1,800 m, it is restricted to mesic broadleaf forest. Recorded from three major satellite islands: Navassa Island, Île de la Gonâve, and Isla Beata.
Status: Formerly considered an abundant breeding resident in many areas, but current populations of this species are greatly reduced because of habitat destruction and excessive shooting, despite legally protected status. Uncommon and threatened even in national parks. Overall, this pigeon has declined widely on Hispaniola and throughout the West Indies.
Comments: Primarily arboreal, it feeds opportunistically on the ground. Usually occurs singly or in small flocks. Most frequently observed flying high over the forest canopy.
Nesting: Lays 2 glossy white eggs. A frail stick nest is usually constructed up to 5 m above ground in a tree, among palm fronds, or even on a bromeliad. On uninhabited islands, said to nest on the ground. Breeds year-round, but principally March to June. There is some suggestion of habitat segregation between this species and White-crowned Pigeon in the breeding season, with Scaly-naped Pigeon nesting in the interior hills and White-crowned Pigeon primarily in the coastal mangroves.
Range: Caribbean region only. Present on almost all major islands in West Indies.
Local names: DR: Paloma Turca, Paloma Morada; **Haiti:** Ramye Kou Wouj, Ramier.

WHITE-CROWNED PIGEON—*Patagioenas leucocephala* **Plate 25**
Breeding Resident—*Threatened*

Description: 30–35 cm long; 245 g. This medium-sized pigeon appears entirely slaty gray except for its distinctive white crown. Male has a clear white crown, but this is grayish white on females. Bill is red with a pale tip; iris is white. Immatures differ from females only in having grayish white crown limited to forecrown.
Similar species: Slightly larger Scaly-naped Pigeon also appears entirely dark gray at a distance but has a dark, rather than white, crown and purple-red head and neck. Plain Pigeon is slightly larger and paler overall, with a white band and a reddish brown patch on wing.
Voice: High-pitched *cruu, cru, cu-cruuu*, which sounds like *Who took two*. This call is a bit faster and less deliberate than that of Scaly-naped Pigeon, with the second syllable having a characteristic rising inflection. Also a distinct low purring sound.
Hispaniola: Principally a lowland and middle-altitude species, often roosting and nesting colonially in coastal mangroves, and flying into adjacent broadleaf forests in the interior to feed on fruits during the day. During the non-breeding season, disperses inland and to higher elevations in the mountains, probably in response to food and cover availability. Reported from nearly all major satellite islands.
Status: Formerly considered an abundant breeding resident nearly islandwide, but the population has been reduced an estimated 50–75%, depending on the locality, because of breeding-habitat destruction, deforestation of feeding areas, and excessive shooting. The species will be at serious risk if this trend continues.
Comments: A highly gregarious, arboreal species typically occurring in flocks of all sizes. Single roosts and breeding colonies formerly supported hundreds of thousands of White-crowned Pigeons, but colonies on Hispaniola are no longer anywhere near this size. An obligate fruit eater, this species is a powerful flyer, often commuting 45 km or more between its roosting and feeding grounds. Regular seasonal movements, sometimes between islands, occur in response to variations in the timing and intensity of local fruit production.
Nesting: Lays 1–2 glossy white eggs. Usually breeds in colonies but may also nest in isolated pairs. A flimsy twig nest is built, typically in mangroves or dry scrub, as all records of breeding are from areas below 75 m elevation. Has been found to nest in all four species of mangrove—black mangrove (*Avicennia germinans*), red mangrove (*Rhizophora mangle*), white mangrove (*Laguncularia racemosa*), and buttonwood mangrove (*Conocarpus erectus*)—as well as a variety of trees at inland localities. Has also nested in pock-holes on 6-m-high coastal cliffs facing to windward on Isla Beata. Timing and success of breeding are closely tied to fruit production. Breeding colonies usually begin to form in early May, coinciding with the onset of the rainy season, but may form as early as March or as late as mid-July. The numbers of birds in the colonies increase for several weeks because of immigration, resulting in asynchronous breeding stages in the colony. Breeding may continue to September or October. Breeding colonies may consist of hundreds or thousands of birds in coastal areas or closely spaced mangrove keys, with nests typically more densely spaced at the core of the colony. Although previous observations of nesting colonies reported trees weighed down with multiple nests, colonies now appear to be less densely packed.
Range: Occurs in s. Florida and Caribbean region only. In West Indies, resident on almost every major island south to Barbuda and Antigua in Lesser Antilles; rare visitor at Guadeloupe; casual south to St. Lucia.
Local names: DR: Casquito Blanco, Paloma Coronita; **Haiti:** Ramye Tèt Blanch, Pigeon à Couronne Blanche.

PLAIN PIGEON—*Patagioenas inornata* **Plate 25**
Breeding Resident—*Threatened*

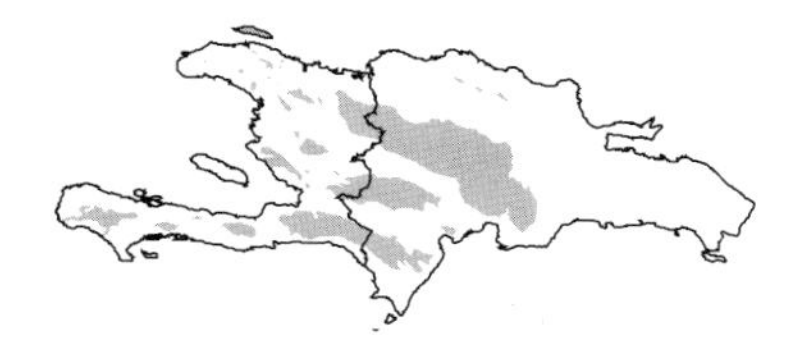

Description: 38–40 cm long; 250 g. A large pigeon, paler than other arboreal species, with more brown in plumage, white edges on upperwing coverts, and reddish brown on wings and breast. Iris is white; eye-ring is red. Immature birds are darker and browner than adults. In flight, shows a thin, white band across wing.
Similar species: Distinguished from Scaly-naped Pigeon when perched by reddish brown color on wings and breast, and white band on leading edge of wing. White-crowned Pigeon is uniformly darker and has a white or gray cap. White-winged Dove is smaller, browner, and has a much more pronounced white wing patch.
Voice: Deep, deliberate *whoo, wo-oo* or *who, oo-oo,* with emphasis on the first syllable, and other variations.
Hispaniola: Formerly found from the lowlands to middle elevations in lowland scrub, cactus thorn woodland, savanna, and primary and secondary evergreen forest. Now mostly restricted to less populated areas, especially in the pine zones. Locally common only in the Sierra de Bahoruco at elevations of 750 m and higher. Recorded in Massif de la Hotte, Haiti, at 1,200 m elevation. Reported from Île de la Tortue.
Status: Once considered among the most common resident pigeons, this species is currently uncommon almost everywhere. A decline of more than 90% in Los Haitises National Park from 1976 to 1996 probably reflects the species' overall trend. It has declined dramatically throughout its Caribbean range (although Puerto Rican populations increased in 1986–1992 and 1997–2001). Considered threatened by habitat loss and hunting.
Comments: Arboreal, typically in flocks. Surprisingly tame.
Nesting: Lays 1 white egg with a somewhat glossy surface. Builds a flimsy nest well above the ground, typically in a tree, bromeliad, or vine tangle. Has been found breeding in a wide variety of habitat types in DR from lowland cactus scrub at Azua, the foothills north and south of Lago Enriquillo, wet limestone karst forest of Los Haitises National Park, upland pine/scrub forest near Constanza, dry karst woodland at Boca de Yuma of Del Este National Park, humid upland deciduous forest at Las Cruces in Sierra de Bahoruco, wet montane forest near Miches, and lowland coastal wet forest at Sabana de la Mar. Breeds year-round, though primarily February to June.
Range: Caribbean region only, including Cuba, Hispaniola, Jamaica, and Puerto Rico.
Local names: DR: Paloma Ceniza; **Haiti:** Ramye Miyèt, Pigeon Simple.

WHITE-WINGED DOVE—*Zenaida asiatica* **Plate 25**
Breeding Resident

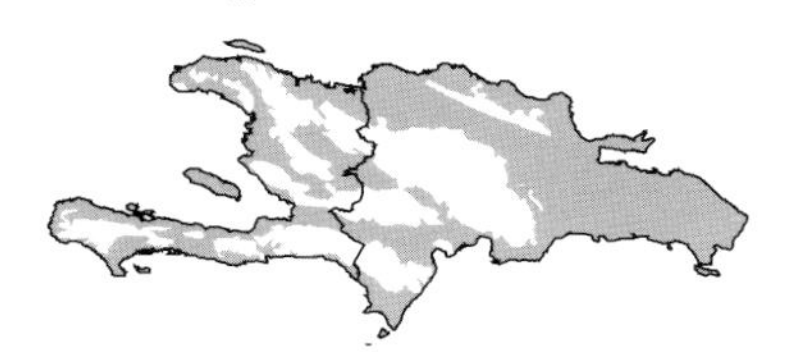

Description: 28–30 cm long; 150 g. A medium-sized, heavy-bodied, grayish brown dove with a large, white central wing patch, visible in flight or along the edge of the folded wing, and a small black crescent below the auricular. Tail is square and tipped white.
Similar species: Zenaida Dove has a white band on wing restricted to trailing edge of secondaries. Mourning Dove has no white in wing and a long, pointed tail. White-winged Dove is more arboreal than either of these species. Plain Pigeon is much larger and has a much less distinct white band transversing its wing.
Voice: Rhythmic *coo-co, co-coo* or *coo-co, co-co-coo* on a single pitch, which sounds like *who cooks for you.* Also a distinctive yodel-like cooing modulating between 2 notes.

Hispaniola: Resident of lowlands and lower hills islandwide, somewhat more numerous in drier woodland and mesquite scrub than in more humid habitats. Not known from higher elevations in the interior, but reported from nearly all major satellite islands.
Status: Moderately common year-round resident.
Comments: Usually flocks. Primarily arboreal in forested areas, but a ground dweller in urban areas with scattered trees where it frequently feeds with Zenaida Doves and Common Ground-Doves.
Nesting: Lays 2 white eggs with a faint tinge of cream and a somewhat glossy sheen. Nest is a frail platform of twigs and grasses usually built within 1–3 m of the ground. Nests year-round in more humid areas, but breeding is concentrated April to July in arid areas.
Range: Resident from c. Arizona and New Mexico, n. Chihuahua, c. Texas, and s. Alabama south to Baja California, Honduras, nw. Costa Rica, and locally in w. Panama. In West Indies, occurs in Bahamas and Greater Antilles east to Virgin Islands; vagrant at Saba.
Local names: DR: Tortola Aliblanca; **Haiti:** Barbarin, Toutrèl Zèl Blanch.

ZENAIDA DOVE—*Zenaida aurita* — Plate 25
Breeding Resident

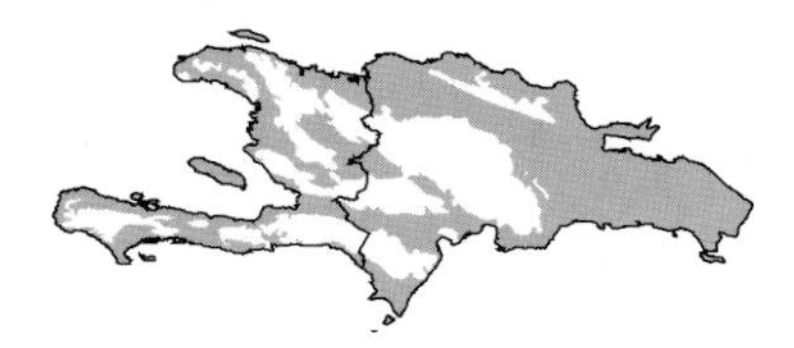

Description: 25–28 cm long; 159 g. A medium-sized, warm brown dove with a white trailing edge to dark secondaries, and a rounded tail with a single black subterminal band and broad white tips. A violet sheen on the neck is visible in good light. Hispaniolan birds are cinnamon brown above, paler below.
Similar species: Mourning Dove has no white in wing and a longer, pointed tail. White-winged Dove has a large, white central wing patch.
Voice: Gentle, mournful cooing, almost identical to that of Mourning Dove, *coo-oo, coo, coo, coo,* the second syllable rising sharply.
Hispaniola: Common resident in the lowlands of open country, second growth, and scrub, and occasionally found in gardens and cultivated areas. Does not occur in rainforest, and rare in the interior mountains, but reported from nearly all major satellite islands.
Status: Common year-round resident. Despite being a popular game species throughout its range, may have become slightly more numerous and widespread since the 1930s because of the clearing of forests and an increase in agricultural acreage.
Comments: Zenaida Dove occurs primarily in rural areas, whereas Mourning Dove occurs more often in settlements. Zenaida Dove may be displaced locally by White-winged Dove, as appears to be the case on other islands. Typically feeds on seeds on the ground, but also takes seeds and fruits from trees.
Nesting: Lays 2 white eggs. The nest is a thin platform of twigs built in a bush or tree, though sometimes on the ground. Nests year-round in some urban and in more humid areas, but primarily April to July in arid habitats.
Range: Caribbean region only; virtually throughout West Indies.
Local names: DR: Rolón, Rolón Turco; **Haiti:** Toutrèl Wouj, Grosse Tourterelle.

MOURNING DOVE—*Zenaida macroura* — Plate 25
Breeding Resident; Non-breeding Visitor

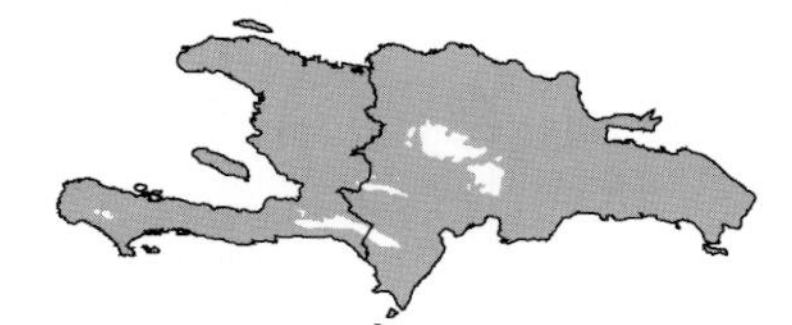

Description: 22.5–34 cm long; 120 g. A medium-sized, slender dove with a small head and a long, tapered tail tipped with white. Upperparts are grayish blue to grayish brown, underparts are buffy. Upperwing coverts contain black spots, but lacks white in wing. Male is identified by a purplish sheen on sides of neck and nape, and a bluish crown. Female has re-

duced purplish iridescence on neck, and a brownish crown. Immatures are browner than adults, heavily spotted with black, and lack iridescence.
Similar species: Zenaida and White-winged doves have white wing markings and rounded tails.
Voice: Mournful cooing almost identical to that of Zenaida Dove, *coo-oo, coo, coo, coo,* the second syllable rising sharply.
Hispaniola: Occurs in all parts of the island except the highest mountains. Primarily found in lowland open country, dry coastal forests, and agricultural lands, often near bodies of freshwater. Also found near agricultural areas in mountains. Reported from nearly all major satellite islands.
Status: Common breeding resident islandwide, particularly in the lowlands where its numbers reflect the disturbed nature of the habitat. Counts of 50–100 per day are not unusual. Numbers and range of this dove are increasing in the West Indies, likely the result of forest clearing and agricultural expansion. Band recoveries indicate that some non-breeding visitors come from the North American mainland.
Comments: A ground feeder, usually found in flocks except when adults pair to breed.
Nesting: Lays 2 pure white eggs with a slight gloss. A flimsy nest of twigs and grasses is typically built 2–6 m above ground in a tree or bush. Has been known to reuse nests of Northern Mockingbird. Primary nesting period is January to August, but possibly nests at other times as well.
Range: Breeds from c. British Colombia east to s. Manitoba, c. Ontario, and Nova Scotia, south to s. Baja California, Oaxaca, Texas, and Florida. Occurs in non-breeding season south to c. Panama. In West Indies, occurs in Bahamas and Greater Antilles.
Local names: DR: Tortola, Fifí, Rabiche; **Haiti:** Toutrèl Ke Fine, Tourterelle Triste.

COMMON GROUND-DOVE—*Columbina passerina* **Plate 25**
Breeding Resident

Description: 15–18 cm long; 30 g. A tiny, stocky dove, by far the smallest columbid on Hispaniola. Sandy brown overall with rufous primaries and a mostly dark tail with white-tipped corners. Male has a bluish gray crown and nape, and pinkish tint on underparts. Female has a sandy gray crown. In flight, stubby wings flash rufous primaries and underwings in quick, shallow wing beats.
Similar species: None; tiny size and bold rufous wing flash are unmistakable.
Voice: Monotonous, often repeated call either of single or double notes, *coo, coo, coo, coo* . . . or *co-coo, co-coo, co-coo* . . . or *hoop, hoop, hoop* . . . in staccato fashion.
Hispaniola: Resident throughout the island in the lowlands, especially in pastures, fields, clearings, and second growth. Becomes scarcer as elevation increases, but still found in open pine forests and at other higher elevation sites with patches of open ground. Also reported as a breeding resident from all of the major satellite islands.
Status: Common to abundant year-round resident. Always considered common, but its numbers have probably increased because of the clearing of forest and an increase of agricultural acreage. Counts of 50 per day are not unusual. The subspecies *C. p. navassae* is endemic to Navassa Island.
Comments: Primarily ground dwelling, it sometimes seeks refuge in trees. Often seen in small flocks. Typically flies only short distances when flushed.
Nesting: Lays 2 white eggs with a distinct gloss. The nest is of rootlets, grasses, or twigs and is built in a bush, tree, or cactus. Average height above ground of 95 nests in desert habitats of Sierra de Bahoruco, DR, was 1.8 m. Of 88 nests with known outcomes, 58 fledged at least 1 young, and 30 failed, most because of predation. Where free from terrestrial predators, such as mongoose, may nest on the ground. Individuals will reuse their own nest; also known to reuse nests of Northern Mockingbird. Believed to nest year-round with a peak in May and June.

Range: Breeds from s. California east to c. Texas and South Carolina, south to e. Brazil and Ecuador. Occurs virtually throughout West Indies.
Local names: DR: Rolita; **Haiti:** Zòtolan, Ortolan, Colombe à Queue Noire.

KEY WEST QUAIL-DOVE—*Geotrygon chrysia* **Plate 25**
Breeding Resident—*Threatened*

Description: 28–30 cm long; 170 g. A strikingly beautiful, reddish brown dove, with feathers suffused with iridescence. Underparts tawny; legs and feet red. All plumages show a bold white line under eye. Males have violet, blue, and green iridescence on the nape, mantle, and sides of breast. Females are duller with less iridescence on nape and upper back. Immature is more uniformly brown and buff, but bold facial stripe and pinkish legs are still present.
Similar species: Ruddy Quail-Dove has more reddish brown underparts and a duller buff streak below eye.
Voice: Low moan on one pitch, *ooo-wooo*, gradually increasing in volume and then fading rapidly. Distinguished only with extreme caution from the very similar call of Ruddy Quail-Dove which fades more gradually. Very ventriloquial.
Hispaniola: Occurs primarily in dry forest habitat in the lowlands islandwide, becoming scarce above 500 m and absent from the highest elevations. Prefers dense woods and scrubby thickets with ample leaf litter primarily in arid and semiarid zones, but also occurs in moist forests with an undisturbed understory. Also reported as a breeding resident of Île de la Gonâve, Île de la Tortue, and Islas Catalina and Saona.
Status: Moderately common resident. Numbers may have declined slightly because of hunting pressure and reduction of forest habitat.
Comments: A ground dweller, often heard deep among dense vegetation where its secretive habits make it difficult to see.
Nesting: Lays 1–2 buff-colored eggs. Nest is a loose platform of twigs, rarely on the ground, often in a low bush, occasionally as high as 5 m up in a tree. Breeds primarily February to August, but young chicks have been found in mid-November, suggesting that some nesting may take place year-round.
Range: Caribbean region only. In West Indies, resident in Bahamas, Cuba, Hispaniola, and Puerto Rico; casual in Florida.
Local names: DR: Perdía, Perdíz Grande; **Haiti:** Perdrix Grise, Pèdri Vant Blanch.

WHITE-FRONTED QUAIL-DOVE—*Geotrygon leucometopia* **Plate 26**
Breeding Resident—*Endangered Endemic*

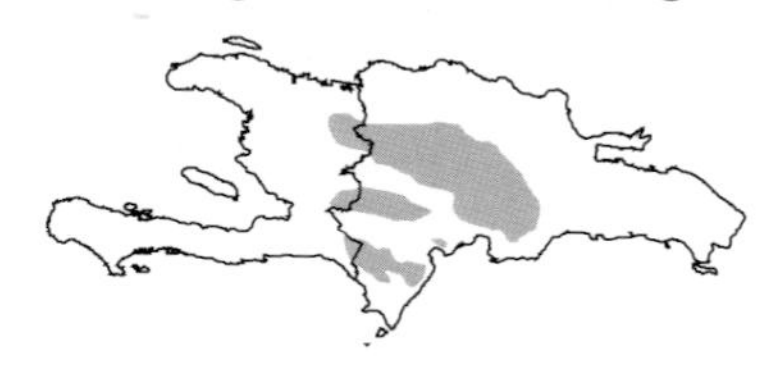

Description: 28 cm long; 171 g. A beautiful pigeon-like, ground-dwelling dove. Crown, nape, and sides of head are slate gray, with sides of neck suffused with reddish purple or violet, sometimes extending onto and across breast. Back is slightly darker and has a metallic purplish blue sheen. Underparts are also gray, becoming rufous on lower belly and undertail coverts. Striking forecrown is white; eye is red. Bill is reddish but becomes paler toward tip, which may be yellowish in adults. Legs are also reddish. Immatures are similar to adults but browner, lacking the reddish purple or bluish sheen on mantle and sides of neck.
Similar species: All other quail-doves on Hispaniola have facial stripes and are browner.

Voice: Continuous, low *uup-uup-uup-uup* without pauses, then changing to a prolonged *coo-o-o*. Typically calls from a low branch.
Hispaniola: Resident of humid mountain forests with plenty of decayed leaves in DR; not yet reliably reported from Haiti though likely to have occurred at least in the Massif de la Selle west of known sites in the Sierra de Bahoruco, and in the Massif du Nord west of the Cordillera Central. Reported from elevations of 745–1,685 m in the Cordillera Central and Sierra de Bahoruco, DR.
Status: Uncommon and local resident, considered endangered. This species has certainly become less widespread and rarer since the 1930s because of hunting pressure and cutting of high-elevation forest habitat. In fact, habitat destruction has nearly extirpated this species from the Cordillera Central, and it appears to have already been extirpated from the Sierra de Neiba.
Comments: Forages for seeds on the ground alone or in pairs. Displays a peculiar balance of the neck and tail while walking or perching. Often occurs with Ruddy Quail-Dove, and known from recaptures of marked individuals to remain in favored localities. Formerly considered a subspecies of Gray-headed Quail-Dove (*G. caniceps*), which occurs in Cuba, but separated based on differences in plumage coloration, shape of the primaries, tail length, and habitat preferences.
Nesting: Lays 1–2 beige eggs. A nest in montane forest of Sierra de Bahoruco was a platform constructed largely of dead pine needles, placed in a tree 6 m above ground. Believed to breed January through August.
Range: Endemic on Hispaniola.
Local names: DR: Perdíz Coquito Blanco; **Haiti:** Pèdri Fron Blanch, Perdrix à Front Blanc.

RUDDY QUAIL-DOVE—*Geotrygon montana* **Plate 25**
Breeding Resident—*Threatened*

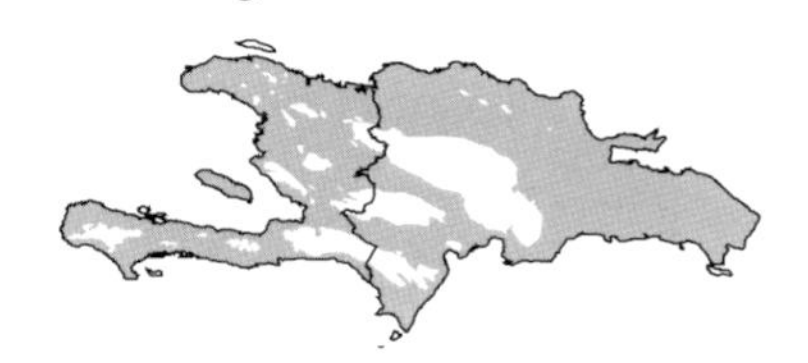

Description: 23–25 cm long; 110 g. Medium-sized, plump, short-tailed dove of forest floors. Male is predominantly rufous brown above, sometimes glossed violet, buffy brown below with conspicuous light buffy auriculars cut by a rufous brown stripe beneath eye. Female is brownish olive, less reddish than male, with a less conspicuous facial stripe. Immatures resemble females but are even more pale, with back feathers extensively edged with cinnamon.
Similar species: Key West Quail-Dove has a more pronounced white stripe beneath eye and lacks overall reddish brown plumage of Ruddy Quail-Dove.
Voice: Simple, mournful *hoooooo*, gradually fading in strength and sometimes in pitch, like blowing across the mouth of a bottle. Distinguished only with extreme caution from very similar call of Key West Quail-Dove which fades more rapidly. Very ventriloquial.
Hispaniola: Resident of dense humid forest, and coffee and cacao plantations at midelevations in the hills and mountains, but also locally on the coast. Also reported to be a breeding resident of Île de la Gonâve and Isla Catalina.
Status: Moderately common resident. Numbers appear to have declined because of habitat loss and hunting pressure.
Comments: Ground dwelling, much more often heard than seen.
Nesting: Lays 2 light buff eggs. Builds a loose nest of twigs and leaves usually in a low bush or on the ground. Breeds February to August, but time of breeding may vary annually.
Range: Caribbean region and from Sinaloa and Veracruz, Mexico, south to ne. Argentina and e. Bolivia. In West Indies, resident of Greater Antilles, and in Lesser Antilles from Guadeloupe south to Grenada, but not on Barbados or Grenadines.
Local names: DR: Perdíz Colorada; **Haiti:** Pèdri Fran, Perdrix Rouge.

PARAKEETS AND PARROTS: FAMILY PSITTACIDAE

The parrots and parakeets form a distinctive family of brightly colored, arboreal birds typical of warmer climates. They are easily recognized by their raucous calls, large head, and extremely heavy bill, which is often used to assist their movements among tree branches. Hispaniolan species are gregarious and primarily green. Parakeets are distinguished by their smaller size and long, pointed tail and wings; parrots are larger with a shorter, squared tail and broader wings. Flight is direct, with rapid, shallow wing beats. All species feed mainly on seeds and fruit, which they often lift to their bill with dextrous feet. Nesting occurs in cavities. Although many species have been introduced to other islands in the West Indies, Hispaniola seems to have as yet largely escaped this potential problem.

HISPANIOLAN PARAKEET—*Aratinga chloroptera* **Plate 27**
Breeding Resident—*Threatened Endemic*

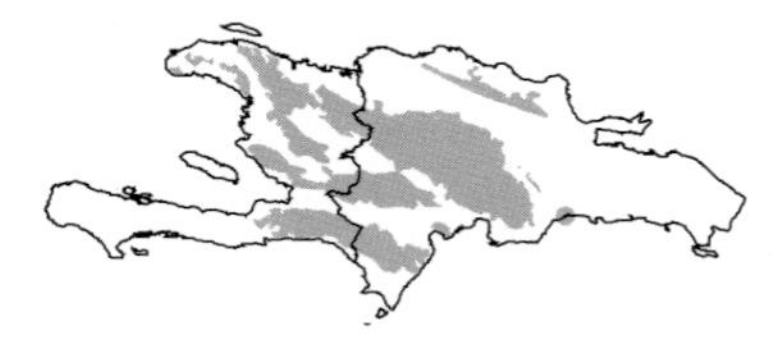

Description: 30–33 cm long; 145g. A large parakeet. Bright green overall, with a long, pointed tail, white eye-ring, and red edge along the bend of wing. In flight, note the red underwing coverts.
Similar species: Hispaniolan Parrot is more robust, larger, and with a shorter, squared tail. See Olive-throated Parakeet.
Voice: Screeching flight and perch calls, similar to those of Olive-throated Parakeet, but more raucous and often repeated in groups of 3–4 notes, such as *creek-creek-creek-creek* or *clack-clack-clack-clack*. Calls are repeated often in flight, and are much simpler than those of Hispaniolan Parrot.
Hispaniola: Locally common in undisturbed habitat, but elsewhere rapidly declining and reduced in distribution. Most often seen in the pine zone, but may also occur in adjacent shade coffee plantations and other agricultural habitats when feeding. In the DR, it still occurs in good numbers in the Sierra de Bahoruco and Sierra de Neiba, particularly at middle elevations between 900 and 1,800 m. Extirpated from Los Haitises National Park about 1986, but still common in urban parks of Santo Domingo. In Haiti, rare and possibly near extirpation in the Massif de la Selle, reportedly still common in the Citadelle area in the Massif du Nord, uncommon elsewhere, and unknown on the Tiburón Peninsula west of the Jacmel-Fauché depression in southern Haiti. Reported from Isla Beata.
Status: Formerly common breeding resident in the high mountains of the interior, especially in the pine belt, and locally distributed elsewhere, though somewhat less numerous in Haiti than DR. Numbers were apparently reduced by 1930 from early flocks which numbered on the order of 1,000 birds and were often pests on maize crops; hundreds were shot to drive them away. From 1930 to 1980, this species declined sharply to the point where it is now found relatively seldom in lowlands, and flocks larger than 50 are unusual anywhere, except at a few known communal roosts. The decline has been the result of habitat loss, shooting, and taking for the pet trade. Although common in Los Haitises National Park, DR, in 1975–1976, it was apparently extirpated from there by 1986. The species is considered threatened and is seriously at risk if present trends continue.
Comments: Feeds mostly on fruits and seeds. Travels and forages in flocks, often feeding on corn and other crops, for which it is persecuted by farmers. As has been the pattern throughout the West Indies, parakeet populations have declined more rapidly than have parrot populations. A subspecies of the Hispaniolan Parakeet (*A. c. maugei*), endemic to Puerto Rico and Mona Island, became extinct at the end of the 19th century, the last specimen being taken in 1892.
Nesting: Lays 3–4 white eggs. Nests in tree cavities and arboreal termite nests up to 26 m

above ground. In the Sierra de Bahoruco, 4 nests were in cavities 10–18 m high in wind trees (*Didymopanax tremulus*). Breeds February to June.
Range: Endemic to Hispaniola.
Local names: DR: Perico; **Haiti:** Perruche, Perich.

OLIVE-THROATED PARAKEET—*Aratinga nana* Plate 27
Introduced(?) Breeding Resident

Description: 24 cm long; 77 g. A small, green, slender parakeet with blue primaries and a long, pointed tail. Underparts are dark brownish olive. Note the grayish to creamy yellow eye-ring and bill.
Similar species: Besides the color of underparts, Olive-throated Parakeet can be distinguished from Hispaniolan Parakeet because the former is smaller, has slightly darker green plumage and blue primaries, and has no red at the bend of wing. Bill of Olive-throated Parakeet is grayish white, that of Hispaniolan Parakeet a mixture of pinks and oranges. Olive-throated Parakeet has a more erratic, rapid flight than the native parakeet.
Voice: Screeching call of paired notes, such as *creek-creek* or *clack-clack*, repeated often in flight, but less raucous than Hispaniolan Parakeet.
Hispaniola: Locally common at lower elevations in the Sierra de Bahoruco where it can be found in pine forests, broadleaf scrub and woodlands, and especially croplands and gardens. Reports indicate two populations, one between Puerto Escondido and Aguacate on the north side of the Sierra de Bahoruco, and the other from Los Arroyos to Las Mercedes on the south side of the Sierra de Bahoruco. Rarely reported in Santo Domingo.
Status: Breeding resident, believed present by local people in the Las Mercedes area of Sierra de Bahoruco, DR, since before 1970, but others believe that the species was introduced, either intentionally or inadvertently, by passengers on ore boats that plied regularly from near Pedernales, DR, to Jamaica. Otherwise, first documented in DR in 1995 in Santo Domingo, and in the Sierra de Bahoruco just west of Barahona and near Aguacate. Populations in the Sierra de Bahoruco seem to be increasing rapidly in numbers and may be competing with the native Hispaniolan Parakeet, possibly displacing that species to some extent.
Comments: Typically forages and roosts in small flocks. Feeds on a wide variety of fruits and seeds, resulting in its periodically becoming a pest to cultivated crops. It has yet to be determined whether the Dominican birds are of Jamaican or Central American stock.
Nesting: Few data from DR, but in Jamaica lays 3–5 white eggs in a nest built in a termite mound or tree cavity. Breeds March to June in Jamaica. In DR, a nest cavity in a termite mound on a rock only about 1.4 m tall was observed at Los Naranjos. The pair initiated the nest in mid-January, and the cavity was complete by mid-February.
Range: Resident on Gulf of Mexico and Caribbean slope of Middle America from s. Tamaulipas to w. Panama. In West Indies, found in Jamaica, DR, and Puerto Rico.
Local names: DR: Perico Amargo, Perico Haitiano; **Haiti:** Perich Doliv.

HISPANIOLAN PARROT—*Amazona ventralis* Plate 27
Breeding Resident—*Threatened Endemic*

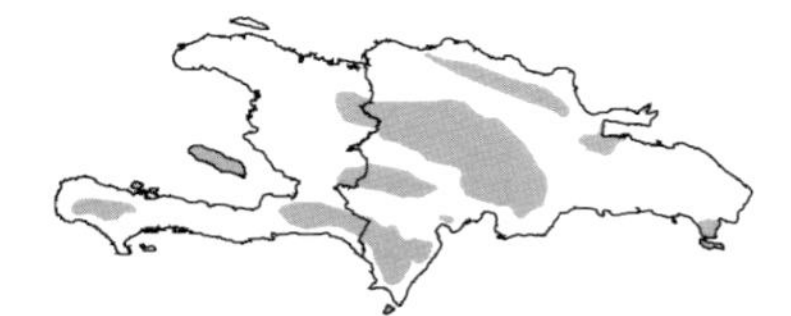

Description: 28–31 cm long; 250–300 g. Distinguished by its large size, chunky shape, and short, squared tail. Bright green overall, with a white forecrown, dark auricular spot, and maroon belly. In flight, shows bright blue primaries and secondaries. Flight is ducklike with rapid wing beats, and wings move below the plane of back.
Similar species: Hispaniolan and Olive-throated parakeets are smaller with long, pointed tails.

Voice: Loud bugling calls in flight; perch calls consist of loud squawkings and screeches.
Hispaniola: Found at all elevations in forests, woodlands, and scrub wherever suitable fruits and seeds are available and it is free from human persecution. Formerly common throughout the main island, the parrot is now much reduced in numbers to the point of being extirpated or uncommon in most areas. This parrot is still locally common only in major forest reserves such as Del Este, Jaragua, Armando Bermúdez, and José del Carmen Ramirez national parks. It occurs sparsely in Santo Domingo. Also recorded from the Cayemites, Île de la Gonâve, and Islas Beata and Saona.
Status: Formerly a locally abundant breeding resident islandwide, with flocks of 500 observed. By 1930, however, this species was common only in the interior montane forests. Since then it has declined alarmingly and is now extirpated or uncommon in most areas. There was at least a 95% decline in Los Haitises National Park, DR, from 1976 to 1996, and only 2 pairs were found in the southern part of the park in June 2000 where formerly there was a large breeding population. Earlier reports mention this species being common in the Massif de la Hotte and on the La Selle ridge, Haiti, but it is now rare and endangered in both La Visite National Park and Macaya Biosphere Reserve. Today it is unusual to see a flock of 30 or more anywhere except in a few major forest reserves. The decline is the result of habitat loss, excessive hunting, and especially the capture of nestlings for the pet trade, despite legal protection which is not enforced. Parrots are frequently seen as pets throughout the DR, but trade is partly driven by the high price these birds command on the international market. This species is seriously at risk if present trends continue.
Comments: Normally feeds on seeds and fruits at middle to high levels in trees. Forms foraging flocks that sometimes depredate crops. Often roosts at higher elevations, descending to lower elevations to feed.
Nesting: Lays 2–4 white eggs. Typically nests in a tree cavity up to 20 m above ground, but 1 record of a nest 1.5 m high. Also reported to nest on rock ledges and in small caves in limestone rock faces in Sierra de Bahoruco and Los Haitises National Park, DR. Breeds February to June.
Range: Endemic to Hispaniola.
Local names: DR: Cotorra; **Haiti:** Jako, Jacquot, Perroquet.

CUCKOOS AND ANIS: FAMILY CUCULIDAE

Cuckoos are sleek, slender birds with a long, graduated tail; long, thin bill that is somewhat down-curved; and feet with two toes pointed forward and two toes pointed back. They are slow and deliberate, moving furtively through the dense foliage of trees and shrubs; their flight is smooth and direct. Many species of this cosmopolitan family, though not those on Hispaniola, are brood parasites, laying their eggs in the nests of other birds. Diets are diverse, consisting of insects, reptiles, fruits, small mammals, eggs, and nestling birds. The **anis**, which are confined to the tropics and subtropics of the Western Hemisphere, form an unusual assemblage of three species within the Cuculidae. All are black with a characteristic heavy bill and share the unique communal nesting habit described for the Smooth-billed Ani.

BLACK-BILLED CUCKOO—*Coccyzus erythropthalmus* **Plate 29**
Passage Migrant

Description: 28–31 cm long; 51 g. A slender cuckoo with a long, white-tipped tail; dull white underparts; long, dark, down-curved bill; reddish or yellowish eye-ring; and upperwing coverts entirely gray-brown.
Similar species: Very similar Yellow-billed Cuckoo lacks red eye-ring and has yellowish lower

bill, prominent reddish brown in primaries, and more conspicuous white markings under tail. Mangrove Cuckoo has black auricular patch and buff-colored lower underparts. Immature Black-billed Cuckoos in September and October may have a bit of yellow in bill, yellowish eye-ring, and some reddish brown in primaries. These birds can be distinguished from Yellow-billed Cuckoo by their pale gray undertail which lacks distinctive white spots.
Voice: Series of 3–4 notes, *cu-cu-cu-cu.*
Hispaniola: Rare spring and fall transient. Only 4 DR records, from Puerto Palenque, Lago Limón, La Descubierta, and Lago Enriquillo. Most often found in April and May, but also recorded September to November.
Status: Regular but rare passage migrant.
Comments: May be expected in scrublands, mangrove forests, and both dry and moist forests in the lowlands.
Range: Breeds in most of s. Canada from Alberta east to Nova Scotia and from e. Colorado east to ne. Oklahoma and w. North Carolina. Non-breeding-season distribution poorly known, but primarily in South America from Colombia east to w. Venezuela and south to c. Peru, possibly to Bolivia. In West Indies, rare passage migrant primarily in Bahamas, Cuba, and DR; a vagrant elsewhere.
Local names: DR: Pájaro Bobo Pico Negro; **Haiti:** Ti Tako Bèk Nwa.

YELLOW-BILLED CUCKOO—*Coccyzus americanus* **Plate 29**
Passage Migrant; Breeding Resident

Description: 26–30 cm long; 64 g. A slender bird distinguished by its clean white underparts and absence of black on auriculars. Tail is long and graduated with broad white tips; long, down-curved bill is yellow below with a black tip. In flight, reddish brown wing patch is diagnostic.
Similar species: Mangrove Cuckoo has black auricular patch and buff-colored underparts. Black-billed Cuckoo lacks yellow in bill and reddish brown in wing, and has less conspicuously white-tipped tail.
Voice: Throaty *ka-ka-ka-ka-ka-kow-kow-kowlp-kowlp-kowlp-kowlp*. Volume increases initially, then remains constant. The call slows substantially during the final syllables.
Hispaniola: Found nearly islandwide in lowland scrub and dry forests at elevations below 700 m, occasionally in pine forest at higher elevations. Also reported from nearly all of the major satellite islands.
Status: Common spring and fall passage migrant and uncommon breeding resident May to August. Breeding residents migrate southward in fall, when joined by birds from North America, and return northward in spring. North American migrant cuckoos sometimes pass through Hispaniola in large concentrations. May be expected mid-March, but peak spring migration period appears to be the last week of April and first 2 weeks of May. Extreme late migrants are indistinguishable from summer residents. The first fall migrants are suspected by early September, and peak fall migration period is October. Unknown on Hispaniola from late November until March.
Nesting: Lays 2–5 blue eggs in a flimsy, cup-shaped nest of twigs and dried grass, usually constructed low in a bush, occasionally up to 5 m high. Breeds April to July. Nesting birds known from Matadero, Baní, and from pine forest at 1,400 m in the Aceitillar sector of Sierra de Bahoruco National Park, DR.
Range: Breeds from n. Minnesota and se. North Dakota east to s. Quebec and New Brunswick, south to Baja California, Tamaulipas, and Florida. In non-breeding season occurs primarily in South America east of Andes south to Bolivia and n. Argentina. In West Indies, breeds in Greater Antilles, possibly in Bahamas, locally eastward to St. Kitts and St. Martin; a passage migrant elsewhere.
Local names: DR: Pájaro Bobo Pico Amarillo; **Haiti:** Ti Tako Bèk Jòn.

MANGROVE CUCKOO—*Coccyzus minor* **Plate 29**
Breeding Resident

Description: 28–32 cm long; 64 g. Slender with a long, white-tipped tail that is blackish below, and a long, down-curved bill with yellow lower mandible. Note the black "mask" through eyes and buff-colored abdomen. Flight is direct with short glides, and wings lack reddish brown.
Similar species: All other cuckoos lack black mask through eye. Yellow-billed and Black-billed cuckoos have dull white underparts. Hispaniolan Lizard-Cuckoo is much larger.
Voice: Call is similar to that of Yellow-billed Cuckoo but slower, lower pitched, and more nasal.
Hispaniola: Found islandwide in mangroves, lowland thickets, dry scrub, and dry forest at elevations up to 365 m. Also reported from nearly all of the major satellite islands.
Status: Generally a fairly common year-round resident. May have become more abundant since the mid-1970s, as it became twice as common in Los Haitises National Park from 1976 to 1996.
Comments: Arboreal and secretive, this inconspicuous bird is usually located by its call.
Nesting: Lays 2–4 blue eggs in a flimsy stick nest in a tree or shrub. Breeds February to July, although the observation of a pair copulating in dry forest in Sierra de Bahoruco in early December suggests breeding may sometimes begin earlier.
Range: Breeds in coastal s. Florida, Caribbean region, and throughout Central America southward from s. Sonora and s. Tamaulipas, Mexico. South American range poorly known, but occurs in Venezuela, Suriname, Guiana, and n. Brazil. Found virtually throughout West Indies, although rare on Cuba. Believed to be non-migratory throughout range.
Local names: DR: Primavera, Pájaro Bobo Menor; **Haiti:** Ti Tako, Coulicou Manioc.

HISPANIOLAN LIZARD-CUCKOO—*Saurothera longirostris* **Plate 28**
Breeding Resident—*Endemic*

Description: 41–46 cm long; 110 g. A large cuckoo identified by its size, pale gray breast, long tail with white tips, and fairly straight and slender bill, prominently hooked at tip. Has a reddish brown wing patch. Throat color varies from whitish on Île de la Gonâve to dull orange on mainland Hispaniola.
Similar species: Bay-breasted Cuckoo is larger and has dark reddish brown breast.
Voice: Prolonged throaty *ka-ka-ka-ka-ka-ka-ka-ka-kau-kau-ko-ko* in descending tones. Also a guttural *tuc wuh-h-h* with the *tuc* being sharp and stacatto.
Hispaniola: Endemic to Hispaniola. Widely distributed in forested and wooded areas, including suburban gardens and plantations of shade coffee, at elevations from sea level to 1,700 and rarely 2,200 m. Also recorded as a breeding resident from several major satellite islands, including Île de la Gonâve, Île de la Tortue, and Islas Beata, Catalina, and Saona.
Status: Generally common breeding resident in appropriate habitat, but this species showed evidence of an almost 90% decline at Los Haitises National Park from 1976 to 1996. The subspecies *S. l. longirostris* is endemic to Hispaniola and Isla Saona; the subspecies *S. l. petersi* is endemic to Île de la Gonâve.
Comments: Forages at all heights from the understory to the forest canopy. Often seen moving deliberately through the vegetation where it pursues lizards and larger insects, or dropping from the subcanopy to trunks of trees. Also chases prey somewhat clumsily along trunks and branches.
Nesting: Lays 2–4 white eggs. Nest is a well-hidden rough structure of twigs, situated at a moderate height. Breeds March to June.

Range: Endemic to Hispaniola and associated satellite islands.
Local names: DR: Pájaro Bobo, Tacot, Tacó; **Haiti:** Tako, Tacco d'Hispaniola.

BAY-BREASTED CUCKOO—*Hyetornis rufigularis* **Plate 28**
Breeding Resident—*Endangered Endemic*

Description: 43–51 cm long; mass unavailable. A large, active cuckoo distinguished by its dark reddish brown throat and breast and thick, somewhat decurved bill. Also has a reddish brown wing patch and very long tail with white tips.
Similar species: Hispaniolan Lizard-Cuckoo has a pale gray breast.
Voice: Strong *cu-aa*, sometimes followed by a guttural, accelerating *u-ak-u-ak-ak-ak-ak-ak-ak-ak-ak*.
Hispaniola: Rare and locally distributed in the DR. In the early 1900s this species was recorded at elevations as high as 900 m at Hondo, DR, and Moustique, Haiti, and in the 1950s and 1960s was found in the Cordillera Central near Constanza. Since the 1990s, found most often on the northern and extreme eastern slopes of the Sierra de Bahoruco, and in the Río Libón valley; also reported to be in the Río Limpio-Carrizal area of northwestern DR. Current status in Haiti is unknown. Preferred habitat appears to be the narrow transition zone between dry forest and moist broadleaf at low to moderate elevations, but also reported in mixed pine and broadleaf, and occasionally in overgrown pasture or agricultural sites.
Status: Breeding resident in the DR, but distribution of this endemic has become very local and the species has declined to the point of endangerment. Numbers have greatly declined since 1930 because of habitat destruction and alteration and hunting pressure, because of the mistaken belief that this species' flesh has medicinal properties. Believed by some local people in the DR to cure arthritic and other body pains. The species is at serious risk if its downward population trend continues.
Comments: Shy and secretive. Feeds on insects, lizards, frogs, small mammals, and bird eggs and nestlings.
Nesting: Lays 2–3 dirty white or light gray eggs in a nest in a bush or dense foliage of tree at low to moderate height. Breeds March to June. Chicks fledge in only 10–11 days.
Range: Endemic to Hispaniola, including Île de la Gonâve.
Local names: DR: Cúa, Tacot, Tacó; **Haiti:** Tako Kabrit, Piaye Cabrite.

SMOOTH-BILLED ANI—*Crotophaga ani* **Plate 29**
Breeding Resident

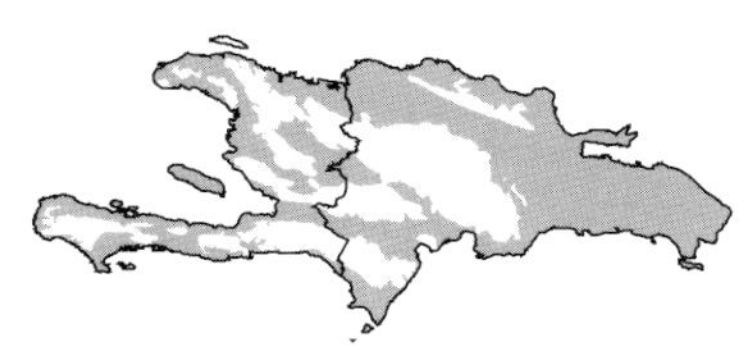

Description: 30–36 cm long; 105 g. A large, entirely black bird with a heavy, parrotlike bill and conspicuously long, flat tail. Flight is straight and slow, with rapid, shallow wing flaps alternating with longer glides. Anis typically fly in single file, calling. Often lands clumsily. Typically occurs in small, noisy flocks.
Similar species: Generally black Greater Antillean Grackle is smaller and has a long tail, but tail is V-shaped and not as long as that of ani.
Voice: Loud and conspicuous, slurred and whining whistled *ah-nee* or *a-leep*. Also a variety of short "growls," "coughs," and "barks."
Hispaniola: Typically found in a wide range of open habitats with scattered trees or bushes. Occurs primarily in agricultural areas, scrub, gardens, and plantations principally at low elevations, but rarely has moved to elevations as high as 1,600 m as dense forests have been cleared. Also reported from almost all of the major satellite islands.
Status: Common to abundant year-round resident on Hispaniola. Although abundance ap-

parently changed little from 1850 to 1930, numbers appear to have increased since 1930 as forest cover has been reduced and the amount of secondary growth and agricultural land has risen. Population in Los Haitises National Park doubled from 1976 to 1996. Daily numbers of 10–20 individuals are now routinely seen in appropriate habitat.
Nesting: Smooth-billed Ani, along with the other two members of its genus, is unique among birds in building a bulky nest which is often used communally by several females. Twenty or more eggs are regularly found in a single nest, with groups of 4–5 eggs laid in layers separated by twigs and leaves. Only the top layer hatches. The eggs are pale blue to greenish blue with a chalky, white outer coating that is easily scratched off. Apparently breeds year-round.
Range: Resident in s. Florida, West Indies, and locally in Central America from Nicaragua and Costa Rica to Panama. Resident in South America south to w. Ecuador and n. Argentina. Present on most islands in West Indies, but locally absent in some areas.
Local names: DR: Judío; **Haiti:** Boustabak, Ani à Bec Lisse.

BARN OWLS: FAMILY TYTONIDAE

The barn owls form a small family of widely distributed birds. Their heart-shaped facial disk and long legs are distinctive. They exhibit habits similar to those of typical owls, but their voices include shrieks and hisses, rather than the hoots of most owls. Ages and sexes are similar, but females are larger and tend to be darker than males. Barn owls are highly nocturnal and primarily depend on their acute sense of hearing to locate prey. More fossil than living species are known, including gigantic flightless forms.

BARN OWL—*Tyto alba* **Plate 29**
Breeding Resident; Non-breeding Visitor

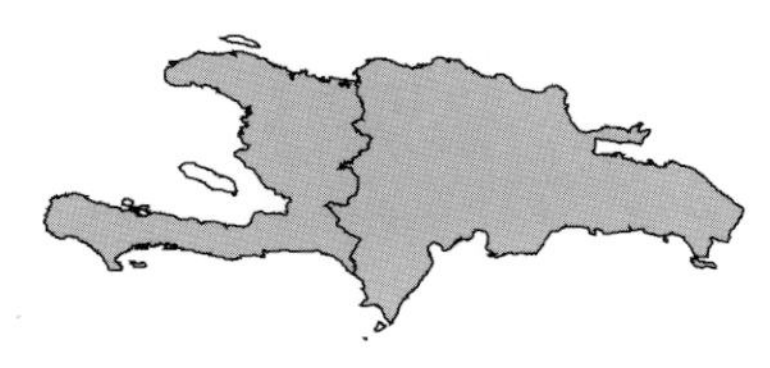

Description: 30–40 cm long; 525 g. A large nocturnal owl with a flat, heart-shaped face and dark eyes. Birds in the Greater Antilles are pale colored, with light orangish brown upperparts and mostly white underparts.
Similar species: Very similar Ashy-faced Owl is darker overall, and its heart-shaped face is silver gray rather than white.
Voice: Drawn-out scream, *karr-r-r-r-r-ick,* is the best-known call, but also emits a harsh, long hissing screech, as well as loud clicking sounds.
Hispaniola: Occurs islandwide over a wide range of elevations in relatively open areas that include rice fields, dry scrub, open woodlands, and human settlements. Also reported from Navassa Island and Isla Saona.
Status: Moderately common breeding resident islandwide from sea level to the interior mountains. This species may be a relatively recent colonist of Hispaniola, possibly having arrived from the Bahamas or North America. It was unrecorded by early ornithologists, and James Bond speculated that it became established in Hispaniola in the mid-1900s. Although the species may have been simply overlooked, the spread of agriculture and deforestation since 1930 has greatly increased the amount of appropriate habitat, and the introduction of exotic rodents has increased its potential food supply. Specimen records indicate that the resident population is joined by at least a few owls from North America in the non-breeding season. Like all other owls in DR and Haiti, it is persecuted.
Comments: A nocturnal species. Sometimes perches on fence posts or utility poles, especially in rice fields. Seen in large numbers at oil-palm plantations, this owl feeds on rats and mice which are attracted by the palm fruit.

Nesting: No nest is built. Typically lays 3–8 white eggs in a cave, tree cavity, or similar site. Breeds in all months but primarily August to April.
Range: Resident from sw. British Columbia east to s. Idaho, s. Ontario, and Massachusetts, south to Tierra del Fuego. One of the most broadly distributed land birds worldwide, it is widespread in Eastern Hemisphere. In West Indies, resident in Bahamas, Cuba, Jamaica, Hispaniola, Cayman Islands, and in Lesser Antilles on Dominica, St. Vincent, Grenadines, and Grenada.
Local names: DR: Lechuza Común, Lechuza Blanca; **Haiti:** Frize, Fresaie.

ASHY-FACED OWL—*Tyto glaucops* **Plate 30**
Breeding Resident—*Endemic*

Description: 35 cm long; mass unavailable. A reddish brown, long-legged owl with a silver gray, heart-shaped face.
Similar species: Barn Owl has a white facial disk and underparts.
Voice: Series of high-pitched, ratchety clicks, sometimes followed by a hissing cry.
Hispaniola: Found in open woodlands and dry and moist broadleaf forest, from sea level to 2,000 m. In the DR, it is common particularly in and around Santo Domingo, Los Haitises National Park, and Samaná Province, the southeastern portion of the island, and throughout the Barahona Peninsula, especially in coastal forests south of Barahona near limestone cliffs where it likely nests. Current distribution in Haiti is unknown.
Status: Endemic to Hispaniola where it is fairly widespread and locally common. Probably has declined somewhat because of habitat loss since 1930.
Nesting: Lays 2–7 unmarked white eggs in a nest in a tree cavity, or on a ledge in a sinkhole or cave. Breeds January to June.
Comments: Nocturnal. Tends to be found in sites which are more heavily forested than those typically frequented by Barn Owl. Frequently roosts in palm trees or pines at higher elevations. Less likely to be found around cities and towns than Barn Owl, but has been seen with Barn Owls at oil-palm plantations feeding on rats and mice which are attracted by the palm fruit. Prey includes rodents, bats, lizards, frogs, and birds.
Range: Endemic to Hispaniola.
Local names: DR: Lechuza Cara Ceniza; **Haiti:** Frize Figi Gri.

TYPICAL OWLS: FAMILY STRIGIDAE

The typical owls are primarily nocturnal birds of prey, although some are active during the day. They are characterized by a distinctive facial disk, a large head with eyes directed forward, and silent flight. Sexes are similar, although females are larger. Owls swallow their food whole and regurgitate bones, feathers, and insect wings in a compact pellet.

BURROWING OWL—*Athene cunicularia* **Plate 29**
Breeding Resident

Description: 19–25 cm long; 155 g. A small, long-legged owl, distinguished by its habit of being active during daylight hours, often standing on open ground or perched on a low post. Conspicuously bobs when approached. Upperparts of adults are brown profusely spotted with white, underparts are white with broad brown barring. Juveniles are similar but underparts are buffy and unbarred. Flight is rapid, direct, and low to ground. Usually flies only a

short distance, but may frequently hover, and sometimes flycatches in aerial pursuit of insects.

Similar species: No other small owls occur on Hispaniola.

Voice: Soft, high-pitched, 2-note *coo-cooo,* given only by the male. When alarmed or agitated, both sexes give a clucking chatter or scream.

Hispaniola: Resident of semiopen dry habitats, scrubby areas, sandy pine savannas, pastures, and limestone ravines throughout much of Haiti, and in the DR west of 70º00′ W longitude; virtually unknown east of that. Recorded from near sea level to as high as 2,200 m at Valle Nuevo, La Vega Province, DR, at an old sawmill site. It is most numerous in and along the sides of the Neiba Valley, DR, and its extension into Haiti, the Cul de Sac Plain, and in parts of Barahona and Pedernales provinces, DR. Also found on Île de la Gonâve and Isla Beata.

Status: Still a locally common year-round resident in the western half of Hispaniola, despite the persecution that all owls receive in both the DR and Haiti. The subspecies *A. c. troglodytes* is endemic to Hispaniola, Île de la Gonâve, and Isla Beata.

Comments: Active day and night. Feeds opportunistically on insects and large spiders, including beetles, locusts, and tarantulas, but also small birds, lizards, and occasionally frogs and small mammals. Insects are taken most often during the day, and mammals at night. Among birds, the most commonly captured species include todies, Bananaquits, hummingbirds, and Green-tailed Ground-Tanager.

Nesting: Lays 4–8 glossy white eggs which are rounded in shape. Nest is a burrow under the ground 1.0–2.5 m deep, typically with a mound at the entrance used as a lookout perch by the adults. Also nests in the chalky layer of soil in road cuts in the Sierra de Bahoruco. Found nesting March to July in northern Sierra de Bahoruco and on the south side of Lago Enriquillo; 4 well-feathered chicks were seen in late May near Jimaní.

Range: Breeds from s. British Columbia east to s. Alberta and s. Manitoba, w. Minnesota, e. Texas, and Florida, south to c. Mexico; also locally in South America to Argentina. In West Indies, resident in Bahamas, Cuba, and Hispaniola.

Local names: DR: Cucú; **Haiti:** Koukou, Chevêche des Terriers.

STYGIAN OWL—*Asio stygius* **Plate 29**

Breeding Resident—*Critically Endangered*

Description: 41–46 cm long; 675 g. Identified by its large size, dark coloration, and conspicuous ear tufts. Upperparts have cream-colored and white spots, and underparts are streaked. Facial disk is dark, and eyes are yellowish orange. Juvenile is pale buff overall with a barred belly.

Similar species: All other Hispaniolan owls are smaller, lighter colored, and lack or have very short ear tufts.

Voice: Generally silent; occasionally gives a single deep, abrupt *hu* or *whu.* During the breeding season, the male calls with a short, low-pitched *fool* repeated at regular intervals, and the female answers with a higher-pitched *niek* or *quick.*

Hispaniola: Rare resident in DR; current distribution in Haiti unknown. Found in dense deciduous and pine forests in remote areas, from semiarid to humid and from sea level to mountains. All the localities from which it has been recorded in recent years are relatively remote old forests, sometimes near caves or in wooded ravines, but never near human dwellings or in second-growth habitats. Since the mid-1980s, this owl has been found sporadically in the pine forests of the Cordillera Central in Armando Bermúdez National Park, on the Samaná Peninsula, in Los Haitises National Park, and in the Sierra de Bahoruco. Also occurs on Île de la Gonâve.

Status: Generally very rare breeding resident, and endangered because of forest destruction and hunting. Stygian Owl has been persecuted by local people who believe it has supernatural powers and can transform itself into a witch. Its call is considered a bad omen. It has also been suggested that introduced predators have resulted in the disappearance of native small mammals on which this species depends. The subspecies *A. s. noctipetens* is endemic to Hispaniola, including Île de la Gonâve.

Nesting: Believed to lay 2 white eggs. Thought to breed April to May, but few data available. A nest believed to be of this species at Hoyoncito, DR, was 4.5–6.0 m above ground in a cana palm and largely composed of grasses placed loosely together. May use old nests of other birds. In Cuba, nests high in trees, including palms, and in cavities, building a platform nest of sticks. May occasionally nest on the ground.

Range: Resident in w. Mexico and in Veracruz, south locally in Middle America to Nicaragua, and locally in South America from Colombia to Argentina. In West Indies, resident of Cuba and Hispaniola.

Local names: DR: Lechuza Orejita, Ciguapa; **Haiti:** Mèt Bwa, Chouette.

SHORT-EARED OWL—*Asio flammeus* **Plate 29**

Breeding Resident; Non-breeding Visitor?

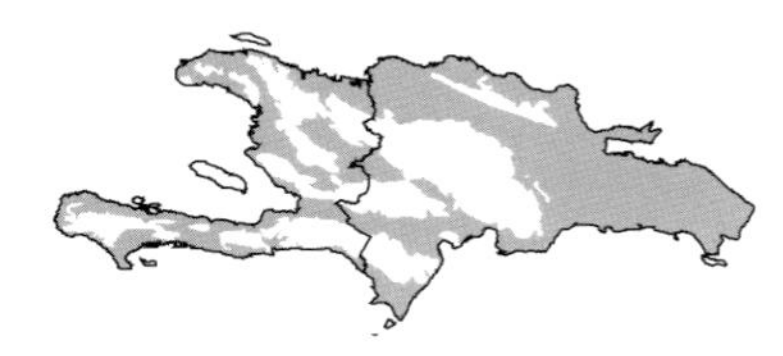

Description: 35–43 cm long; 345 g. An owl of open-country habitats, active day and night, hunting low above ground, often quartering or hovering. Tan or buffy below; heavy dark streaking on breast is reduced on lower belly. Whitish facial disk is distinct and round; eyes are yellow. Buoyant flight consists of slow, deliberate flaps and alternating glides, often in a characteristic erratic pattern. In flight, shows conspicuous black wrist patches on whitish underwings and large buff patches on upperwings.

Voice: Short, emphatic barking call, *bow-wow* or *uh-uh*. Also a distinct wing-clap during courtship flights and nest defense, produced by clapping wings beneath the body.

Hispaniola: Local resident from sea level to about 490 m elevation. Inhabits open country of the lowlands, including pastures, short-grass marshlands, savannas, rice fields, and citrus plantations. Probably most numerous in eastern DR (Laguna Redondo, Laguna Limón, Sabana de la Mar, San Pedro de Macorís, Santo Domingo), but clearly occurs farther west with reports from Moca, Jarabacoa, San Juan de la Maguana, and Cabral, DR. There is some evidence too that the species may be more abundant in the nonmontane uplands than in the lowlands. The present status of this species in Haiti is unknown, as the only documented record for Haiti is a specimen collected in 1928 in the Plateau Central near Saint-Michel de L'Atalaye.

Status: Locally common year-round resident on Hispaniola, and has increased dramatically since the 1930s. Resident birds are likely joined by individuals that breed on the North American mainland, but this is unconfirmed. Individuals from Cuba are also to be expected on Hispaniola, as the invasion of southern Florida since 1970 through postfledging dispersal of Cuban birds has been well documented. West Indian birds differ from continental forms in plumage and vocalizations. This has led to the suggestion that the West Indian form be considered a separate species. Other authors have recently assigned full species rank (*A. domingensis*) to the subspecies *A. f. domingensis* which is endemic to Hispaniola.

Comments: Most active at dawn and dusk, when it has the distinctive habit of flying low over open areas in search of prey. Sometimes observed perched on fence posts or in low bushes, where it takes cover during the day.

Nesting: Lays 3–4 white eggs in a crude ground nest under a bush or thick clump of grass. Breeds primarily April to June, but possibly throughout year, as nesting has been recorded as early as December.

Range: Breeds from n. Alaska east to n. Quebec, n. Labrador, and Newfoundland, south to c. California, n. Nevada, Kansas, c. Ohio, and n. Virginia. In non-breeding season occurs south to Oaxaca and Veracruz. Nearly cosmopolitan in South America and Eastern Hemisphere. In West Indies, breeds on Cuba, Hispaniola, and Puerto Rico.

Local names: DR: Lechuza de Sabana; **Haiti:** Chat-huant, Chwèt Savann.

GOATSUCKERS (NIGHTJARS): FAMILY CAPRIMULGIDAE

The nightjars are sometimes called "goatsuckers," from the ancient myth that at night these birds use their gaping mouths to rob goats of their milk, causing the udders to dry up and the animals to go blind. This myth was long ago shown to have no foundation. Rather, the birds use their huge, bristled mouths to engulf nocturnal insects on the wing. During the day, most species are inactive and rest on the ground or lengthwise on a branch, where their mottled plumage serves as perfect camouflage.

COMMON NIGHTHAWK—*Chordeiles minor* **Plate 31**
Passage Migrant

Description: 22–24 cm long; 61 g. A cryptically plumaged caprimulgid with a large, flattened head and large eyes; small bill and enormous mouth; slender, pointed wings; and conspicuous white bar across primaries. Wings project beyond tail tip at rest. Flight is graceful but erratic, almost batlike, with continual flaps interspersed with sporadic periods of gliding.
Similar species: Virtually identical to Antillean Nighthawk which has tan, rather than blackish, underwing coverts, and tends to be paler above and more buffy below, with a warmer brown color on face. These are not diagnostic field marks, and the two species are identified with certainty only by voice. Exhibits similar behavior and occurs in same habitats as Antillean Nighthawk, whose flight is more "fluttery" and which has a similar white wing band that is typically more mottled. Common Nighthawk is also about 10% larger overall than Antillean Nighthawk.
Voice: Distinctive, single-noted, very nasal *neet* or *peent*.
Hispaniola: Certainly a regular spring and fall transient in moderate numbers, although little documentation of its status is available. This species is a migrant throughout the West Indies, moving southward in September and October and northward in April and May. Its true abundance is little known on Hispaniola, because of its being distinguishable in the field from the more common Antillean Nighthawk only by voice, and because both species are nearly always silent in September and October. It is likely that the presence of this nighthawk tends to be underestimated.
Status: Regular passage migrant, but abundance is poorly documented.
Comments: Typically seen at dawn or dusk foraging for flying insects high over open areas such as fields, pastures, savannas, and coastal fringes.
Range: Breeds from s. Yukon east to n. Manitoba, c. Quebec, s. Labrador, and Nova Scotia, south virtually throughout North America to s. California and east to ne. Sonora, Tamaulipas, and s. Florida. Non-breeding-season distribution poorly known, but primary range appears to encompass lowlands of e. Ecuador and e. Peru, s. Brazil, Paraguay, Uruguay, and n. Argentina. A regular migrant nearly throughout West Indies, but apparently more numerous in western part of region.
Local names: DR: Querebebé Migratorio; **Haiti:** Petonvwa Etranje.

ANTILLEAN NIGHTHAWK—*Chordeiles gundlachii* **Plate 31**
Breeding Visitor

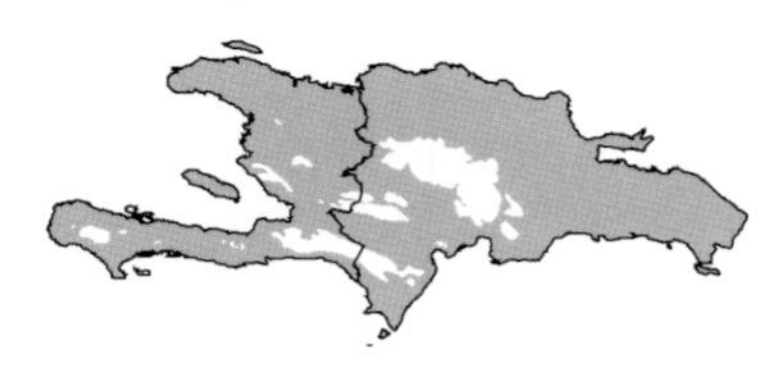

Description: 20–22 cm long; 50 g. A dark hawklike bird, recognizable as a caprimulgid by long, slender wings, medium-length notched tail, and cryptic, mottled dark grayish brown plumage. Also distinguished by its conspicuous white wing band and erratic, but graceful, flight. Wings are the same length as tail tip at rest. Male has a narrow white band near

tail tip and a white patch across throat; female lacks white tail band and has a buffy throat patch. Juvenile is similar to female but lacks the throat patch.
Similar species: Nearly identical to Common Nighthawk which has blackish, rather than tan, underwing coverts and tends to be whiter below and darker above. Antillean Nighthawk averages 10% smaller than Common Nighthawk, with shorter wings and wing tips appearing more rounded and blunt. The two are distinguished with certainty only by voice. All other species of this family on Hispaniola, such as nightjars and poorwills, lack the white wing patch.
Voice: Loud, raspy, distinctive rendition of its local name, *querebebé*.
Hispaniola: Seasonal resident of open country such as fields, pastures, savannas, and coastal fringes, from sea level to 1,165 m elevation. Has also occurred on Île de la Gonâve, Île de la Tortue, and Islas Saona and Beata.
Status: Common breeding visitor on Hispaniola from April to August, but occurring as early as March and as late as October. Some breeding birds may remain through the non-breeding season, but current information is inadequate to determine this.
Comments: Typically seen in numbers foraging for flying insects high over open areas at dawn or dusk. It may also be abroad during daylight when the weather is overcast. Sometimes seen sitting on country roads at night.
Nesting: Lays 1–2 slightly glossy white, bluish, or greenish tinted eggs, speckled evenly with fine spots of fuscous black, neutral gray, or brown. Eggs are unusually variable in size. Nest is invariably on bare ground in dry forested or stony localities, sometimes on pebbly sea beaches. Of 9 nests monitored from early April to early July in Sierra de Bahoruco, DR, 8 clutches held 1 egg, 1 clutch consisted of 2 eggs.
Range: Breeds in Florida Keys, Bahamas, Cayman Islands, and throughout Greater Antilles; only locally common on Puerto Rico and Virgin Islands. May be a rare and local breeder on Florida mainland. Non-breeding range poorly known but presumed to be South America. Migratory routes also little known, but populations appear to migrate from throughout Caribbean region via a more western route.
Local names: DR: Querebebé; **Haiti:** Petonvwa Peyi, Peut-on-voir.

LEAST PAURAQUE—*Siphonorhis brewsteri* **Plate 32**
Breeding Resident—*Threatened Endemic*

Description: 17–20 cm long; mass unavailable. A very small, darkly mottled nightjar. Adult has a distinct white throat band and narrow white terminal band on tail. Juvenile is more buffy than adult. Flight is erratic and floppy, with a mothlike appearance.
Similar species: Greater Antillean Nightjar is larger and darker. Chuck-will's-widow is much larger and more reddish brown in color. Both lack the white terminal tail band.
Voice: Guttural repetition of its local name, *torico, torico,* heard at dawn and dusk. Also a rising whistle.
Hispaniola: Most numerous in dry habitats and semiarid areas of cactus and thorn scrub from sea level up to about 300 m elevation, but may also occur as high as 800 m. In the DR found at Monte Cristi; near Galindo in the Sierra de Martín García; in the Sierra de Bahoruco at Puerto Escondido, Boca de Cachón near Cabral, and Santa Elena near Barahona; in Sierra de Neiba at La Descubierta Canyon and near Hondo Valle; at Manzanillo and Laguna Saladilla; in the Cordillera Central at La Leonor and Jarabacoa; near La Romana at Pedro Santana; and near Loma de Cabrera at Río Limpio. In Haiti, occurs (or formerly occurred) between Arcahaie and Montruis north of Port-au-Prince. Formerly numerous in scrubby woods on Île de la Gonâve.
Status: Formerly believed extinct, Least Pauraque is a local but increasingly scarce resident, threatened as a result of habitat destruction and introduced predators. The true status and local distribution of the species are poorly known. Surveys at Las Cruces, Sierra de Bahoruco,

in March 1996 revealed 4.5 birds/km, essentially unchanged from 4.8 birds/km on 1976 surveys in the same area. However, in most of its range it has surely declined since the 1930s because of habitat loss, primarily from slash-and-burn agricultural techniques. Its present status in Haiti is unknown.
Comments: Entirely nocturnal, but most active just after sunset and just before sunrise. The downy young look like a fluffy ball of white cotton and appear to mimic a round, whitish cactus which grows on the nesting grounds. In fact, the first described nestling was found by a botanist collecting cacti and was thought to be a cactus until it moved.
Nesting: Lays 2 dull white eggs with rather evenly distributed spots of pale violet-gray, gray, or brown and numerous spots or scrawls of buff and pale brown, in a scrape or depression in leaf litter on the ground, often at the base of a tree. Breeds April to July.
Range: Endemic to Hispaniola, including Île de la Gonâve.
Local names: DR: Torico; **Haiti:** Grouiller-corps, Gwouye-kò.

CHUCK-WILL'S-WIDOW—*Caprimulgus carolinensis* **Plate 31**
Non-breeding Visitor

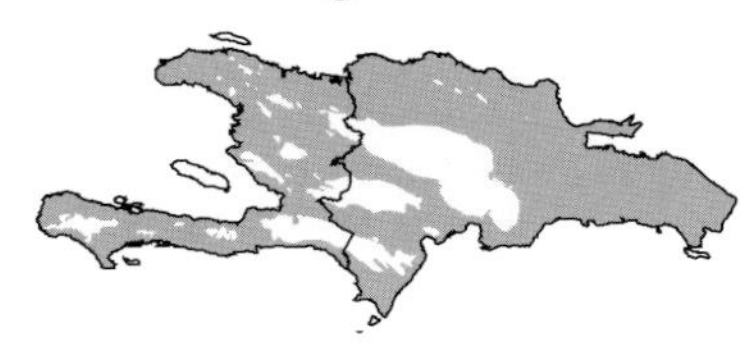

Description: 28–32 cm long; 120 g. A large nightjar, cryptically patterned cinnamon brown overall. Upperparts are mottled brown, buff, and black; breast is mottled blackish; rest of underparts are cinnamon to pale buff, barred dark brown. A whitish collar on throat is seen on male; females and immatures have a pale buff collar. Males have white on inner webs of the three outer tail feathers; females lack white in tail. Both sexes have buffy tips on these outer three rectrices.
Similar species: Greater Antillean Nightjar is smaller, darker, less reddish, and has white breast spots. Antillean Nighthawk has a white band on outer wing. Least Pauraque is much smaller, less reddish brown, and has full terminal white tail band.
Voice: Distinctive, clear, whistled rendition of its name with the first syllable lower and weaker, *chuck- will's-WID-ow*. Very vocal on breeding grounds, but seldom calls on Hispaniola.
Hispaniola: A moderately common non-breeding visitor nearly islandwide in lowland dry forest and transitional broadleaf forest at least up to 750 m elevation and possibly higher. Apparently absent from pine habitat. Exact distribution and abundance uncertain because of the species' secretive habits. Numbers of non-breeding visitors may vary from year to year. Also reported from Isla Alto Velo.
Status: Common non-breeding visitor, with most birds believed to arrive in mid-September and remain to March or April. Some individuals may arrive in August. Greater Antillean birds thought to originate from eastern part of the breeding range.
Comments: Entirely nocturnal; rarely flushed during the day from among dense thickets. Sometimes seen at night along rural roads. Forges aerially for insects, including moths and beetles.
Range: Breeds from e. Iowa east to nw. Indiana, Maryland, and north along coastal plain to Massachusetts, south to c. Texas, sc. Louisiana, and Florida. In non-breeding season found from s. Florida and se. Mexico south through Central America to Colombia east of Andes. In West Indies, occurs in non-breeding season in Bahamas, Greater Antilles, and northern Lesser Antilles.
Local names: DR: Don Juán; **Haiti:** Petonvwa Karolin.

HISPANIOLAN NIGHTJAR—*Caprimulgus eckmani* **Plate 32**
Breeding Resident—*Endemic*

Description: 28 cm, mass unavailable. Mottled dark grayish brown overall; narrow buffy band below blackish throat; breast and belly irregularly spotted with white. Male has outer tail feathers broadly tipped with white; female has buff tail tips.
Similar species: Smaller, darker, and less reddish brown than Chuck-will's-widow, which

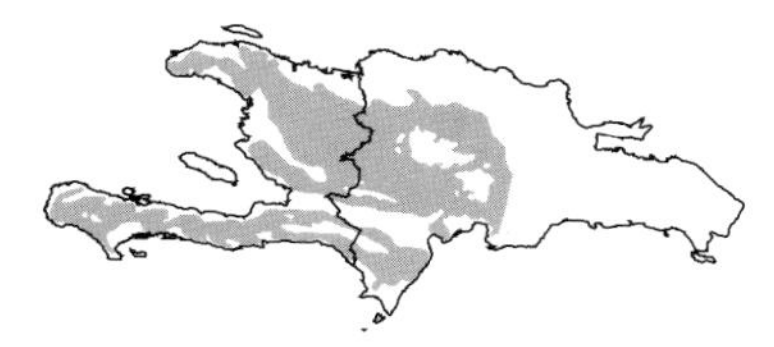

lacks white breast spots. Antillean Nighthawk, much more likely to be seen, is characterized by a bold white band on outer wing. Least Pauraque is smaller and paler.

Voice: Plaintive, frequently repeated *pi-tan-guaaaa,* reflecting its local name.

Hispaniola: Rare and local in eastern two-thirds of DR and known primarily from middle altitudes in the foothills and mountains of the western third of DR where it is found in Cordillera Central, Loma de Cabrera, the foothills of the Sierra de Neiba, and regularly in the Sierra de Bahoruco. In the Sierra de Bahoruco its distribution appears to coincide with broadleaf tree distribution. It is common from 300 to 750 m elevation, and is rarely reported from broadleaf cloud forest at 1,200 m, but does not appear in lower elevation thorn scrub or intermediate-elevation pine forest.

Status: Fairly common breeding resident in the western DR, but current status in Haiti is unknown. Has certainly declined in numbers islandwide since European settlement, and particularly since 1900, because of the introduction of non-native mammal predators and the destruction of forest habitat. Hispaniolan Nightjar was recently split from Cuban Nightjar (*C. cubanensis*) based on pronounced differences in vocalizations and morphological characters, but this split has not been formalized by the Committee on Classification and Nomenclature of the AOU.

Comments: Nocturnal. At night, perches at the edge of roads in wooded areas. Feeds primarily on insects that it captures in flight.

Nesting: Lays 2 light greenish gray to grayish white eggs spotted with light brown in a shallow scrape on the ground. Nests April to July.

Range: Endemic to Hispaniola.

Local names: DR: Pitanguá; **Haiti:** Petonvwa Peyi.

POTOOS: FAMILY NYCTIBIIDAE

The potoos are chunky, neotropical relatives of nightjars, which they superficially resemble. The plumage is soft and fluffy, with cheek feathers modified as bristles on the sides of the enormous gape. These odd-looking birds hawk flying insects, often sallying from and returning to the same perch, and using favored perches regularly. Potoos are solitary, nocturnal, and found in forested and semiopen habitats. During the day they typically roost atop a stump or on a tree branch, where their cryptic coloration and motionless, upright posture give them the appearance of branches.

NORTHERN POTOO—*Nyctibius jamaicensis* **Plate 31**
Breeding Resident—*Threatened*

Description: 43–46 cm long; mass unavailable. A large, brownish, cryptically patterned, long-tailed nocturnal bird best identified by its habit of perching nearly upright and motionless on the end of a stump or post. Upperparts are mottled brown, gray, and black; underparts are grayish brown to pale grayish cinnamon with dark streaking. Iris is yellow but appears reddish in a beam of light. Juvenile is paler than adult.

Similar species: Distinguished from nightjars and poor-wills by its conspicuously longer tail, more rounded wings, and larger size.

Voice: Deep, guttural call, often followed by 1–2 short notes, *kwah, waugh, waugh, waugh, kwaah* or *wahhrr wah-wah*. Also a hoarse *waark-cu-cu*.

Hispaniola: Occurs in arid and humid forests, and in scrublands adjacent to open areas from near sea level to at least 900 m elevation and probably higher. Also seen in palm groves, pastures, and cattle corrals. Reports prior to the 1930s include Morne La Visite, Port-à-Piment, Jérémie, and the Cul de Sac region of Haiti; and Puerto Plata, Hondo in the Cordillera Central, and Padre Las Casas, DR. Since then found to be an uncommon resident of the Sierra de Bahoruco foothills in Independencia, Pedernales, and Barahona provinces, and in the Neiba Valley and near La Descubierta, DR. Also found locally in the eastern interior of DR near Hato Mayor, Almirante, and Hoyón, and in Los Haitises National Park. Also resident on Île de la Gonâve.
Status: Local and uncommon to rare resident with a spotty distribution. Probably because of the species' harsh call and a general aversion to night birds, local people consistently persecute it to the point that it is considered threatened. The subspecies *N. j. abbotti* is endemic to Hispaniola, including Île de la Gonâve.
Comments: Cryptically plumaged, and often allows a close approach. Arboreal. Feeds on moths, beetles, and other insects, sallying from a perch in a manner similar to flycatchers. May use the same perch for months. Seems to be more active at night than most caprimulgids, which tend to favor the hours around dusk and dawn for feeding.
Nesting: The nest is a mere indentation atop a stump or broken branch at a moderate height (average 7 m) in which a single white egg is laid. Believed to breed April to July on Hispaniola.
Range: Resident from s. Sinaloa and s. Tamaulipas, Mexico, south to e. Honduras and c. Costa Rica. In West Indies, found on Jamaica, Cuba, and Hispaniola.
Local names: DR: Don Juan Grande, Bruja; **Haiti:** Chanwan Ke Long.

SWIFTS: FAMILY APODIDAE

Swifts are the supreme aerialists among land birds, rarely seen other than in flight. They pursue flying insects, often high overhead, throughout the day, seldom landing to rest, and even copulate in flight. Swifts are propelled rapidly by shallow flaps of their long, stiff, bow-shaped wings. One species is reported to hold the avian speed record, a remarkable 322 km per hour. Nests are typically constructed of saliva and twigs or other plant matter.

BLACK SWIFT—*Cypseloides niger* **Plate 31**
Breeding Visitor

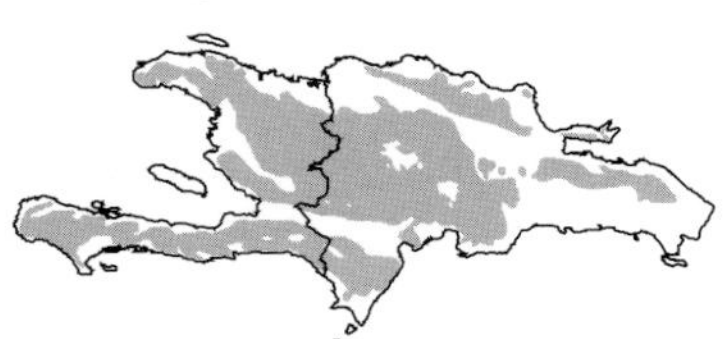

Description: 15–18 cm long; 46 g. A fairly large, black swift with long, pointed wings and a slightly notched tail. At close range, white is visible on forecrown and supercilium. Flight is less erratic than in other swifts; when gliding, wings are held below horizontal.
Similar species: White-collared Swift is larger, with pronounced white on neck. Chimney Swift and Antillean Palm-Swift are markedly smaller, with shorter and more squared tails, more darting flight, and quicker wing beats.
Voice: Mostly silent, but occasionally gives a soft *tchip, tchip* when flying.
Hispaniola: Recorded from sea level to 1,800 m elevation. Typically seen over forest or adjacent open areas in the mountains, less frequently in lowlands and coastal areas. Expected in the Cordilleras Septentrional, Oriental, and Central, and Sierra de Bahoruco, though this species is known to descend frequently into the lowlands, especially at the leading edge of rainstorms. Also occurs along the extension of Bahoruco and Cordillera Central ranges into Haiti, such as a flock of more than 50 seen at Trou Caïman in January 2004.
Status: Locally common breeding visitor, with some part of the population remaining through

the non-breeding season, but the portion doing so is uncertain and may vary from year to year. Appears that migrants from the south usually return in April or May. Southbound migration is thought to occur in late September to November, though little documentation is available. Migrating birds may follow the Lesser Antilles island chain to South America or fly directly to South America from Hispaniola.
Comments: Aerial. Often in flocks, feeding on insects high in the air. Reported to feed largely on winged ants. It is believed that during the breeding season this species usually ranges to higher elevations on average than White-collared Swifts.
Nesting: First found to breed in canyons of the Sierra de Bahoruco in 2002, though it almost certainly breeds elsewhere as well. As at other West Indian sites, nests in small colonies, laying a single white egg in a shallow cup nest of moss in a crevice or steep rock face in the mountains, often near or behind a waterfall. Probably breeds March to September.
Range: Disjunct breeding range from se. Alaska and sw. Alberta south through Pacific states to s. California and from n. Idaho and nw. Montana south to nc. New Mexico and Costa Rica. Non-breeding-season range poorly known but thought to be primarily in n. and w. South America. In West Indies, breeds locally on Cuba, Jamaica, Cayman Islands, Hispaniola, Puerto Rico, and Virgin Islands, and known on all Lesser Antilles islands from St. Barthélemy to Grenada.
Local names: DR: Vencejo Negro; **Haiti:** Chiksòl, Zwazo Lapli Nwa, Martinet Sombre.

WHITE-COLLARED SWIFT—*Streptoprocne zonaris* **Plate 31**
Breeding Resident

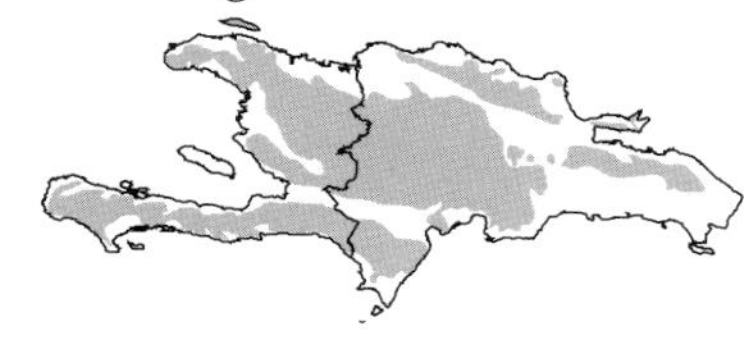

Description: 20–22 cm long; 98 g. A large, black swift with a prominent white collar and slightly forked tail. Immatures are similar, with narrower and paler collar. Flight is rapid and powerful, with deep, slow wing beats, spectacular swoops, and aerobatics; often soars without flapping.
Similar species: All other regularly occurring swifts are smaller and lack white collar.
Voice: High-pitched, screeching *screee-screee* or rapid, chattering *chip-chip-chip-chip.*
Hispaniola: Principally found at middle and upper elevations in foothills and mountains, occasionally down to sea level. Highest known elevation is 2,280 m in La Visite National Park, Haiti. Primarily seen over foothills, mountain valleys, and forests, and less regularly over lowlands, particularly during bad weather. However, this strong-flying, wide-ranging species could appear virtually anywhere. Also found on Île de la Tortue.
Status: Moderately common breeding resident.
Comments: Aerial. Typically in small flocks, but sometimes up to 200 birds. Forages for flying insects at all heights above the forest.
Nesting: Lays 2 white eggs in a shallow nest of mud and moss attached to a cliff, often behind a waterfall, in a cave, or in a dead palm. May nest colonially. Breeds April to June.
Range: Resident from Guerrero, San Luis Potosí, and Tamaulipas, Mexico, south through Middle America to South America east of Andes to se. Brazil and nw. Argentina. In West Indies, resident on Cuba, Jamaica, and Hispaniola.
Local names: DR: Vencejo de Collar; **Haiti:** Zouazo Lapli Kou Blanch.

CHIMNEY SWIFT—*Chaetura pelagica* **Plate 31**
Passage Migrant

Description: 13–14 cm long; 24 g. A medium-sized, dark, uniformly colored swift, sooty olive or brown above, plain grayish brown below, but noticeably paler on throat, chin, and auriculars. Plumage is slightly glossy, especially on wings. Very short, rounded tail is barely visible in flight.

Similar species: White-collared and Black swifts are larger; the former has a distinct white band around the neck, the latter a longer, slightly notched tail.
Voice: Loud, rapid, high-pitched twittering in flight, seldom heard during migration.
Hispaniola: Rare transient northbound in spring; very rare in fall. Documented spring records from mid-April to mid-May; only recorded twice in autumn. Early records are from Port-au-Prince, Morne La Selle, Hinche, and Balladère, Haiti; and from Comendador and Barahona, DR. Available reports suggest the species may occur more often in western Hispaniola than farther east. Also recorded from Île de la Tortue.
Status: Rare passage migrant, but may be concentrated largely at high elevations and thus go undetected. No documented records since 1973.
Comments: Forages for flying insects above cities and towns. Also seen above open fields and woodlands.
Range: Breeds from e. Saskatchewan east to c. Ontario, New Brunswick, and Nova Scotia, south through e. Colorado to s. Texas, Gulf Coast, and Florida. Found in non-breeding season in nw. Brazil, upper Amazon basin of e. Peru, and n. Chile. In West Indies, transient in Bahamas, Greater Antilles, Virgin Islands, Swan Island, and Cayman Islands.
Local names: DR: Vencejo de Chimenea; **Haiti:** Ti Irondèl Etranje.

ANTILLEAN PALM-SWIFT—*Tachornis phoenicobia* **Plate 31**
Breeding Resident

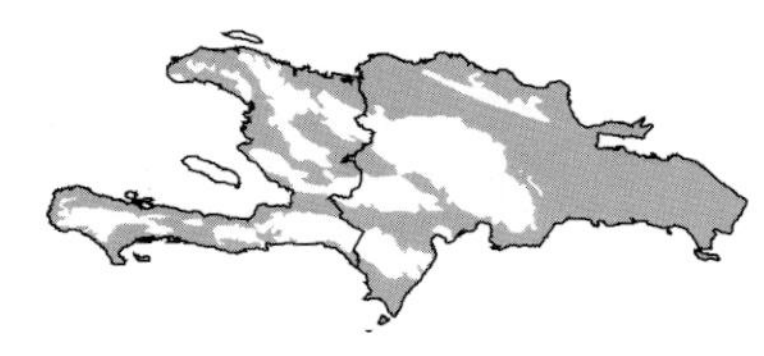

Description: 10–11 cm long; 9 g. A very small swift, dark brown above, with a conspicuous white rump. Adults have a blackish breast band. Males are dark with a white throat; females are paler below, with a grayish white throat. Immatures are similar to females but darker and more buffy below. Flight is erratic and darting.
Similar species: No other swift is as small and boldly patterned. Bank Swallow is somewhat similar but lacks white rump and has shorter, broader wings with a slower, less agile flight.
Voice: Constantly emitted, faint, high-pitched twitter.
Hispaniola: Found islandwide in a variety of habitats at lower elevations. Highest elevation known is 1,120 m at a water source in the Sierra de Bahoruco, but most reports are from below 450 m. Reported in both moist and dry habitats, over forest, scrub, cultivated areas, seashores, and savannas, and in towns and cities. Reported from several major satellite islands, including Île à Vache and Islas Beata, Catalina, and Saona.
Status: Locally common resident at lower elevations islandwide. Daily counts of 100 are not unusual. May have declined slightly since 1930, though found regularly in Los Haitises National Park in 1996 where it was less numerous in 1976.
Comments: Aerial. Forages in flocks for flying insects, its sole food, with a rapid and erratic, batlike flight, gliding between flapping spurts. Usually forages low over ground, at no higher than 20 m.
Nesting: Lays 2–5 white eggs in a globular nest of plant fibers, feathers, and other materials glued with saliva to a frond of the royal palm (*Roystonea hispaniolana*). Typically nests in colonies of 10–30 nests. Has also been reported nesting under a sea cliff overhang in association with Cave Swallows and Caribbean Martins. Breeds March to May.
Range: Resident on Cuba, Hispaniola, and Jamaica; vagrant in Florida Keys and Puerto Rico.
Local names: DR: Vencejito Palmar; **Haiti:** Jolle-jolle, Ti Irondèl Palmis.

HUMMINGBIRDS: FAMILY TROCHILIDAE

This large, Western Hemisphere family is particularly diverse in the tropics. Best known for their small size, remarkable flight powers, and brilliant colors, most species show iridescent green above but appear black in poor light. They characteristically have a long, slender, pointed bill for probing into flower corollas from which they obtain insects and suck nectar with their long, tubular tongue. Hummingbirds feed by hovering before blossoms with wings appearing blurred as they beat nearly 50 times per second. The second smallest bird in the world, the Vervain Hummingbird, is found on Hispaniola. Hummingbirds are very aggressive, particularly around their feeding territories. Nests are small cups of lichen, plant fiber, spider silk, hair, and other fine material.

ANTILLEAN MANGO—*Anthracothorax dominicus* **Plate 34**
Breeding Resident

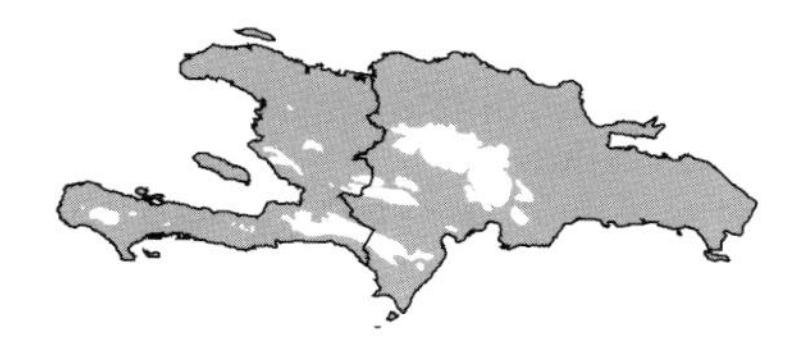

Description: 11–12 cm long; 5.4 g. A large hummingbird with pale, yellowish green upperparts and a long, black, down-curved bill. Adult males are primarily black below with a green throat. Females are whitish below, with whitish tips to reddish brown tail. Immature males are similar to females, but with a black stripe down the center of whitish underparts.
Similar species: Female Hispaniolan Emerald is smaller, with a straighter and shorter bill, paler lower mandible, and more green on outer tail feathers.
Voice: Rather loud, unmusical, thin trill. Also sharp, chipping notes.
Hispaniola: Occurs from sea level well up into the mountains islandwide; highest reported elevation is 2,600 m in the Cordillera Central, although reports from elevations above 1,500 m may indicate that the native forest has been degraded. Occurs in both moist and arid habitats but more abundant in semiarid regions. It is the most abundant hummingbird in dry forest and desert habitats, is common in broadleaf forests above 300 m elevation, but is unusual in pine forest above 1,100 m. It is also found in shade coffee plantations in the Cordillera Central, but it is not as numerous there as other hummingbird species. Recorded as a breeding resident on nearly all major satellite islands.
Status: Common resident throughout its range. The subspecies *A. d. dominicus* is endemic to Hispaniola and associated islands.
Nesting: Lays 2 white eggs in a deep cup nest made of cottony plant fibers often coated with lichens, moss, and spider webs, placed on a branch of a tree or shrub, or a cactus pad 1.5–10 m above ground. Average height of 43 nests in desert scrub habitat of Sierra de Bahoruco was 3.1 m (range 1.0–7.0). May breed year-round, but known to nest December to August.
Range: Resident on Hispaniola and Puerto Rico. Formerly rare in Virgin Islands on Anegada, St. Croix, and St. John, but now possibly extirpated.
Local names: DR: Zumbador Grande; **Haiti:** Ouanga Négresse, Oiseau-mouche, Wanga-nègès, Mango Doré.

HISPANIOLAN EMERALD—*Chlorostilbon swainsonii* **Plate 33**
Breeding Resident—*Endemic*

Description: 10 cm long; 4.9 g. A small, straight-billed hummingbird. Male is green overall, with dull black breast patch, deeply forked tail, and lower mandible mostly pinkish. Female is green above, dull grayish below with metallic green sides, a conspicuous white spot behind eye, and whitish outer tail tips.
Similar species: Female Antillean Mango is larger and has reddish brown outer tail feathers

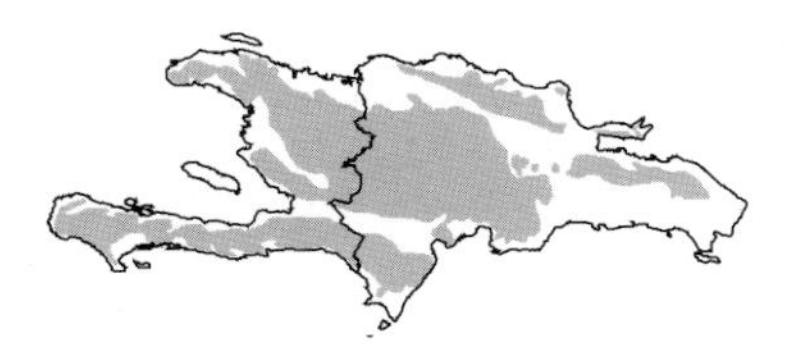

and a darker bill. Vervain Hummingbird is similar to female but is much smaller and has a shorter, darker bill. Female Ruby-throated Hummingbird, a vagrant on Hispaniola, has a darker bill and less white behind eye than female Hispaniolan Emerald.
Voice: Series of sharp, metallic chips, *tic-tic-tic*.
Hispaniola: Resident islandwide of moist forest in the mountains and foothills of the interior, as well as in the limestone karst region. May also be found in clearings at higher elevations. Common to abundant in shade coffee plantations in the Cordillera Central, DR. In the Sierra de Bahoruco, common in broadleaf habitats above 300 m elevation, but abundant in pine forests above 1,100 m. Seldom recorded below 200 m, but populations at higher elevations may descend to lower sites after the breeding season. Not recorded from any major satellite island.
Status: Generally common breeding resident, but overall numbers have likely declined since 1930 because of significant reductions in extent of moist forest. Although formerly considered to be very common in forested habitats above 1,500 m, such as in Haiti's La Visite National Park and Macaya Biosphere Reserve, it is now considered threatened in Massif de la Selle and Massif de la Hotte because of destruction of its preferred moist forest habitat.
Nesting: Lays 2 white eggs in a cup-shaped nest of moss, plant materials, lichens, and cobwebs, usually built in a low tree or shrub 0.5–2.0 m above ground. Of 30 nests in montane forests of Sierra de Bahoruco, average height was 2.8 m (range 1–10 m). Breeds at least January to August. Average fledging date of 10 Bahoruco nests was July16.
Range: Endemic to Hispaniola.
Local names: DR: Zumbador Mediano, Zumbador Verde; **Haiti:** Wanga-nègès Mòn, Ouanga Négresse, Emeraude d'Hispaniola.

RUBY-THROATED HUMMINGBIRD—*Archilochus colubris* **Plate 34**
Vagrant

Description: 8–10 cm long; 3.2 g. A small hummingbird with metallic bronze-green upperparts. Male has a brilliant red throat, moderately forked tail, and whitish underparts with dull greenish sides. Female's throat is whitish, and rest of underparts are dull grayish white, with some buff coloration on sides. Tail is rounded and broadly tipped with white. Both sexes have a small white spot behind eye. Bill of adults is dark.
Similar species: Female Ruby-throated Hummingbird is similar to Hispaniolan Emerald which has a paler bill, more conspicuous white stripe behind eye, and greenish sides.
Voice: High, squeaky, rapid chips, *tchi-tchi-tchi-chit* or *tic-tic*.
Hispaniola: Three records of single males. One seen at Bizoton, Haiti; 1 in dry scrub forest near Duvergé, DR; and 1 at Hotel Gran Bahía on the Samaná Peninsula, DR.
Status: Vagrant.
Habits: May be looked for at flowers in large gardens, wood edges, and clusters of trees.
Range: Breeds from c. Alberta east to Quebec, New Brunswick, and Nova Scotia, and south to s. Texas, Gulf Coast, and s. Florida. Occurs in non-breeding season from s. Florida and s. Mexico south to c. Costa Rica and w. Panama. In West Indies, a regular transient in w. Cuba and a vagrant in Hispaniola.
Local names: DR: Zumbador Migratorio; **Haiti:** Wanga-nègès Gòj Wouj.

VERVAIN HUMMINGBIRD—*Mellisuga minima* **Plate 34**
Breeding Resident

Description: 6 cm long; 2.4 g. A tiny hummingbird, green above and predominantly whitish below with a straight black bill. Chin and throat are sometimes flecked greenish; sides and flanks are dull green. Adult male has a deeply notched tail, and adult female has a rounded and white-tipped tail. Immature male is similar to adult female, except chin

and throat are flecked with gray, and white tips on outer tail feathers are less sharply defined.
Similar species: Much smaller than any other hummingbird in its range. Female Hispaniolan Emerald has a longer, paler bill.
Voice: Loud, rhythmic song of high-pitched, metallic squeaks. Also an extended throaty buzz.
Hispaniola: Found in all habitat types except dense, moist forest, including scrub, second growth, broadleaf, and pine from sea level to a maximum known elevation of 1,600 m. Known from Île à Vache, Île de la Gonâve, Île de la Tortue, the Cayemites, and Islas Catalina and Saona.
Status: Locally common and widespread resident. The subspecies *M. m. vieilloti* is endemic to Hispaniola and associated islands.
Comments: Often hovers at food source with tail cocked up, but more often heard than seen. This species may at least occasionally descend to lower elevations in non-breeding season. Common to abundant in shade coffee plantations in Cordillera Central, DR, especially when the guama (*Inga vera*) canopy is in flower. The world's second smallest bird; its egg weighs only 0.4 g, and the hatchling weighs 0.3 g.
Nesting: Lays 2 white eggs in a deep cup nest of plant fibers covered with lichens, bark, and cobwebs. Nests are usually built 0.5–3.0 m above ground in a bush, attached to the side of a branch. Breeds year-round but primarily December to May.
Range: Resident on Jamaica and Hispaniola; vagrant to Puerto Rico.
Local names: DR: Zumbadorcito; **Haiti:** Ouanga Négresse, Sousaflè, Zwazo-mouch, Wanga-nègès.

TROGONS: FAMILY TROGONIDAE

Trogons are pantropical, brightly colored birds that spend much time perched motionless in the forest and may be easily overlooked. All are short necked with a short, broad bill; small, weak feet on very short legs; short, rounded wings; and a long, squared tail that hangs straight down. They are strictly arboreal, usually solitary or in pairs, and inhabit forests characterized by low levels of human impact. The flight of trogons is undulating and rapid, and rarely extends more than a short distance. The birds feed by darting from their perch and snatching insects from the air or from foliage, or plucking small fruits or flowers. Nests are in natural tree cavities.

HISPANIOLAN TROGON—*Priotelus roseigaster* **Plate 36**
Breeding Resident—*Threatened Endemic*

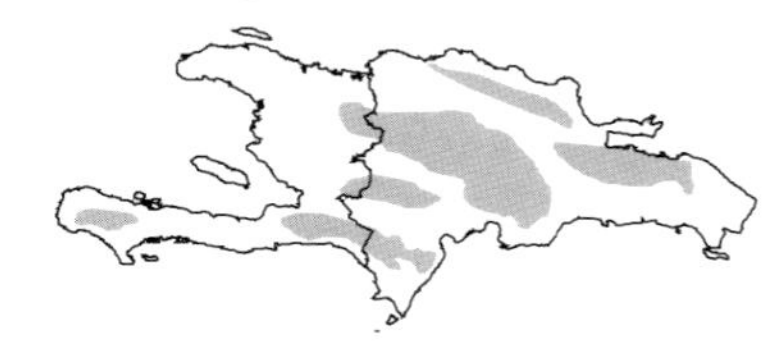

Description: 27–30 cm long; 73–81 g. Identified by its glossy green crown and upperparts, bright red belly, yellow bill, and gray chin, throat, and breast. Also note the blackish facial mask. Long, dark blue tail is conspicuously tipped with white below. Male has wings with fine black and white barring, and prominent white notches on outer edges of primaries. Female has a gray-green wing and lacks the fine white barring; red on belly is also less vivid.
Similar species: Unmistakable.
Voice: Hollow *toca-loro; coc, ca-rao;* or *cock-craow*, repeated several times, especially early in the morning during breeding season, but may call throughout day. Also cooing and whimpering sounds.
Hispaniola: Occurs in undisturbed habitat in foothills and mountains, rarely down to sea

level. Highest elevation recorded is 3,000 m. Previous reports that this species may descend to lower elevations during the non-breeding season appear to be incorrect as it is found year-round at Rabo de Gato and La Placa (380–400 m, near Puerto Escondido) and at Ebano Verde. Most often found in humid forest, both deciduous broadleaf and pine, though once reported in mangrove swamps at sea level at Miragoâne, Haiti. Reported from Sierra de Bahoruco, Sierra de Neiba, Cordillera Central, Cordillera Septentrional, and their associated foothills. This species also presumably still occurs where appropriate habitat exists in the mountain ranges that extend from the DR into Haiti, including the Massif des Montagnes du Noires, Montagnes du Nord Ouest, and their associated foothills, but few recent data exist. Still common in Macaya Biosphere Reserve but uncommon and threatened in La Visite National Park, Haiti, because of habitat loss. Not recorded from any major satellite island.
Status: Locally common breeding resident, but declining overall and threatened with extinction. In DR still quite common in appropriate, undisturbed habitat. Still abundant in parts of Haiti (Massif de la Hotte), but declining and considered threatened there. Populations have been considerably reduced throughout their range because of habitat destruction through loss of forests for logging and agriculture. Loss of appropriate nest sites (old decayed trees) is of particular concern.
Comments: Relatively inactive, trogons are strictly arboreal, and actually incapable of walking or even turning around on a perch without using their wings. They are most often seen perched upright with the tail pointing straight down on a horizontal branch between the foliage and the trunk. Feed mainly on fruit and insects, but will complement their diet with invertebrates and even small anole lizards. Usually forage in the subcanopy or middle strata of interior forests. Typically occur in pairs, the two individuals calling regularly to one another. The call's ventriloquial nature frequently makes the birds difficult to locate.
Nesting: Nests in tree cavities 2–10 m above ground, exceptionally up to 15 m, often in the abandoned nest of a Hispaniolan Woodpecker, or in cavities the trogon excavates in very decayed trees. Lays 2–4 pale green, unmarked eggs. Breeds March to July. Eight of 12 nests in montane forests of the Sierra de Bahoruco failed because of predation; 2 fledged on June 25.
Range: Endemic to Hispaniola.
Local names: DR: Papagayo, Cotorrita de Sierra, Piragua, Loro; **Haiti:** Caleçon Rouge, Pic de Montagne, Dame Anglaise, Demoiselle Anglaise, Kanson Wouj.

TODIES: FAMILY TODIDAE

The todies are a group of five species most closely related to the neotropical motmots, and more distantly to kingfishers. Although the family is now confined to the Greater Antilles, it is believed to have Central American origins. Fossil evidence indicates that the family once extended well into North America, and possibly Europe. Todies are opportunistic feeders with a voracious appetite; a captive specimen ate about 40% of its weight in insects daily. Although todies often perch quietly in the forest, they are attentive parents, with one of the highest rates of feeding young ever recorded for insectivorous birds. Nests are in long, narrow, curved burrows in earthen banks, which the birds excavate with their bills and feet.

BROAD-BILLED TODY—*Todus subulatus* **Plate 33**
Breeding Resident—*Endemic*

Description: 11–12 cm long; 9.8 g. Bright green above with a red throat. Underparts are grayish white washed with yellow and pink, creating a dirty appearance; sides are reddish pink. Underside of lower mandible is entirely reddish. Iris is brown. Briefly held juvenal plumage lacks red on throat, and breast is streaked with pale green.
Similar species: Narrow-billed Tody is slightly smaller, cleaner white below, with a pale blue

iris, and underside of lower mandible is usually red with a black tip. The two species are best distinguished by voice.

Voice: Monotonous, often repeated *terp, terp, terp,* uttered in a complaining tone. A single call note in the same tone contrasts with Narrow-billed Tody's 2-note call. May also be located by a unique *burrrrrrr* made by the wings in short flights.

Hispaniola: Occurs islandwide from sea level to a maximum known elevation of 1,700 m. Favored habitats (in approximate order of population density) include dry limestone forests; shade coffee plantations; lowland and montane second-growth of any kind, including low bracken; deciduous understory of pine forest; and wet lowland forest and mangroves. Not found in dense rainforests, and unusual in the driest habitats. Often occupies vegetated ravines and areas with earthen embankments. Also reported from Île de la Gonâve.

Status: Common to abundant resident. Of the two todies, Broad-billed Tody is thought to be the more abundant and general in habitat preferences, suggesting that its population has probably not substantially declined in recent years and may have even increased in some areas where forest has been removed.

Comments: The distribution of this species overlaps with that of Narrow-billed Tody between elevations of 435 and 1,700 m across three major habitat types in the Sierra de Bahoruco: very humid montane forest (including stands of pine), dry lower montane forest (humid forest in ravines), and dry forest. In the Cordillera Central this overlap is possible because the two species occupy slightly different microhabitats, with Broad-billed Todies tending to forage higher in the vegetation, in less dense foliage, and at slower rates.

Nesting: Typically constructs a nest chamber at the end of a 30- to 60- cm tunnel dug into the bank of a stream, path, road, or cliff face, often 0.6–1.8 m above ground. An atypical nest dug into flat ground was documented in 2003. Lays 3–4 white eggs with a distinct gloss. Breeds March to June. May nest twice annually.

Range: Endemic to Hispaniola, including Île de la Gonâve.

Local names: DR: Barrancolí, Barranquero; **Haiti:** Perroquet de Terre, Kolibri.

NARROW-BILLED TODY—*Todus angustirostris* **Plate 33**
Breeding Resident—*Endemic*

Description: 11 cm long; 6.7 g. A small, chunky bird with brilliant green upperparts, red throat, whitish underparts tinted with yellow, and reddish pink sides. Iris is pale blue. Underside of lower mandible is reddish, usually with a black tip. Juveniles briefly lack red on throat, and breast is streaked with pale green.

Similar species: Broad-billed Tody is slightly larger, grayish white below, with entirely red underside of lower mandible, and a brown iris. However, Narrow-billed Tody sometimes lacks the black tip on lower mandible, and the two species are most reliably distinguished by voice.

Voice: Call note a frequently repeated, 2-part *chick-kweee,* accented on the second syllable. Also a chattering *chippy-chippy-chippy-chip,* dropping in pitch but not in tone.

Hispaniola: Most numerous in mature forests, including wet montane forest (including pines), moist ravines, and coffee plantations between 900 and 2,400 m elevation, but occurs up to 3,000 m, and is a locally common resident at lower elevations such as at Los Haitises National Park. However, this species favors wet, high-elevation broadleaf or pine forests where dense vegetation is carpeted with epiphytic mosses, leafy liverworts, orchids, Spanish moss, ferns, and climbing bamboo. Primarily occurs in ravines and near earthen embankments. Reported from all the mountain ranges of both countries and the Samaná Peninsula, DR. Not recorded from any major satellite island.

Status: Common resident, but has surely declined in overall numbers since 1930 because of

the great reduction of its preferred habitat types, particularly in Haiti where forest destruction has been greatest. Recent DNA work has suggested that two populations of *T. angustirostris* differ enough genetically to be considered separate species; one population occurs north of the Neiba Valley/Cul de Sac Plain, and the other south of it in the Sierra de Bahoruco and Haiti's southern peninsula.
Comments: Both Narrow-billed and Broad-billed todies are broadly sympatric over an extensive altitudinal range of more than 1,000 m, but at higher elevations and near ravines Narrow-billed is typically the only tody present. Perches with tail pointed down and broad bill pointed up at an angle of 45 degrees, actively turning its head to watch for insects. It captures its prey in quick sallying flight on whirring wings with an audible snap of its bill. It is more active than Broad-billed Tody and forages lower in the vegetation among the inner, more dense foliage.
Nesting: Lays 3–4 unmarked, glossy white eggs in a nest in a chamber at the end of a tunnel in a soil bank up to 30 cm deep. Occasionally nests in tree cavities, including one discovered 9 m above ground in montane forest of Sierra de Bahoruco. Eggs are distinctly smaller than those of Broad-billed Tody. Breeds April to June.
Range: Endemic to Hispaniola.
Local names: DR: Chi-cuí; **Haiti:** Tèt-sèch, Kolibri Mòn, Chikorèt.

KINGFISHERS: FAMILY ALCEDINIDAE

The kingfishers are a cosmopolitan family with only a single representative on Hispaniola. Typically associated with water, kingfishers often perch quietly, overlooking rivers and pools. As a group they are characterized by a large, often crested head; long, fairly heavy, pointed bill; and brilliant coloration. Most species feed on fish and may hunt by hovering.

BELTED KINGFISHER—*Ceryle alcyon* **Plate 34**
Non-breeding Visitor; Passage Migrant

Description: 28–35 cm long; 148 g. An aquatic bird distinguished by its large, crested head; stout, sharply pointed bill; bluish gray upperparts and head; and nearly complete white collar. Legs are very short. Male has a single blue-gray breast band; female has one blue-gray and one rufous breast band, with rufous on flanks. Immature has rusty spotting in a slaty breast band.
Similar species: Unmistakable.
Voice: Loud, reverberating, dry rattle.
Hispaniola: Occurs along all parts of the coastline, the edges of both fresh and saline lakes and ponds in the lowlands, and the courses of rivers and streams, sometimes far into the interior. Highest known recorded elevation is 1,085 m at Aceitillar, Sierra de Bahoruco, DR. Reported from several major satellite islands, including Navassa Island, Île de la Gonâve, Île à Vache, and Islas Beata, Catalina, and Saona.
Status: Regular and moderately common non-breeding visitor from North America, and spring and fall passage migrant, with documented records from September to May. Periods of migrant passage are poorly defined, because of the difficulty of distinguishing transients from residents, but are thought to be October–November and March–April.
Comments: Frequents conspicuous perches, usually near or over water, and often hovers before plunging into the water after prey. Feeds mainly on fish, but may take insects, reptiles, and amphibians.

Range: Breeds from Alaska east to n. Ontario, Labrador, and Newfoundland, south to s. California, sc. Texas, and c. Florida. Occurs in non-breeding season from coastal s. Alaska through most of continental U.S. south to Colombia, Venezuela, and Guyana. Common throughout West Indies from fall to spring, a few birds remaining year-round, but not known to breed.
Local names: DR: Martín Pescador; **Haiti:** Pipirit Rivyè.

WOODPECKERS AND ALLIES: FAMILY PICIDAE

Members of this nearly cosmopolitan family (absent only from Australia, New Zealand, and New Guinea) use their chisel-like bill to bore into trees in search of insects and to excavate nest cavities. Woodpeckers have an unusually long, tubular tongue with barblike tips for extracting prey from deep cavities. Most species have toes arranged with two behind and two forward. Central tail feathers are stiffened and serve as supports during vertical climbing on trunks and branches. Drumming on hollow trees or other resonant surfaces serves to advertise territories. Cavities excavated by woodpeckers provide important nest sites for other birds, including parrots, trogons, and the endangered Golden Swallow.

ANTILLEAN PICULET—*Nesoctites micromegas* **Plate 35**
Breeding Resident—*Endemic*

Description: 13–16 cm long; 33 g. A very small, chunky, un-woodpeckerlike bird that taps tree trunks or branches occasionally and crisscrosses its way along twigs and vines. Bill is fairly small, chisel-like, and sharply pointed. Adults are olive above and pale yellowish below with heavy dark spots and streaks. Pale auriculars are streaked. Male has a bright red patch in center of a yellow crown; female is slightly larger than the male and lacks red in the yellow crown patch. Juvenile is duller green above with a duller yellow crown, and abdomen is more heavily barred. Flight is direct and even, not undulating as in typical woodpeckers.
Similar species: Unmistakable. Smaller than other woodpeckers and lacking in typical barring patterns.
Voice: Staccato, woodpeckerlike *tu-tu-lu-feo,* accented on the last note, surprisingly loud for this diminutive bird. This is sometimes heard as *Al pícaro, no le fía,* which translates roughly as "Don't trust the rascal." Both sexes frequently call to one another. Drumming apparently does not signal territorial status.
Hispaniola: Locally common in both dry and humid forest, semiarid scrub, pines mixed with some broadleaved trees, and rarely in coastal mangroves. Found from sea level to at least 1,770 m in Sierra de Bahoruco, DR, but is most numerous in undisturbed dry forest at intermediate elevations (e.g., 400–800 m) and only occasionally descends into bushier portions of desert thorn scrub. Inhabits the highest elevations in Sierra de Bahoruco, but in Cordillera Central occurs primarily in pines down to 450 m. Common in mesic broadleaf forest as high as 1,700 m in Macaya Biosphere Reserve, but not found in La Visite National Park, Haiti. A slightly paler form is resident on Île de la Gonâve and is more common there than the species is on the main island, possibly because of the lack of competition from Hispaniolan Woodpecker.
Status: Breeding resident. The population islandwide is certainly reduced from the 1930s because of habitat destruction, especially in Haiti where the species is now threatened. The population in Los Haitises National Park fell sharply from 1976 to 2000 because of extensive habitat destruction. Downward trends may be expected to continue as more habitat is disturbed. The subspecies *N. m. micromegas* is endemic to Hispaniola; the subspecies *N. m. abbotti* is endemic to Île de la Gonâve.

Comments: This species mostly gleans insects and other arthropods, but takes some fruit as well. Forages in dense clusters of vines, branches, and leaves. Does not hammer like a woodpecker, but may peck weakly or probe fruit, flowers, and leaf clusters. Although widespread, it is often shy and difficult to see, but it is also noisy and thus easy to hear. The only piculet in the West Indies, it was originally described in error from Brazil. The earliest known New World fossil of Picidae was identified by 15- to 20-million-year-old feathers preserved in amber and discovered in the DR. Comparisons of feather structure showed this fossil to be closely related to *Nesoctites*.
Nesting: Lays 2–4 white eggs with a distinct gloss. The nest is a natural cavity or abandoned woodpecker hole in a palm, tree, or post 1–5 m above ground. Breeds March to July.
Range: Endemic to Hispaniola, including Île de la Gonâve.
Local names: DR: Carpintero de Sierra, Carpintero Bolo, Tutulufeo; **Haiti:** Sèpantye Bwa, Sèpantye Mòn, Ti Sèpantye, Picumne des Antilles.

HISPANIOLAN WOODPECKER—*Melanerpes striatus* — Plate 35
Breeding Resident—*Endemic*

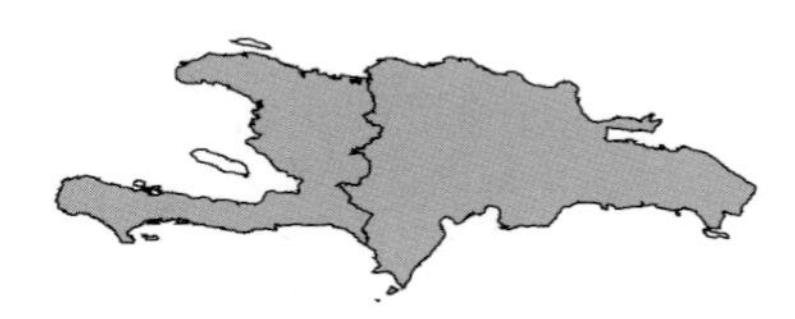

Description: 22–25 cm long; 65 g. A conspicuous, medium-sized woodpecker. Upperparts are barred blackish and greenish yellow; underparts are unmarked dark buffy olive. Forecrown and face are grayish, and there are white and black patches on nape. Uppertail coverts are red, and tail is black. Eye is whitish to yellow. Male is distinctly larger and longer-billed than female, with crown entirely red. Female has a black crown, with red limited to nape.
Similar species: Yellow-bellied Sapsucker is smaller, with distinct black and white facial markings and barred upperparts, and a conspicuous white wing patch.
Voice: Strong, variable vocalizations including a loud, rolling call interrupted with throaty noises. Call notes include *wup* (alarm-aggression) and *ta-a* (defensive); also a short *b-d-d-d-t* with 3–5 distinct notes. Drumming signals given in the vicinity of the nest also serve to define territory.
Hispaniola: Common resident in virtually all habitats containing some trees from sea level to 2,400 m. Habitats used include both dry and humid forest, pine and broadleaf, thorn scrub, coffee plantations, palms, cacti, and coastal mangroves. Most numerous in hilly, partly cultivated, and partly wooded areas, especially where palms are scattered among cultivated fields. Absent only where trees are lacking. Among satellite islands, known only from Isla Beata.
Status: Common and widespread breeding resident; this is possibly the most widely distributed of Hispaniola's endemic birds. Considered a pest in some agricultural areas; its use of crops such as corn, oranges, and cacao has resulted in conflict with humans. The species was historically subject to a bounty paid on its tongue.
Comments: Hispaniolan Woodpecker has several distinctive traits, including the large difference in bill size between sexes (male's bill is 20% longer), its gregariousness, and its habit of nesting colonially. Nest sites excavated by this woodpecker are critical to other birds, including parrots, parakeets, trogons, flycatchers, and swallows, which use abandoned cavities for their own nests. Consumes insects, fruit, and seeds taken through gleaning, probing, and pecking on trees, bushes, epiphytes, and cacti. Forages at all levels but not on the ground. Diet also includes tree sap.
Nesting: Nests in cavities excavated by both sexes in living or dead palms, trees, cacti, and telephone poles. Reputed to nest even in cliff burrows. Nesting often clustered in loose colonies of 5–20 or more pairs. Lays 3–6 white eggs. Breeds mainly February to July but to some extent throughout the year.
Range: Endemic to Hispaniola.
Local names: DR: Carpintero; **Haiti:** Sèpantye, Charpentier, Pic d'Hispaniola.

YELLOW-BELLIED SAPSUCKER—*Sphyrapicus varius* **Plate 34**
Non-breeding Visitor

Description: 20–23 cm long; 50 g. A medium-sized woodpecker, identified in all plumages by its large, white upperwing covert patch, especially visible in flight. Adult has red forecrown and crown, with black border at rear, prominent black and white striped facial pattern, and broad black breast band. Chin and throat are red in adult male but white in adult female, which may also have crown partially or wholly black. Immature resembles adult but with a brownish head showing indistinct facial stripes, usually with faint red mottling on crown; chest and sides are pale brown with dusky barring.
Similar species: Smaller than common Hispaniolan Woodpecker, from which it is readily distinguished by its large white wing patch and striking black and white facial pattern.
Voice: Mostly silent in the West Indies, but occasionally gives a soft, nasal *mew*. Most often detected by its tapping on tree trunks.
Hispaniola: Found in forests, forest edges, woodlands, and gardens, virtually always in mountainous areas above 400 m. Aside from a single report near La Romana, DR, and specimens taken at Île de la Gonâve and Île de la Tortue in March and April when the birds were presumably migrating northward, all other records are from the montane areas of Massif de la Hotte and Massif de la Selle, Haiti; Cordillera Central, Sierra de Neiba, and Sierra de Bahoruco, DR. Most often recorded in pine forest, once at active sap wells in pine trees; also occasionally found in deciduous broadleaf forest. Reported from November to April but arrival as early as mid-October is probable.
Status: Uncommon to rare but regular non-breeding visitor. Typically rather quiet and solitary, it probably occurs more frequently than recorded. Common on nearby Cuba, where it also occurs routinely at low elevations in the non-breeding season. The species' apparent rarity on Hispaniola is puzzling.
Comments: Drills a series of horizontal holes in a live tree and feeds on the exuded sap and insects attracted to it. Individuals often return to the same feeding trees.
Range: Breeds from ec. Alaska, n. Saskatchewan, c. Quebec, and c. Newfoundland south to ne. British Columbia, s. Alberta, e. North Dakota, ne. Ohio, w. Pennsylvania, and New Hampshire, and locally in Appalachian Mountains to w. North Carolina. In non-breeding season occurs from n. Kansas, c. Ohio, and Long Island south to c. Panama. In West Indies, occurs mainly in Bahamas and Greater Antilles; rare in Puerto Rico and Virgin Islands; vagrant to St. Martin, St. Barthélemy, and Dominica.
Local names: DR: Carpintero Migratorio, Carpintero de Paso; **Haiti:** Chapentier à Bec Jaune, Sèpantye Etranje, Pic Maculé.

TYRANT FLYCATCHERS: FAMILY TYRANNIDAE

The tyrant flycatchers are a large Western Hemisphere family that reaches its greatest diversity in Central and South America. Widespread and common in nearly all habitats, the family is characterized by a broad, somewhat flattened bill that is often slightly hooked, with rictal bristles at the base. Most species are dully plumaged, some with a colorful crown patch that is usually concealed. Many species typically sit upright on exposed perches from which they sally forth to capture flying insects. Some also feed on fruit. Vocalizations are simple and unmusical but are often the best means of detection or identification. Some flycatchers, particularly the *Tyrannus* kingbirds, are noisy and aggressive, attacking birds of all sizes that intrude into their breeding territories.

GREATER ANTILLEAN ELAENIA—*Elaenia fallax* **Plate 37**
Breeding Resident

Description: 14–15 cm long; 14 g. A small, fairly chunky, rather nondescript flycatcher. Diagnostic field marks include a faint dark eye-line, two distinct wing bars, and a small bill with a pinkish base. Head and upperparts are grayish olive, underparts are pale gray washed with yellow. Neck and breast are faintly streaked with gray. White crown patch is usually concealed, and erectile crest is usually held flat. Immature lacks crown patch.
Similar species: Hispaniolan Pewee is slightly darker below and lacks wing bars. Stolid Flycatcher is much larger, yellow of underparts is more distinct, and it is a lowland bird.
Voice: Harsh *pwee-chi-chi-chiup, see-ere, chewit-chewit.* Also a trill sung at dawn.
Hispaniola: Occurs primarily in pine forests, but also in open country with scattered trees and wet broadleaf forests at high elevations from 500 m to a maximum recorded elevation of at least 2,500 m on Pico Duarte. Reported from the Sierra de Bahoruco, Sierra de Neiba, Cordillera Central, and Los Haitises National Park, DR, and Massif de la Hotte, Massif de la Selle, and Massif des Montagnes Noires, Haiti. Not recorded from any major satellite island.
Status: Locally common year-round resident. May inhabit recently disturbed areas dominated by pine, if the forests are left to regenerate the fruit-bearing shrubs favored by the elaenia. Populations in the DR have likely declined since 1930, and the species' current status in Haiti is uncertain because of the degree of sustained deforestation and habitat loss that have occurred. The subspecies *E. f. cherriei* is endemic to Hispaniola.
Comments: Sallies from its perch to snatch prey off leaves and twigs. Frequently feeds on fruit as well as insects. Forages from near ground level to high in the canopy, often in pairs, but also is a regular participant in mixed-species flocks in pine forests on Hispaniola.
Nesting: Lays 2–3 pale pinkish and lightly spotted eggs in a cup nest of moss lined with feathers. Nest may be built high in a tree or near the ground in a bush or thicket. Breeds May to July.
Range: Resident on Jamaica and Hispaniola; vagrant to Mona Island and Caicos Islands.
Local names: DR: Maroíta Canosa; **Haiti:** Chitte Sara, Ti Tchit Sara, Pipirit Sara.

HISPANIOLAN PEWEE—*Contopus hispaniolensis* **Plates 37, 41**
Breeding Resident—*Endemic*

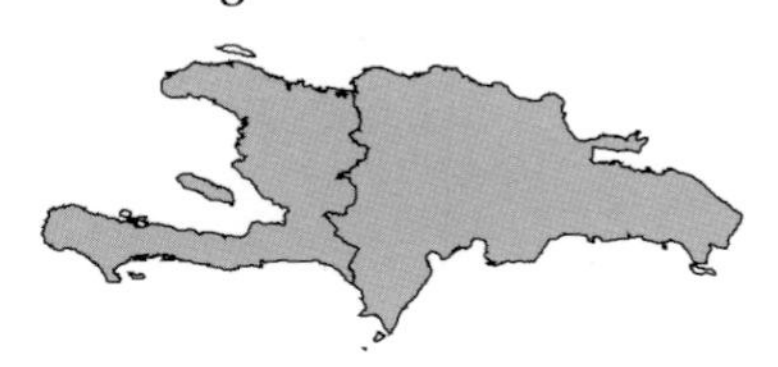

Description: 15–16 cm long; 11 g. A small flycatcher with drab grayish olive upperparts, somewhat darker on head. Pale, light grayish to buffy wing bars are inconspicuous or absent. Underparts are lighter gray with an olive, yellow, or brown wash. Bill is relatively long and broad, blackish above, pinkish yellow or pale orangish below. Often flicks tail when perched.
Similar species: Stolid Flycatcher is substantially larger, browner, and has white wing bars. Greater Antillean Elaenia has smaller bill, is paler below, and has two distinct wing bars.
Voice: Strong, mournful *purr, pip-pip-pip-pip.* The distinctive dawn song is a loud, rapid-fire volley of short notes, *shurr, pet-pet, pit-pit, peet-peet,* with the paired syllables successively rising in pitch.
Hispaniola: Occupies a variety of habitats, including both dry and humid broadleaf forest, scrubby woodlands, shade coffee plantations, orchards, and limestone karst broadleaf forest at all elevations from sea level to at least 2,000 m, but occurs most commonly in pine forest in the mountains and foothills.
Status: Common and widespread breeding resident. The subspecies *C. h. hispaniolensis* is endemic to Hispaniola; the subspecies *C. h. tacitus* is endemic to Île de la Gonâve.

Comments: A tame bird. It flies from a low perch to catch small insects on the wing, frequently returning to the same perch. Although the pewee also feeds on small fruits, it is more strictly insectivorous than the elaenia. In pine forests, this pewee regularly forages in mixed-species flocks.
Nesting: Eggs are creamy white with a pinkish tint, heavily marked with reddish brown and grayish spots and blotches which are concentrated around the broad end or middle of the egg. The nest is a cup of lichens, moss, and rootlets placed in the fork of a branch 3–5 m above ground. A clutch of 2–4 eggs is typically laid March to June, but date probably varies to some degree with elevation.
Range: Endemic to Hispaniola, including Île de la Gonâve. Has been recorded as a vagrant on Mona Island and on Providenciales, Bahamas.
Local names: DR: Maroíta; **Haiti:** Pipirit Tèt Fou; Moucherolle Tête-fou.

GREAT CRESTED FLYCATCHER—*Myiarchus crinitus* **Plate 37**
Vagrant

Description: 17–21 cm long; 33 g. A large, noisy flycatcher. Upperparts olive-brown, wings reddish brown with whitish wing bars, tail also reddish brown. Throat and breast gray, belly bright yellow. Often raises crown feathers to form crest when excited.
Similar species: Stolid Flycatcher is darker above, especially on head, with wings and tail less rufous. Belly is paler yellow.
Voice: Loud, harsh *wheeep* with a rising inflection, heard throughout the year.
Hispaniola: One identified at Bayahibe, DR, in November 1987 by an experienced observer, and another bird reported near Puerto Escondido in April 2001. A specimen in the Museo Nacional de Historia Natural (Santo Domingo) with no data is of uncertain origin.
Status: Vagrant.
Range: Breeds from ec. Alberta, s. Ontario, and c. Nova Scotia south to c. North Dakota, w. Kansas, s. Texas, Gulf Coast, and Florida Keys. In non-breeding season occurs from s. Florida and s. Mexico south to Colombia and n. Venezuela. Rare in Cuba; vagrant in Bahamas, DR, and Puerto Rico.
Local names: DR: Maroíta de Cresta; **Haiti:** Pipirit Bwòdè, Tyran Huppé.

STOLID FLYCATCHER—*Myiarchus stolidus* **Plate 37**
Breeding Resident

Description: 20 cm long; 23 g. A medium-sized, large-headed flycatcher with two pale white wing bars and primaries strongly fringed with white. Upperparts are olive-brown, darker on head, and tail is brownish with reddish brown inner webs. Throat and breast are grayish white, belly is pale yellow, and bill is black and moderately heavy. A raised crest is often noticeable.
Similar species: Pewee and elaenia are substantially smaller; kingbirds are larger and more robust.
Voice: Prolonged, rolling *whee-ee-ee, swee-ip, bzzrt*. Also a plaintive *jui* on Hispaniola; often snaps its bill loudly and repeatedly.
Hispaniola: Occurs most commonly in dry forest and desert scrub habitats from sea level to 750 m, but is also regular in more humid habitats such as pine and moist broadleaf forest and shade coffee plantations to 1,800 m elevation. Present islandwide though probably absent from the highest elevations in the mountains. There is some indication of seasonal altitudinal migration in Sierra de Bahoruco, DR, with birds moving to higher elevations in the breeding season. Also has occurred on the adjacent islands of Île à Vache, Île de la Gonâve, Île de la Tortue, the Cayemites, and Islas Beata, Catalina, and Saona.

Status: Common resident islandwide. Its current status in Haiti is poorly known, though populations there are surely reduced since 1930 because of widespread habitat destruction. The subspecies *M. s. dominicensis* is endemic to Hispaniola and associated islands.
Comments: An active bird, easily approached. Feeds on insects, which it typically captures on the wing by snatching them from a twig or leaf. Also feeds on fruits, which are plucked while hovering.
Nesting: Lays 2–5 eggs in a nest in a cavity of a tree, cactus, post, or even house, often only 0.5–3 m above ground. The nest itself is lined with moss, hair, cottony plant material, and feathers. Eggs are pale ivory yellow marked with brown, most heavily at the broad end. Breeds April to July.
Range: Resident on Jamaica and Hispaniola, including satellite islands.
Local names: DR: Manuelito; **Haiti:** Louis, Pipirite Gros-tête, Pipirit Gwo-tèt.

GRAY KINGBIRD—*Tyrannus dominicensis* **Plate 37**
Breeding Resident

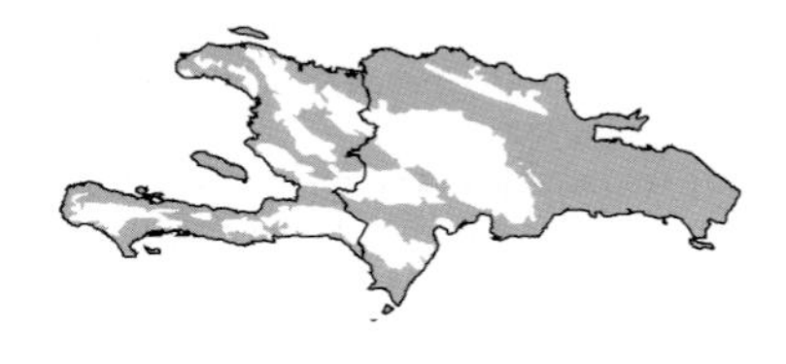

Description: 22–25 cm long; 44 g. A large, conspicuous tyrant flycatcher. Gray above, pale grayish white below, with a distinct blackish mask from the base of bill through auriculars. Faint whitish wing bars are present, and tail is notched. A small yellowish orange or reddish orange patch on crown is usually concealed. Immature has pale brownish wing bars and lacks crown patch.
Similar species: Loggerhead Kingbird has an entirely dark crown extending below eye, squared tail, and distinctly different voice. Northern Mockingbird is also gray and white but cocks its longer tail upward and has pronounced white wing and tail patches.
Voice: Call is a loud, emphatic *pe-ti-gre* or *pe-cheer-ry*. The song is a more complex, musical, chattered variation of the call, *pe-ti-gréee, pe-ti-grro*. Also note a harsh *peet*, and *burr*.
Hispaniola: A species of open areas. Most numerous in lowland dry forest, cactus scrub woodland, and agricultural areas below 450 m, but also occurs in coastal mangroves. Rarely occurs in cleared lands in interior valleys in the mountains such as near Constanza, DR, at about 1,100 m, and in the outskirts of towns, but avoids dense forest. Also reported from almost all of the major satellite islands.
Status: One of the most conspicuous and common species in appropriate habitat throughout Hispaniola. Primarily a year-round breeding resident on Hispaniola. Migratory activity off the island is possible, as numbers present are much lower from the end of November through January than at other seasons. Some birds present in Hispaniola during the non-breeding season may be from mainland North American breeding populations, though this remains to be documented. The species is still widespread in Haiti, where the extensive clearing of forest has likely provided additional habitat.
Comments: Usually seen on exposed perches such as telephone lines and bare treetops. Feeds primarily on large insects captured on the wing by sallying from its perch; often returns to the same perch and batters the insect before eating it. Some birds take advantage of streetlights to feed at night on the attracted insects. Also takes lizards and small fruits. Outside the breeding season this species typically roosts communally at night. Such gatherings can be large, such as 300+ in an area less than 0.6 km long at Cumayasa, DR. It is a very aggressive bird, sometimes attacking much larger species such as hawks and frigatebirds, and even dogs and cats.
Nesting: Lays 2–4 elliptical oval eggs, which vary in color from white tinged with ivory yellow to pale pinkish buff, and spotted boldly with chestnut red, purplish brown, and paler brown and gray, the markings often concentrated on the larger end. The nest is an open structure loosely built of small twigs with an inner cup lined with fine rootlets, built in a tree or shrub 2–12 m above ground. Individuals will reuse their own nests. Breeds May to July.

Range: Breeds on Atlantic and Gulf coasts from Georgia south to Florida Keys, parts of interior s. Florida, west to s. Mississippi, locally in n. Colombia and n. Venezuela east to Tobago. In non-breeding season, occurs south to Guianas and n. Brazil. Breeds throughout West Indies. Populations from Florida and Bahamas south to Cuba and Jamaica are wholly or partially migratory. Migratory behavior and patterns in this species are poorly understood.
Local names: DR: Petigre; **Haiti:** Pipirit Gri, Titiri.

LOGGERHEAD KINGBIRD—*Tyrannus caudifasciatus* **Plate 37**
Breeding Resident—*Endangered*

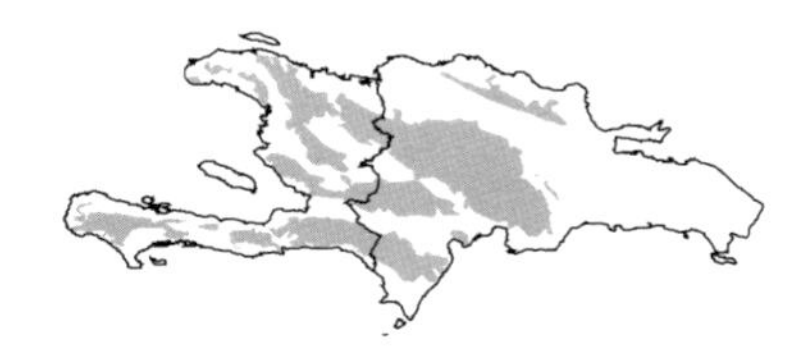

Description: 24–26 cm long; 44 g. A large, handsome flycatcher with a large bill. Crown and facial mask are entirely blackish but contrast with dark brown back and tail, and wings and tail which are edged in rufous. Yellow or pale orange crown patch is usually concealed. Underparts are whitish, lightly washed with gray or brown. Tail is squared. Juvenile is gray above, buffy white below, with buffy or brownish upperwing coverts; lacks crown patch.
Similar species: Gray Kingbird has a lighter crown and lighter gray upperparts which contrast with a prominent black facial mask. Stolid Flycatcher has a much smaller bill and yellowish wash on underparts.
Voice: Calls variable, but usually a long, rolling trill, br-r-r-r-r-r, often terminating with several explosive notes. Mostly silent outside of breeding season. Also snaps its bill loudly.
Hispaniola: Resident islandwide from sea level to at least 2,000 m elevation but primarily found at mid- to upper elevations. This species essentially replaces Gray Kingbird in densely forested areas and at higher elevations, though it is also found in broadleaf and pine habitats and in coffee plantations where Gray Kingbird occurs only sparsely. Loggerhead Kingbird has been found occasionally in lowland cactus scrub habitat in Del Este National Park, and near Azua, Monte Cristi, and Barahona, DR, but this is unusual on Hispaniola, although the species routinely occupies such low-elevation habitats in Cuba. Since the early 1990s, records include single birds or pairs from Aquacate, Mencia, and Aceitillar in the Sierra de Bahoruco, and Río Limpio, El Rubio near Monción, and Manabao in the Cordillera Central. Also recently reported from La Visite National Park in the Massif de la Selle of Haiti. Not known from any major satellite island.
Status: Very uncommon and local breeding resident. This species appears to have declined dramatically since the 1970s, as it was formerly considered abundant and widespread. Once thought to be more common in the southwestern peninsula of Haiti than elsewhere, but the extensive deforestation that has occurred in that region and elsewhere in Haiti, as well as in the DR, has surely reduced this species' numbers. The subspecies *T. c. gabbii* is endemic to Hispaniola.
Comments: Commonly sallies from its perch low to the ground to capture insects, usually by plucking them off a leaf or twig rather than in midair as Gray Kingbird does. Also eats fruits and small lizards. Most often seen on natural exposed perches, such as posts and tree branches.
Nesting: Lays 2–4 reddish brown eggs with darker brown and violet markings at the broad end. Nest is an unlined cup of twigs, stems, and grasses built in a tree or shrub, typically in the fork on the outer part of a branch. Breeds principally February to July, but has been recorded in November to January as well. Breeding season may vary with altitude, beginning earlier at lower elevations.
Range: Resident in Bahamas, Greater Antilles east to Puerto Rico and Vieques, and Cayman Islands; vagrant to s. Florida.
Local names: DR: Manjuila; **Haiti:** Tête Police, Pipirit Chandèl, Pipirit Tèt Lapolis.

SCISSOR-TAILED FLYCATCHER—*Tyrannus forficatus* **Plate 37**
Vagrant

Description: Female 22–30 cm long, male 25–37 cm long; 43 g. A pale grayish flycatcher with a conspicuously long, forked tail. Upperparts are pearl gray, wings and tail are dusky brown, and underparts are whitish with a salmon pink wash on sides, flanks, and undertail coverts. Axillaries are reddish. Females and immatures are generally paler with a less intense salmon and reddish coloration than in males, and with shorter tail.
Voice: Variable, chattering. Seldom heard in non-breeding season.
Similar species: Fork-tailed Flycatcher is somewhat darker, has a black head, and lacks salmon pink wash.
Hispaniola: Only 1 record; 1 seen and photographed about 20 km west of Oviedo, Pedernales Province, in late October 1996.
Status: Vagrant.
Range: Breeds from se. Colorado and se. New Mexico east to w. Louisiana and south through s. Texas and ne. Mexico. Occurs in non-breeding season from s. Mexico south to c. Costa Rica, rarely to w. Panama. In West Indies, a vagrant to Bahamas, w. Cuba, DR, and Puerto Rico.
Local names: DR: Cola de Tijeras; **Haiti:** Pipirit Ke Long, Tyran à Longue Queue.

FORK-TAILED FLYCATCHER—*Tyrannus savana* **Plate 37**
Vagrant

Description: 33–41 cm long; 29 g. A grayish flycatcher with a dark head and a conspicuously long, forked tail. Adult male has a gray back, blackish brown wings, black crown and nape, and entirely white underparts. Long tail is black and edged white but may be shorter during molt. Females and immatures are similar but duller, head is browner, and tail is shorter.
Voice: Variable, thin tittering notes, and a high *jeek*, but not very vocal.
Similar species: Scissor-tailed Flycatcher is paler gray, including head, and has a salmon pink wash on sides, flanks, and undertail coverts.
Hispaniola: Only 1 record; a well-described individual seen by several observers at the airport in Punta Cana in January 2005.
Status: Vagrant.
Comments: Caribbean vagrants of this species are thought to be migrants from South America that have overshot their usual wintering grounds in Venezuela and Trinidad. The presence of austral migrants from the south is unusual this far north.
Range: Breeds on Caribbean slope of Mexico and Central America south through n. South America, e. Amazonia, and c. Brazil to s. Argentina. Occurs widely in non-breeding season east of Andes through n. South America, Trinidad, and Tobago. In West Indies, a vagrant to Lesser Antilles.
Local names: DR: Cola Ahorquillada; **Haiti:** Pipirit Ke Fann, Tyran des Savanes.

VIREOS: FAMILY VIREONIDAE

Vireos are a New World family of fairly small, drab birds that resemble heavily built warblers but have a thicker bill that is hooked at the tip. Most have dull olive green plumage above and are whitish or yellowish below. Songs vary from rich warbles to simple, repeated phrases; calls typically consist of scolding chatters and mews. Vireos are sluggish and deliberate in their movements, carefully inspecting twigs and the undersides of leaves for insects as they move among the branches. Nests are open cups of woven plant materials suspended in shallow forks of horizontal branches.

WHITE-EYED VIREO—*Vireo griseus* **Plate 38**
Vagrant

Description: 11–13 cm long; 11 g. A small vireo, grayish green tinged with yellow above, and whitish below with yellowish sides and flanks. Neck is more grayish than the rest of upperparts. Note the broad yellow spectacles, dusky lores, and two white to yellowish white wing bars. Adult has a white iris; immature a duller, brownish iris. Legs are bluish gray; bill is black.
Similar species: Thick-billed and Yellow-throated vireos also have yellow spectacles, but both have a dark iris. Thick-billed Vireo is entirely pale yellow below; Yellow-throated Vireo has a bright yellow chin, throat, and breast. Flat-billed Vireo lacks yellow spectacles, has underparts washed with pale yellow, and has white-tipped outer tail feathers. Pine Warbler has a thinner bill, lacks distinct spectacles, and has finely streaked sides.
Voice: Loud, slurred 3- to 7-syllabled song, often beginning and ending with a sharp chip, such as *chip-a-tee-weeo-chip,* repeated with variations. More rapid and less emphatic than that of Thick-billed Vireo. Alarm call is a series of nasal, churring notes. Adults also give soft *pick* and *mew* calls in the non-breeding season.
Hispaniola: Six known records, all from low-elevation sites, including individual birds in xeric scrub near Punta Rucia, DR; in second-growth scrub in Los Haitises National Park; at Jaragua National Park; and at the National Botanical Garden, Santo Domingo, DR. Two birds seen on Île à Vache, Haiti.
Status: Vagrant.
Comments: Solitary. Usually in dense, scrubby vegetation low to the ground where it moves about slowly. These traits, plus the bird's drab appearance and similarity to more common species, result in it being easily overlooked. Primarily insectivorous, mainly gleaning lepidopteran larvae, but may take fruit, especially from the gumbo limbo tree (*Bursera simaruba*), in the non-breeding period.
Range: Breeds from c. Iowa east to s. Michigan, s. New York, and s. Massachusetts, south to wc. Texas and Atlantic slope of Mexico, east to Florida. A sedentary race breeds on Bermuda. Occurs in non-breeding season from se. Virginia south through Florida and west to s. Texas, and Atlantic slope of Mexico south to n. Guatemala and nw. Honduras. In West Indies, common on Bahamas and Cuba, uncommon in Cayman Islands, rare on Jamaica and Puerto Rico, and vagrant on Hispaniola and U.S. Virgin Islands.
Local names: DR: Vireo de Ojo Blanco; **Haiti:** Ti Panach Je Blanch.

THICK-BILLED VIREO—*Vireo crassirostris* **Plate 38**
Breeding Resident

Description: 14 cm long; 13 g. A medium-sized vireo, with generally olive upperparts, grayish green crown and nape, blackish lores, dark iris, two white wing bars, and bright yellow spectacles, sometimes broken around eye. Underparts are dingy whitish to buffy yellow. Immature has poorly defined wing bars and lacks blackish lores.
Similar species: White-eyed Vireo has a white iris, smaller and thinner bill, and is whitish on throat, breast, and belly. Yellow-throated Vireo has a bright yellow chin, throat, and breast, with a white belly; it also has an olive green, rather than grayish, crown and lacks black lores.
Voice: Bubbly and variable *chik-didle-wer-chip,* very similar to song of White-eyed Vireo but slower and less emphatic. Calls include buzzy, low, and nasal notes.
Hispaniola: Moderately common resident confined to Île de la Tortue, Haiti, but common throughout this island. Occurs in dry broadleaf forest, lowland scrub, second growth, and mangroves.
Status: Common year-round resident on Île de la Tortue, where the subspecies *V. c. tortugae* is endemic.
Comments: More often heard than seen, it is usually located by its song or call. Often perches motionless in dense vegetation while singing, making visual detection difficult.
Nesting: Lays 2–3 whitish or pale buffy eggs with heavy spotting in a deep, round cup nest made of bark strips with a lining of leaves and other soft plant materials. The nest is suspended

by its margins in a fork near the end of a branch and is typically built only 35 cm–1 m above ground. Breeding chronology is not well known, but probably April to July.
Range: Resident in Bahamas, certain cays north of Cuba, Cayman Islands, and Haiti; vagrant to s. Florida.
Local names: DR: Vireo de Pico Ancho; **Haiti:** Ti Panach Gwo Bèk, Oiseau Canne, Zwazo-kann.

FLAT-BILLED VIREO—*Vireo nanus* **Plates 38, 43**

Breeding Resident—*Endemic*

Description: 12–13 cm long; 11 g. A small vireo, grayish green above and dull grayish white below, washed with pale yellow. Diagnostic features include outer tail feathers with narrow white tips, two white wing bars, and white iris.
Similar species: White-eyed Vireo has yellow spectacles and a clean white chin, throat, and breast. Black-whiskered Vireo has a narrow malar stripe, white supercilium, dark line through eye, and lacks wing bars. Female Antillean Siskin is finely streaked below and has a paler, heavier bill.
Voice: Chattering, high-pitched *weet-weet-weet-weet-weet-weet-weet*. Also a slower version, often repeated.
Hispaniola: Locally distributed in both the DR and Haiti, principally in lowlands, but recorded from sea level to 1,200 m elevation. Occurs primarily in dry scrub habitats, often at the base of limestone hills, but occasionally in more humid situations such as mesic forest near more typical habitat. The known islandwide historical range of this species includes coastal eastern DR from Guaraguao at the easternmost entrance to Del Este National Park then west along the coast nearly to San Pedro de Macorís, including Isla Saona; the Samaná Peninsula and Los Haitises National Park area; the northern coast of the DR from Río San Juan west to at least Port-de-Paix, Haiti; the valley of Río Yaque del Norte from the river mouth at least as far upstream as Mao; the north slope of the Sierra de Bahoruco, DR, from Duvergé to near Aguacate, but very rare on the south slope at Aceitillar; parts of the south slope of Sierra de Neiba, especially near La Descubierta, DR; parts of the coast of Port-au-Prince Bay, especially near Montrouis, Haiti; and on Île de la Gonâve. It seems likely that this species occurred in much of the Neiba Valley/Cul de Sac Plain all the way from Cabral to Montrouis before large-scale human colonization of this area. Reported in 2003 from the Sierra de Martín García. Its current distribution needs to be clarified.
Status: Uncommon breeding resident, local in distribution.
Comments: Moves deliberately in low vegetation, sometimes foraging on the ground. Feeds on fruits and insects, occasionally pursuing insects in flight.
Nesting: Lays 2 eggs which are unmarked white or lightly marked with gray at one end, in a cup-shaped nest usually built in a low bush. Breeds February to June, but nesting biology poorly known.
Range: Endemic to Hispaniola, including Île de la Gonâve. Formerly on Isla Saona, but current status there is unknown.
Local names: DR: Ciguíta Juliana; **Haiti:** Ti Panach Bèk Plat, Viréo d'Hispaniola.

YELLOW-THROATED VIREO—*Vireo flavifrons* **Plate 38**

Non-breeding Visitor

Description: 13–15 cm long; 18 g. A fairly large vireo. Upperparts olive green with a contrasting gray rump; bright yellow chin, throat, and breast; and white belly and undertail coverts. Bold yellow spectacles, two white wing bars, and dark iris and lores are diagnostic.
Similar species: White-eyed Vireo has a white chin, throat, and breast. Thick-billed Vireo is entirely buffy below, with more distinct black lores and a grayish crown. Flat-billed Vireo is

dull whitish on breast, lacks yellow spectacles, and has a white iris. Pine Warbler has streaked underparts, a slimmer bill, and lacks yellow spectacles.
Voice: Song is a variable series of wheezy, short phrases separated by pauses, *chee-wee, chee-woo, u-wee, chee-wee*. Scolding call is a harsh, descending series, *chi-chi-chi-chi*, similar to those of other vireo species. May occasionally sing in non-breeding season.
Hispaniola: Recorded several times in Santo Domingo, at Punta Cana, near La Romana, and in Jaragua National Park. Also observed several times in Mencia, Pedernales Province, in the non-breeding period of 2003–2004, suggesting individuals may overwinter. One Haitian record from near Pignon.
Status: Regular non-breeding visitor, but in low numbers.
Comments: A bird of the mid- and upper-forest canopy which may be expected in a wide variety of forest types; prefers open semideciduous forest on Cuba. This vireo may be more common in the West Indies than records indicate, because of difficulty of detection and its similarity to other species.
Range: Breeds from se. Saskatchewan and se. Manitoba east to n. Michigan, se. Ontario, and se. Maine, south to c. Florida and se. Texas. Occurs in non-breeding season from ec. Mexico south through Central America to mountains of n. Colombia and nw. Venezuela. In West Indies, occurs regularly in non-breeding season in Bahamas, casually to Cuba, Jamaica, Cayman Islands, DR, and Virgin Islands, and as a vagrant to Puerto Rico and Lesser Antilles south to Grenada.
Local names: DR: Vireo Garganta Amarilla; **Haiti:** Ti Panach Gòj Jòn.

WARBLING VIREO—*Vireo gilvus* — Plate 38

Vagrant

Description: 12–13 cm long; 13 g. A medium-sized, pale, unmarked vireo. Upperparts grayish green, with crown slightly grayer; underparts dingy white, with a cleaner white breast and belly. Sides of neck, sides, and flanks lightly washed yellow, but buffier on neck. Whitish supercilium, pale lores, and indistinct dusky eye-line. No wing bars.
Similar species: Similar to Red-eyed and Black-whiskered vireos but smaller, always paler and grayer, and lacks distinct white supercilium bordered narrowly above and below by black.
Voice: Variety of harsh, nasal mews; also a short, dry *git* or *vit*. Song is a rapid, run-on warble.
Hispaniola: One DR record of a probable northbound migrant seen in early April 2001 near Puerto Escondido on the north side of Sierra de Bahoruco.
Status: Vagrant.
Range: Breeds from se. Alaska, sw. Northwest Territories, c. Alberta, s. Manitoba, sw. Quebec, and New Brunswick south to s. California, mountains of c. Mexico, sw. Texas, and Virginia. Occurs in non-breeding season from s. Baja California and Pacific slope of Mexico south to Honduras and Nicaragua. In West Indies, vagrant to Cuba and DR.
Local names: DR: Vireo Cantor; **Haiti:** Ti Panach Chantè , Viréo Mélodieux.

RED-EYED VIREO—*Vireo olivaceus* — Plate 38

Vagrant

Description: 14–15 cm long; 17 g. A large vireo, distinguished by a gray crown that contrasts with olive green upperparts. Underparts are white. Note prominent white supercilium, bordered by a blackish eye-line below and a distinct lateral crown stripe above. Adult has a red iris; immature has a dull brown iris and yellower tints on flanks and undertail coverts.
Similar species: Black-whiskered Vireo has a larger bill, black malar (or whisker) stripe, more buffy yellow underparts, and is duller green on back and paler gray on crown.
Voice: Usually silent in the West Indies except during its northward migration in April. Call is

a nasal mew, *myaahh*. Song consists of abrupt phrases separated by deliberate pauses, repeated persistently throughout the day, *cherr-o-wit, cheree, sissy-a-wit, tee-oo* or *here-I-am, up-here, see-me? see me?* Black-whiskered Vireo has a similar song but with longer phrases, more repetitious and warbling.
Hispaniola: Four DR records from September to early February: 1 identified in a shade coffee plantation near Jumunucú, Cordillera Central; twice reported from National Botanical Garden, Santo Domingo; and once from near Laguna de Oviedo. One seen and heard calling during migration in late August in acacia trees at Sabanita de Pallavo, Don Juan de Monte Plata. Also reported from Navassa Island.
Status: Vagrant. Possibly occurs more regularly as a passage migrant but is easily confused with juvenile Black-whiskered Vireo which has not yet developed a clear malar stripe. Although most of the North American population is believed to migrate through Middle America, recent reports show that the form *V. o. olivaceus* is also a regular fall transient southbound in small numbers and a rare spring transient northbound through the Lesser Antilles. More work is needed to clarify its exact status on Hispaniola.
Comments: A relatively sedentary canopy species, easily overlooked because of its similarity to common Black-whiskered Vireo.
Range: Breeds from se. Alaska and wc. MacKenzie east to n. Alberta and s. Newfoundland, south to n. Oregon, sc. Texas, Gulf Coast, and s. Florida. Occurs in non-breeding season from n. South America south to e. Peru and c. Brazil. In West Indies, uncommon migrant in Bahamas, Cuba, Jamaica, and Cayman Islands; vagrant in Hispaniola, Puerto Rico, Virgin Islands, and Lesser Antilles south to Barbados.
Local names: DR: Vireo de Ojo Rojo; **Haiti:** Ti Panach Je Wouj.

BLACK-WHISKERED VIREO—*Vireo altiloquus* **Plate 38**
Breeding Visitor; Non-breeding Visitor?; Breeding Resident?

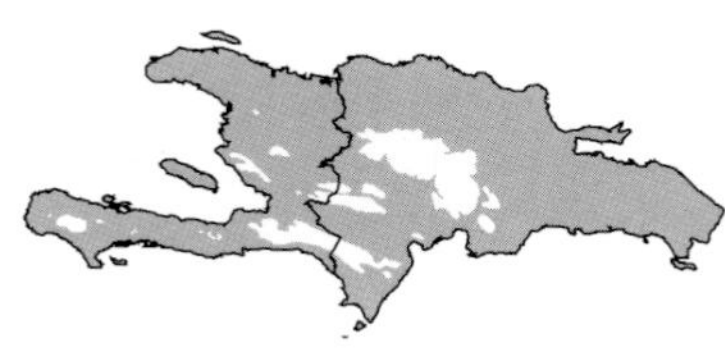

Description: 15–16.5 cm long; 18 g. A large vireo, with olive brownish upperparts and whitish underparts, washed pale olive on sides and flanks, with yellowish undertail coverts. White supercilium contrasts with dusky eye-line and grayish crown. Note narrow black malar stripe, or "whisker." Adult has red iris; immature has brown iris, duller upperparts, more buffy underparts, and may show a faint wing bar.
Similar species: All other vireos in the West Indies lack the black malar stripe. Very similar Red-eyed Vireo has a smaller bill, whiter underparts, greener back, and grayer crown with more distinct black border below.
Voice: Most easily identified by its repetitive, monotonous song heard throughout the day, consisting of short, melodious, 3-syllabled phrases, each slightly different and separated by a distinct pause, *julián chi-ví*. The local name for this bird is a rendition of its song. Calls include a thin mewing *tsit* and a sharp, nasal *yeeea*.
Hispaniola: Found from sea level to at least 1,700 m elevation, though typically not above 900 m, in a wide variety of habitats from semiarid lowland scrub, coastal mangroves, shade coffee plantations, dry and moist broadleaf forest, to humid rainforest on the Samaná Peninsula, DR. Absent from pine and broadleaf forest high in mountains of both countries. Reported from most major satellite islands.
Status: Generally a common to abundant breeding visitor. Whereas a few individuals of this species are present all year, most of the population is believed to migrate to South America, leaving in late August or September and returning in late January or early February. Since the birds rarely sing during this season, the number present in September to February is unclear but may possibly include both individuals that are permanent residents and those that are

non-breeding visitors from Florida or Bahamas populations. Movements and seasonal status of this species on Hispaniola need further study.

Comments: Normally remains nearly motionless in dense canopy foliage where it goes unnoticed except when singing. Feeds mainly on insects which it typically gleans from leaves but occasionally takes while hovering. This species also appears to be more frugivorous than other vireos in the West Indies. Nonmigratory individuals, though year-round residents, may undergo altitudinal migrations.

Nesting: Eggs are white, sparingly marked with minute to moderate spots of black and blackish brown mainly at the larger end. The nest is a deep cup, typically suspended by its margin, 1.5–5.0 m above ground in a fork near the end of a branch or twig. The nest is built of rootlets, moss, and strips of bark and is often lined with cottony plant fibers. Clutch size is 2–3 eggs. Usually breeds April to July.

Range: Breeds in central and south-coastal Florida and throughout Caribbean region. Occurs in non-breeding season on Hispaniola, throughout Lesser Antilles, and in South America from e. Colombia and Guianas south to n. Peru and c. Brazil.

Local names: DR: Julián Chiví; **Haiti:** Ti Panach Pyas-kòlèt, Viréo à Moustaches.

CROWS: FAMILY CORVIDAE

Crows, the largest passerines, are found nearly worldwide, being absent only from New Zealand and most other islands of the Pacific Ocean. They are invariably black, with a fairly robust bill and legs, and strong, grasping feet. Crows are typically gregarious, often noisy and aggressive, and lack a musical song. Diets and feeding behaviors are variable. Food items include insects, fruit, seeds, nestling birds and eggs, reptiles, amphibians, and carrion. Nests are bulky platforms of sticks and twigs, usually at mid- to upper levels in trees. Some species breed cooperatively, with several birds attending a single nest; others nest as separate pairs. Some species may exhibit both breeding behaviors.

HISPANIOLAN PALM CROW—*Corvus palmarum* **Plate 39**
Breeding Resident—*Threatened Endemic*

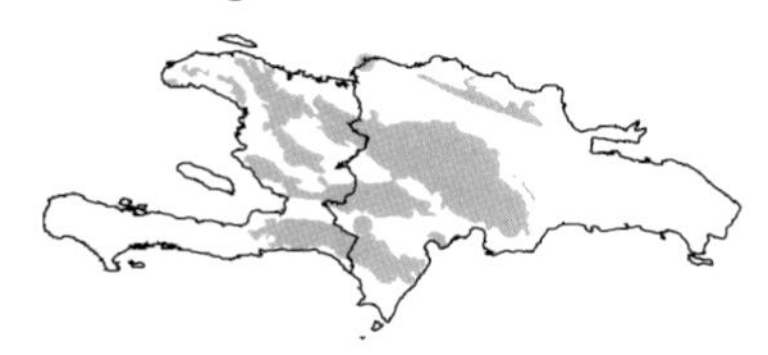

Description: 43 cm long; 289 g. A large, completely black bird, with a purplish blue sheen on back and upperwing coverts, fading to dull brownish black in worn plumage. Best identified by its distinctive voice.

Similar species: Considerably smaller than White-necked Crow, which is a much more robust bird. Hispaniolan Palm Crow has a steadier flapping flight than White-necked Crow, and wings appear shorter. Vocalizations are less variable and more distinctly nasal than those of White-necked Crow.

Voice: Harsh and nasal *aaar*, often with a complaining quality. Also described as *cao cao*, hence its local name. Usually given in pairs or in a series. May flick tail downward, with wing tips lowered, while calling.

Hispaniola: Reported islandwide from sea level to a maximum known elevation of 3,000 m near the summit of Pico Duarte, Cordillera Central, DR, but seldom found above 2,000 m. In the DR, known especially from pine forest, but also recorded irregularly in dry scrub, such as at Monte Cristi and Lago Enriquillo, and in dry forest and humid broadleaf forest. Most common in pine forest habitat above 750 m elevation in the Sierra de Bahoruco, but not found at lower elevations or in populated areas. In Haiti, commonly found on the plains and abundant in the northern pine belt, but has never been recorded west of the Jacmel-Fauché depression on the Tiburón Peninsula. Not reported from any major satellite islands.

Status: Locally common, endemic resident; may occur in flocks as large as 200+ individuals, but small groups of less than 20 are more common. Although previously abundant, and still fairly numerous in remote areas, it is rare to absent from areas of DR inhabited by many people. Excessive shooting of the birds for sport and as game (their flesh being considered a great delicacy) has clearly affected the population. Still reported to be fairly common in much of Haiti, since few people own guns. This species was recently split from Cuban Palm Crow (*C. minutus*) based on differences in vocalizations and morphology, but this split has not been formerly recognized by the Committee on Classification and Nomenclature of the AOU.
Comments: Species is omnivorous and feeds on fruit, seeds, insects, snails, and lizards.
Nesting: Eggs are pale green, marked evenly with moderately large, diffuse spots of clove brown and dark olive. Lays 3–4 eggs in a loosely constructed platform of twigs 10 m or higher in a pine tree or palm. Breeds March to July.
Range: Endemic to Hispaniola.
Local names: DR: Cao; **Haiti:** Cao, Ti Kaw, Ti Gragra, Corneille Palmiste.

WHITE-NECKED CROW—*Corvus leucognaphalus* **Plate 39**
Breeding Resident—*Threatened Endemic*

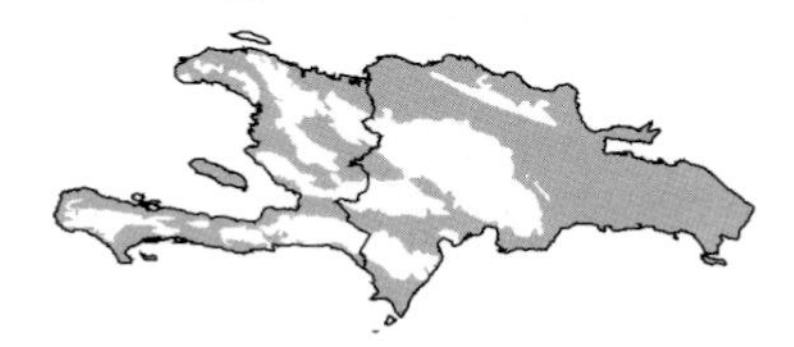

Description: 48–51 cm long; mass unavailable. A large, entirely black corvid with a large bill and red-orange eye. Upperparts have a violet sheen, and base of neck feathers are white, but these features are rarely visible. Often flicks tail downward.
Similar species: Best distinguished from smaller Hispaniolan Palm Crow by its reddish eye and more variable voice. Flight tends to be less direct, and it occasionally soars.
Voice: Wide variety of vocalizations, including a *caw* and clucking, gurgling, bubbling, and laughlike calls and squawks; described as a comical *culic calao calao*.
Hispaniola: An unevenly distributed species resident islandwide from sea level to at least 2,650 m, but uncommon above 1,500 m. Reported in pine woodlands, but most often occurs in lowland swamps, cactus forest, coastal mangroves, and broadleaf forests. In the early 1990s reported as still very common around Port-au-Prince where there are trees, and at an agricultural station at Damian. Recorded from several major satellite islands, including Navassa Island, Île à Vache, Île de la Gonâve, and Isla Saona.
Status: Endemic and locally common to uncommon breeding resident on Hispaniola, but rapidly declining and considered threatened because of habitat alteration and hunting. Although flocks of several hundred birds were reported prior to the 1930s as common from many areas of the island, the species has now disappeared from most of its range because of excessive hunting. By 1982 it had become rare in DR pine forest habitat and at high elevations, and it was only locally distributed in the lowlands. Its current DR stronghold is the Los Haitises National Park area, where its abundance changed little from 1976 to 1996. The species is still regarded as a game bird and considered excellent eating, especially in the DR. Since guns are rare in the Haitian countryside, the species may be more abundant in rural areas.
Comments: Feeds mainly on fruits and seeds, but also takes vertebrates and large insects. Forms large foraging flocks which sometimes leave the forest to raid crops. Sometimes kept as a pet because of its ability to mimic humans and animals alike. The species' rapid decline on Hispaniola, combined with habitat recovery and improved wildlife conservation measures on Puerto Rico, warrants consideration of a reintroduction program for the White-necked Crow in its former range.
Nesting: Lays 3–4 greenish blue eggs with dark markings in a crudely built stick nest in a tree or palm. Breeds February to June.
Range: Endemic to Hispaniola and associated islands; extirpated on Puerto Rico about 1963.
Local names: DR: Cuervo; **Haiti:** Kaw, Kònèy, Corneille d'Hispaniola.

SWALLOWS: FAMILY HIRUNDINIDAE

Swallows are a cosmopolitan family of aerially feeding birds, characterized by their short and broad bill, long, pointed wings, and broad, often forked tail. Swallows are typically gregarious and can often be seen in large numbers over fields and marshes or perched on wires. They drink by dipping into rivers, ponds, and other water bodies while flying. Voices are typically unmusical high twitters and buzzes. Food consists mainly of flying insects captured on the wing. Swallows superficially resemble swifts, but the two families are not closely related. Swifts are distinguished by their much longer wings, shallower wing strokes, and more rapid flight.

PURPLE MARTIN—*Progne subis* **Plate 40**
Vagrant

Description: 18–20 cm long; 49 g. A large swallow with a forked tail. Adult male is entirely glossy bluish black; adult female and immature have scaled pattern on grayish brown breast, light gray forecrown and collar, and an indistinct border between the darker breast and dingy whitish belly.
Similar species: Female Caribbean Martin has a brown wash on breast rather than a scaled pattern; male has a white belly.
Voice: Male's song is a rich, liquid gurgling, which includes a high *twick-twick*. Calls, similar to those of other martins, include high, melodious warbles and low whistles.
Hispaniola: Only 1 unequivocal record of an individual collected in the 19th century. Four sight records from the 1970s and 1980s from Lago Enriquillo, and from Islas Saona and Beata. Because this species migrates south primarily through mainland Middle America, it may be more likely to occur in western Hispaniola than in the eastern part of the island and should be sought in western Haiti.
Status: Vagrant.
Range: Breeds from sc. British Columbia east to c. Alberta and nw. Nova Scotia, south through U.S. to s. Baja California, c. Mexico, Gulf Coast and s. Florida; distribution west of 102 parallel is patchy and local. Occurs in non-breeding season in South America in lowlands east of Andes south to n. Argentina and s. Brazil. In West Indies, a regular transient in Bahamas, Cuba, and Cayman Islands; vagrant on Jamaica, DR, Puerto Rico, and Virgin Islands.
Local names: DR: Golondrina Migratoria Grande; **Haiti:** Irondèl Vyolèt.

CARIBBEAN MARTIN—*Progne dominicensis* **Plate 40**
Breeding Visitor

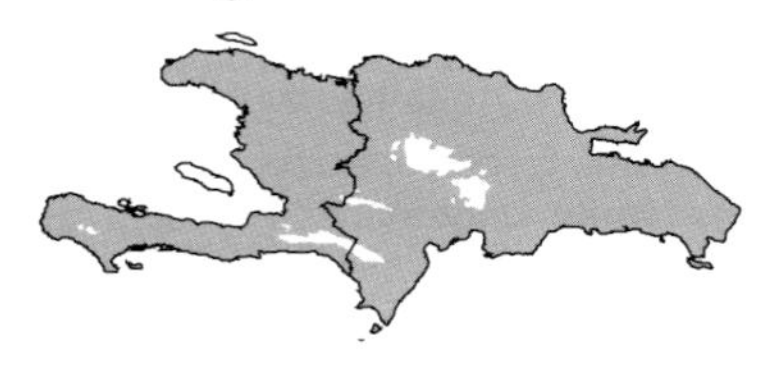

Description: 20 cm long; 40 g. A large, bicolored swallow with a forked tail. Male has upperparts, head, and throat metallic blue; belly is white with a dark band across vent. Female and immature are similar to adult male, but blue of underparts is replaced by a brownish wash that blends gradually into white of belly.
Similar species: Female Purple Martin has a scaled pattern on breast rather than a brownish wash; male Purple Martin is entirely bluish black.
Voice: Rich, liquid gurgling, which includes a high *twick-twick*. Also a melodious warble and a gritty *churr*. Vocalizations are similar to those of Purple Martin.
Hispaniola: Resident from sea level to a maximum known elevation of 1,500 m. Occurs along coastlines, in towns, open country, and in the mountains wherever suitable dead trees

with old Hispaniolan Woodpecker holes are found. Believed to migrate off the island by late September, returning in late January and early February. Never reported in large flocks. Recorded from several major satellite islands, including Île à Vache and Islas Beata and Saona.
Status: Fairly common breeding visitor from January to September.
Nesting: Lays 2–6 white eggs in a simple nest of twigs, leaves, and plant material placed in a cliff crevice, on a building, or typically in an abandoned woodpecker hole in a coconut palm. Breeds February to August.
Range: Restricted to Caribbean region, occurring on Jamaica and Hispaniola, then east and south through Lesser Antilles, Tobago, and Curaçao. Casual in Bahamas, Cayman Islands, and Cozumel Island. Non-breeding-season range is unknown but presumed to be in South America.
Local names: DR: Golondrina Grande; **Haiti:** Irondèl Nwa, Hirondelle à Ventre Blanc.

TREE SWALLOW—*Tachycineta bicolor* **Plate 40**
Passage Migrant; Non-breeding Visitor

Description: 13–15 cm long; 20 g. A medium-sized swallow with a moderately forked tail. Adult male and older females are iridescent greenish blue above with entirely white underparts. Yearling females and immatures have brownish upperparts, and underwing coverts are pale gray.
Similar species: Golden Swallow has a golden sheen above, more forked tail, longer wings, and more graceful flight.
Voice: Mostly silent in the West Indies, but sometimes gives a high, liquid twittering.
Hispaniola: DR records are from El Limón, Santo Domingo, Laguna de Rincón at Cabral, Laguna de Oviedo, Ebano Verde, El Valle, the Aceitillar area of Sierra de Bahoruco, Cabo Rojo, and Aguacate. The only known Haitian report is of approximately 20 birds at Savane Desolée near Gonaïves.
Status: Regular spring and fall passage migrant and non-breeding visitor in small numbers from late September to late April.
Range: Breeds from w. Alaska east to n. Saskatchewan, c. Labrador, and Newfoundland, south to s. California, ne. Texas, and n. Georgia. Occurs in non-breeding season from s. California, Gulf of Mexico, and Florida south to w. Panama and northern coast of South America. In West Indies, a regular non-breeding visitor in Bahamas, Cuba, Hispaniola, and Jamaica; rare on Puerto Rico and St. Croix.
Local names: DR: Golondrina de Árboles; **Haiti:** Irondèl Pye Bwa.

GOLDEN SWALLOW—*Tachycineta euchrysea* **Plate 40**
Breeding Resident—*Endangered*

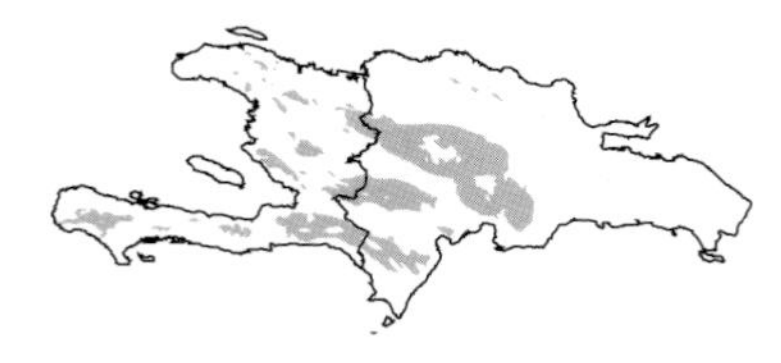

Description: 12.5 cm long; mass unavailable. A small swallow with a moderately forked tail. Adults have iridescent bluish green upperparts with a golden sheen and white underparts. Females are duller above than males, with a grayish wash on breast. Immatures are similar to females but duller above, with a gray breast band.
Similar species: Tree Swallow has a shallower tail notch, relatively shorter wings, and is less graceful in flight.
Voice: Soft twittering *chi-weet* during the breeding season.
Hispaniola: Recorded from sea level to a maximum elevation of 2,000 m, but usually found only above 750 m and only rarely recorded foraging at lower elevations. Occurs primarily over pine forest, occasionally over open fields and humid broadleaf forest. Reported from the

Cordillera Central, Sierra de Neiba, and Sierra de Bahoruco, DR, and in Haiti in the Massif de la Hotte and Massif de la Selle. Several reports support postbreeding dispersal of both adults and immatures from higher elevation breeding areas to lower elevations in late summer. Not recorded from any major satellite island.
Status: Uncommon and increasingly rare and local permanent resident on Hispaniola. Formerly common on Jamaica, but last reported with certainty there in 1989. This swallow's decline on Hispaniola appears to be the result of habitat destruction. Heavy cutting of pine forests in the mountains of Hispaniola, especially in Haiti, has caused this species to decline sharply since the mid-1970s. The subspecies *T. e. sclateri* is endemic to Hispaniola.
Comments: Graceful, often seen foraging over pine forests or perched in small groups in large, dead pine snags.
Nesting: Lays 3 white eggs in a nest in an old woodpecker hole or other cavity in a dead tree, or occasionally in the mouth of a cave. In 2002–2004 found to frequently nest in the earthen wall of abandoned open-pit bauxite mines in the Sierra de Bahoruco where natural tree cavities are thought to be lacking, but these nests are often depredated by mongoose. Sometimes nests in colonies at favorable locations. Breeds April to June.
Range: Formerly resident on Jamaica (where now probably extirpated) and Hispaniola.
Local names: DR: Golondrina Verde; **Haiti:** Oiseau de la Pluie, Jolle-Jolle, Irondèl Vèt, Hirondelle Verte.

NORTHERN ROUGH-WINGED SWALLOW—*Stelgidopteryx serripennis*
Non-breeding Visitor **Plate 40**

Description: 12–14 cm long; 16 g. A small, somewhat stocky swallow, entirely warm brown above, white below, with a pale brownish wash on chin, throat, and chest. Tail is square. Immature has cinnamon wing bars.
Similar species: Bank Swallow has a well-defined brown breast band and shallower and faster wing beats.
Voice: Repetitive series of low, coarse *prriit* or *brrrtt* notes, slightly rising in inflection.
Hispaniola: All known records are from the DR and include La Ciénaga, San Luis, Laguna de Reyes, Cabo Rojo, Puerto Plata, Lago Limón, San Pedro de Macorís, El Valle, Santo Domingo, and Las Salinas.
Status: Probably of regular occurrence from November to April in small numbers, but may be overlooked.
Comments: Often in the company of other swallows; has been recorded in flocks of 30–100 individuals.
Range: Breeds from sw. Alaska east to c. Saskatchewan and sw. Nova Scotia, south to Costa Rica, Gulf Coast, and sw. Florida. Occurs in non-breeding season from s. California, Gulf Coast, and s. Florida south to Panama. In West Indies, rare in Bahamas, Cuba, Jamaica, DR, Cayman Islands, and Virgin Islands; vagrant in Guadeloupe, St. Lucia, and Barbados.
Local names: DR: Golondrina Parda; **Haiti:** Irondèl Gòj Mawon.

BANK SWALLOW—*Riparia riparia* **Plate 40**
Passage Migrant

Description: 12–14 cm long; 15 g. A small, slender swallow with grayish brown upperparts and white underparts, with a diagnostic dark band across breast. Forecrown is pale, and tail is slightly notched. Immature has buffy wing bars.
Similar species: Northern Rough-winged Swallow has a pale brownish wash across breast

and a squared tail. Antillean Palm-Swift has a white rump and longer, narrower wings; its flight is more rapid and darting.

Voice: Rolling, gravelly or buzzy series of repeated, short notes, *chirr* or *bzrrrt*.

Hispaniola: Records are mostly from late April to early May, and from October, and include Laguna de Rincón at Cabral, DR, and Île à Vache and Isla Beata.

Status: Rare passage migrant from North America that is probably overlooked on occasion.

Comments: Typically found in association with other swallows; often seen perched on utility wires.

Range: Breeds from c. Alaska east to n. Manitoba and sw. Newfoundland, south to c. California, n. New Mexico, n. Alabama, and w. North Carolina; also s. Texas and ne. Mexico. In non-breeding season occurs primarily in South America south to n. Chile and n. Argentina, also along Pacific slope of s. Mexico. Widespread breeder in Eurasia. In West Indies, uncommon in Bahamas, Cuba, Puerto Rico, Cayman Islands, and Virgin Islands; rare in DR and in Lesser Antilles south to Martinique; vagrant south to Grenada.

Local names: DR: Golondrina de Collar; **Haiti:** Irondèl Kolye Senp.

CLIFF SWALLOW—*Petrochelidon pyrrhonota* **Plate 40**
Vagrant

Description: 12–15 cm long; 22 g. A stocky, pale-rumped swallow, distinguished by its dark reddish brown chin, throat, and auriculars, along with buff-colored forecrown, collar, and rump. Rest of upperparts are metallic bluish black; underparts are dull white. Chestnut throat patch often contains a dark patch of greenish or bluish black. Tail is short and squared. Immatures are variable but show a much paler throat and auriculars, and have duller and browner upperparts.

Similar species: Cave Swallow has a darker, more extensive patch on forecrown and a much paler buffy throat and auriculars.

Voice: Calls include a burry, nasal *vrrrt* and drier *chrri-chrri* or *chreh*, repeated while pursuing insects in flight.

Hispaniola: Two records: 1 at Cap Haïtien, Haiti, in April 1985, and 2 near Barahona, DR, in September 2000 in a migrating flock of swallows that included several Barn Swallows and Caribbean Martins.

Status: Vagrant. Almost certainly overlooked because of similarity to the common Cave Swallow. A southbound migrant from late August to early December, returning northward from late March to early May.

Range: Breeds from w. Alaska east to n. Manitoba, s. Quebec, and Nova Scotia, south to n. Baja California, c. Mexico, Tennessee, se. Pennsylvania, and ne. Massachusetts; rare and local breeder on Gulf Coast. Non-breeding-season range not well known but apparently extends from s. Brazil south to c. Argentina. In West Indies, vagrant or rare in Bahamas, Cuba, DR, Virgin Islands, and in Lesser Antilles south to St. Vincent.

Local names: DR: Golondrina de Farallón; **Haiti:** Irondèl Fwon Blanch.

CAVE SWALLOW—*Petrochelidon fulva* **Plate 40**
Breeding Resident

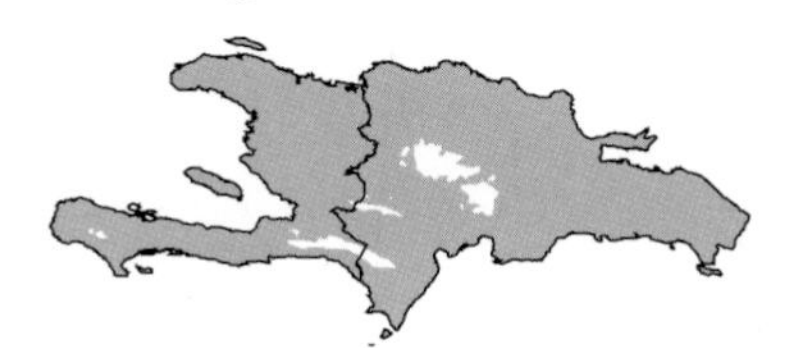

Description: 12–14 cm long; 20 g. A small, stocky, pale-rumped swallow. Identified by its dark rufous buff rump and forecrown and pale reddish brown auriculars, throat, breast, and sides. Tail is short and square to slightly notched.

Similar species: Cliff Swallow has a dark reddish brown throat and auriculars and a lighter forecrown.

Voice: Song is a mix of chattering or twittering notes and nasal buzzes. Common flight call is a soft, rising *twit* or *pwid*.

Hispaniola: Found nearly islandwide from sea level to a maximum known elevation of about 1,640 m. Occurs near cliffs along the coast; inland over fields, open areas, and wetlands; and occasionally in towns. Birds are present year-round, though some of the breeding population may migrate off the island from September to January, as in Cuba. Occasionally gathers in flocks of 100–500 birds. Recorded from most major satellite islands.
Status: Common breeding resident. The subspecies *P. f. fulva* is endemic to Hispaniola and associated islands.
Comments: Highly gregarious; typically seen in flocks, often perched on utility wires or foraging on the wing for flying insects.
Nesting: Lays 2–5 eggs that are white and finely spotted with brown. The nest is constructed primarily of mud, sometimes mixed with bat guano, in a cave mouth, on a building, in a culvert, or under a ledge or bridge. The nest cup is lined with plant materials. Breeds in colonies March to July.
Range: Patchy breeding distribution includes se. New Mexico east to coastal Texas, adjacent ne. and c. Mexico, extreme s. Mexico, s. Florida, Greater Antilles, and coastal lowlands of sw. Ecuador and se. Peru. In Greater Antilles, occurs from Cuba east to Puerto Rico; rare in Cayman and Virgin islands; vagrant in Lesser Antilles south to St. Vincent.
Local names: DR: Golondrina de Cuevas; **Haiti:** Irondèl Falèz, Hirondelle à Front Brun.

BARN SWALLOW—*Hirundo rustica* **Plate 40**

Passage Migrant; Non-breeding Visitor

Description: 15–19 cm long; 16 g. An elegant swallow, with long, slender, pointed wings and long, deeply forked tail. Adults have dark bluish black upperparts, tan to orangish underparts, and dark chestnut throat and forecrown. Tail has white spots on inner webs. Females average paler below with a shorter tail than males. Immature has throat and upper breast tan, remainder of underparts white, and tail less deeply forked.
Similar species: Deeply forked tail and uniformly tan or orangish underparts in all plumages distinguish this species from all other swallows.
Voice: Thin, short *chit* or *chit-chit*.
Hispaniola: Primarily inhabits coastal areas, but also found inland over agricultural lands and wetlands in the lowlands. The highest recorded elevation is one bird at 1,085 m in the Sierra de Bahoruco, DR. Peak of southbound migration appears to be August and early September; counts of 10–30 birds are common during this period. Numbers in the non-breeding season are variable, but flocks of 5–10 are regularly observed from December to March. Peak spring migration period is April. There are few records for May to July. Recorded from several major satellite islands, including Navassa Island, Île à Vache, and Islas Alto Velo, Beata, and Saona.
Status: Common spring and fall passage migrant and regular non-breeding visitor.
Comments: Typically seen in flocks zigzagging over fields, or perched on utility wires. Feeds on small flying insects.
Range: The most widely distributed and abundant swallow worldwide. In New World, breeds from se. Alaska east to n. Saskatchewan and s. Newfoundland, south to s. Mexico, Gulf Coast, and n. Florida. Occurs in non-breeding season from s. Mexico south to Tierra del Fuego. Primarily a transient throughout West Indies but occurs regularly in non-breeding season.
Local names: DR: Golondrina Cola de Tijera; **Haiti:** Irondèl Ke Long, Hirondelle Rustique.

KINGLETS: FAMILY REGULIDAE

Kinglets are a primarily Old World family of very small, delicately built insectivores, with a fine, slender bill and legs. They forage actively in trees and shrubs, often with short, hovering flights to glean food from beneath leaves. Large clutches of up to 12 eggs are laid in deep cup nests of moss, lichen, and plant fibers, usually built at mid- to upper levels in coniferous trees.

RUBY-CROWNED KINGLET—*Regulus calendula* **Plate 44**
Vagrant

Description: 9–11 cm long; 6 g. Distinguished by its tiny size, olive green to gray upperparts, dusky white underparts, bold white eye-ring, and two whitish wing bars. Male has a red crown patch which is usually concealed; female lacks crown patch.
Similar species: Vireos are considerably larger, more heavily built, and larger headed, with a stout bill that has a hooked tip.
Voice: Short, dry *chet* notes, often strung together in prolonged chatter.
Hispaniola: One record of a single bird observed at close range at San Francisco de Macorís, DR, in October 1976.
Status: Vagrant.
Comments: At other Caribbean sites, generally occurs in low, scrubby vegetation through which it actively hops and flits in search of insects, often with conspicuous flicking of the wings.
Range: Breeds from nw. Alaska east to n. Saskatchewan, Labrador, and Newfoundland, south to sc. California, s. New Mexico, n. Wisconsin, n. New York, and s. Vermont. Occurs in non-breeding season through much of nw. and c. U.S. south to Guatemala. In West Indies, rare in Bahamas; vagrant in Cuba, Jamaica, and DR.
Local names: DR: Reyezuelo; **Haiti:** Ti Kouwòn Wouj.

GNATCATCHERS: FAMILY SYLVIIDAE

Gnatcatchers are very small, arboreal birds with a long, graduated tail and generally grayish plumage above and whitish plumage below. The tail is often cocked upward and swung from side to side. Vocalizations typically consist of a series of thin, high-pitched nasal notes. Nests are neat cups of plant fibers, lichens, and bark strips, at low to middle levels in trees or shrubs.

BLUE-GRAY GNATCATCHER—*Polioptila caerulea* **Plate 44**
Vagrant

Description: 10–12 cm long; 6 g. A tiny, active bird with a long, thin tail with white outer feathers, often cocked upward or fanned. Upperparts are bluish gray, underparts are white. Note prominent white eye-ring. Bill is fine tipped and narrow. Sexes are similar in non-breeding season, although females are slightly paler above. In breeding plumage, male distinguished by fine black eyebrow stripe.
Similar species: Smaller, slimmer, and longer-tailed than warblers.
Voice: Often detected by its voice, a mewing and nasal call of 2–6 syllables, *zee-zeet*. Thin, soft, complex song of mews, chips, and whistles, seldom given during non-breeding season.
Hispaniola: One record of an individual seen in broadleaf montane forest in Ebano Verde Reserve, Cordillera Central, DR, in January 1987 at an elevation of 2,745 m. Occurs in scrub-

lands on other Caribbean islands, varied lowland and midelevation habitats from forests to gardens in Cuba, and in mangroves in Cayman Islands.
Status: Vagrant.
Range: Breeds from s. Oregon east to sw. Wyoming, c. Minnesota, sw. Quebec, and s. Maine, south through Baja California, s. Mexico, Belize, Gulf Coast, and s. Florida. Occurs in non-breeding season through much of southern portion of breeding range south to Honduras and Cuba. In West Indies, common year-round resident on larger islands of Bahamas, common non-breeding visitor on Cuba; uncommon non-breeding visitor to Cayman Islands; vagrant to DR.
Local names: DR: Rabuita; **Haiti:** Ti Chwichwi, Gobemoucheron Gris-bleu.

THRUSHES: FAMILY TURDIDAE

The thrushes are a cosmopolitan family, reaching their greatest diversity in temperate regions. They are characterized by a slender bill, fairly long legs, and an erect posture. The wings are relatively long and pointed, and the tail is moderately long and squared. Most feed on insects and fruits, and often forage on the ground. Many thrushes are exceptional singers. Nests are open cups of plant materials, often twigs and grasses, sometimes moss, in trees or shrubs, on the ground, or in cavities. Eggs vary from solid bluish to whitish and are mottled brownish in some species.

RUFOUS-THROATED SOLITAIRE—*Myadestes genibarbis* **Plate 41**
Breeding Resident

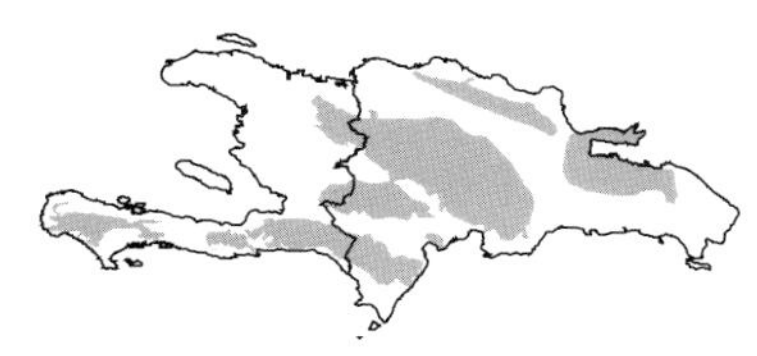

Description: 19 cm long; 27 g. A slender, arboreal thrush, mostly gray above with a white chin, reddish brown throat and undertail coverts, and rest of underparts pale gray. Also note a white crescent below eye. Tail is fairly long with white outer feathers which become visible when fanned in flight. Feet are yellow.
Similar species: Unmistakable.
Voice: One of the most beautiful songs on Hispaniola. A haunting, minor-key whistle of different tones, typically ending in a higher-pitched trill. Most often heard at dawn and dusk.
Hispaniola: Resident in foothills and interior mountains from sea level to 1,800 m elevation, but seldom reported below 400 m. Most common in moist broadleaf forest from 750 to 1,500 m, usually not found in pure pine forest or in shade coffee plantations except when associated with humid riverside habitat. In the DR, reported from Sierra de Bahoruco, Sierra de Neiba, Cordillera Central, Cordillera Septentrional, the Samaná Peninsula, and Los Haitises National Park area, and the Cordillera Oriental. In Haiti, recorded in the Massif de la Hotte, Massif des Montagnes Noires, and the mountains near Dondon in northeastern Haiti, but is now uncommon in La Visite National Park of the Massif de la Selle where forest loss is high. There is some evidence of birds moving to lower elevations in the non-breeding season. Reported to have formerly occurred on Île de la Gonâve but apparently extirpated there before 1928.
Status: Locally common resident. This species has declined substantially islandwide because of destruction of montane forest habitat since 1930, particularly in Haiti. The subspecies *M. g. montanus* is endemic to Hispaniola.
Nesting: Lays 2–3 eggs which are bluish white or blue with white spots. Nest is cup shaped, often mossy, and placed in a crevice, tree fern, or bromeliad, sometimes on a horizontal branch. Seven nests in montane broadleaf forests of Sierra de Bahoruco were 2–7 m above ground. Breeds April to August.
Range: Resident on Jamaica, Hispaniola, Dominica, Martinique, St. Lucia, and St. Vincent.
Local names: DR: Jilguero; **Haiti:** Mizisyen, Oiseau Musicien, Solitaire Siffleur.

VEERY—*Catharus fuscescens* **Plate 42**
Vagrant

Description: 17–19 cm long; 31 g. A medium-sized thrush, generally with uniformly reddish brown upperparts, but may rarely be olive-brown. Throat and chest buffy with indistinct brownish spots; rest of underparts white. Has inconspicuous pale eye-ring.
Similar species: Warmer reddish brown above and more indistinctly spotted below than other *Catharus* thrushes.
Voice: Low, nasal *pheu* or *veer*, rarely heard in migration.
Hispaniola: Three records from the DR: recorded twice at Ángel Félix in Sierra de Neiba, and 1 collected near Bonao.
Status: Vagrant.
Comments: A retiring ground dweller. May be expected in forest understory, especially in second-growth woodlands. Feeds on insects and fruits.
Range: Breeds from interior s. British Columbia east to s. Manitoba and sw. Newfoundland, south to e. Oregon, s. Colorado, n. Indiana, w. Maryland, n. New Jersey, and in Appalachian Mountains to nw. Georgia. Occurs in non-breeding season from n. South America south to s. Brazil. In West Indies, rare in Bahamas, Cuba, and Jamaica; vagrant in DR and Cayman and Virgin islands.
Local names: DR: Zorzal Migratorio Colorado; **Haiti:** Griv Pal, Grive Fauve.

BICKNELL'S THRUSH—*Catharus bicknelli* **Plate 42**
Non-breeding Visitor—*Threatened*

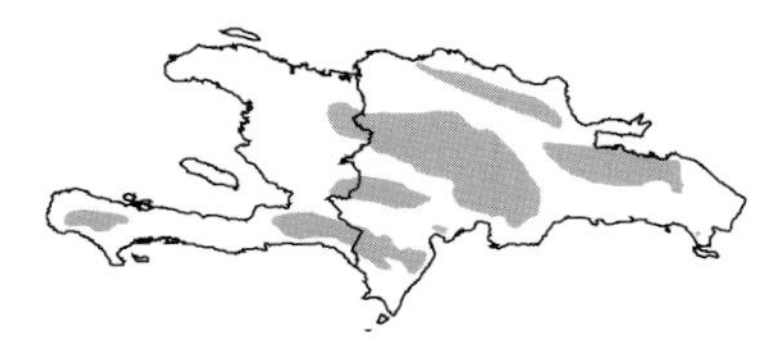

Description: 16–17 cm long; 28 g. A smallish, slender *Catharus* thrush, with olive-brown upperparts and contrasting chestnut-tinged tail, whitish underparts, and sides of throat and breast creamy buff prominently spotted with black. Also distinguished by its grayish auriculars and lores.
Similar species: Although not recorded on Hispaniola, the nearly identical Gray-cheeked Thrush is slightly larger and more olive-gray above with darker lores, a whiter breast, and less yellow on lower mandible; lacks chestnut in tail. Veery is more finely spotted below and more reddish brown above.
Voice: Generally silent in the West Indies except for brief periods of dawn and dusk calling. Calls vary in intensity and pitch, typically a downward slurred whistle, *beer* or *pee-irt*, occasionally a softer *chook* or *chuck*. Song is a high-pitched *chook-chook, wee-o, wee-o, wee-o-tee-t-ter-ee*, slurring downward, infrequently given in subdued form in non-breeding season.
Hispaniola: Prefers dense understory of moist broadleaf forest, or broadleaf forest mixed with relatively few pines. Surveys in the 1990s in the DR found this species in wet and mesic broadleaf forests (75%), mixed pine/broadleaf forests (19%), and pine-dominated forests (6%). Birds were found at all elevations from sea level to 2,200 m, but the majority (62%) of occupied sites were in primary montane forests greater than 1,000 m in elevation. In migration it has been recorded from such lowland DR localities as Puerto Plata and Santo Domingo. In intensive surveys in the 1990s and early 2000s, found at high elevations in Sierras de Bahoruco, Neiba, and Martín García, and Cordilleras Central, Septentrional, and Oriental, as well as Los Haitises National Park and scattered locations in Del Este National Park. Confirmed in February 2004 in Haiti's Massif de la Hotte, where found in wet karst limestone forest at 1,175–1,250 m elevation and wet montane forest with scattered emergent pines at 1,825–1,915 m. Also found in broadleaf forest fragments at 1,575–2,025 m elevation in the Massif de la Selle in February 2005. General arrival in fall is thought to be late October and early November. Peak northbound migration period in spring is probably early to mid-April.
Status: Uncommon to rare, locally distributed non-breeding visitor, seldom observed because of its retiring habits. Most abundant in the Sierra de Bahoruco and Cordillera Central of

the DR; current status in Haiti incompletely known, but probably restricted to remnant patches of mid- to high-elevation forest.
Comments: Typically shy and wary. Searches for invertebrates and fruits on the ground or subcanopy. Formerly considered a subspecies of Gray-cheeked Thrush, from which it can be distinguished mainly by differences in song, or in-hand measurements. Unfortunately, these characters are of little use in the West Indies. There is evidence of habitat segregation by sex, with males predominating in undisturbed montane forests and females and yearlings in younger, more disturbed forest.
Range: Breeds in s. Quebec, maritime provinces except Newfoundland, e. New York, Vermont, c. and n. New Hampshire, and w. and c. Maine. Non-breeding-season range restricted to Greater Antilles, including Cuba, Hispaniola, Jamaica, and Puerto Rico, although most individuals occur in DR.
Local names: DR: Zorzal de Bicknell; **Haiti:** Griv Biknel, Grive de Bicknell.

WOOD THRUSH—*Hylocichla mustelina* **Plate 42**
Vagrant

Description: 19–21 cm long; 45 g. Largest of the spotted thrushes, distinguished by its bright rufous brown crown and nape, slightly duller on back, wings, and tail; conspicuous white eye-ring; and white underparts with heavy blackish spots on breast, sides, and flanks.
Similar species: No other thrush is as heavily spotted below or as distinctly reddish brown on upperparts. Ovenbird is smaller, and its orangish crown is bordered by black stripes.
Voice: Emits short, rapid, staccato *pit-pit-pit* notes; also a lower, clucking *bup-bup* or *tut-tut*. During migration, rarely sings its clear, flutelike song of 3–5 syllables, ending with a trill.
Hispaniola: Two DR records of birds mist netted: 1 near Las Mercedes, Pedernales Province, in November 1996 at about 330 m elevation in dry broadleaf forest; and 1 at Punta Cana in January 2004 in dry forest, suggesting the possibility of non-breeding-season residency.
Status: Vagrant.
Comments: Forages on the forest floor for insects and fruits. Generally inhabits understory of both primary broadleaf and older second-growth forest in non-breeding season.
Range: Breeds from s. North Dakota east to n. Minnesota, s. Quebec, and sw. Nova Scotia, south to ec. Texas, Gulf Coast, and n. Florida. Occurs in non-breeding season from se. Mexico south along Atlantic slope to Panama and nw. Colombia. In West Indies, rare in Cuba, vagrant to Bahamas, Cayman Islands, Jamaica, DR, and Puerto Rico.
Local names: DR: Zorzal Migratorio Pecoso; **Haiti:** Griv Bwa, Grive des Bois.

AMERICAN ROBIN—*Turdus migratorius* **Plate 42**
Vagrant

Description: 23–26 cm; 77 g. A large, conspicuous thrush, with slaty gray upperparts and dull orangish red underparts. Throat is white streaked with black, and undertail coverts are white. Bill is yellow. Males have a blackish head and tail; females are paler overall.
Similar species: Red-legged Thrush has a bright reddish bill and legs and gray breast. La Selle Thrush has darker slaty head and upperparts, dark gray breast, and white streaks on belly.
Voice: Staccato, clucking *puk-puk-puk*; also a low, mellow *pup*. In flight, a thin, descending *see-lip*. Song is a series of short, happy, warbled phrases, followed by pauses, *cheer-up, cheer-ey*.
Hispaniola: One record of a bird observed at close range by an experienced observer at Puerto Plata, DR, in November 1985.
Status: Vagrant.
Comments: Forages primarily on the ground for invertebrates and fruits. Favors open habitats.
Range: Breeds from nw. Alaska, n. Manitoba, n. Quebec, and Newfoundland south to n. Baja California, interior Mexico, Gulf Coast, and c. Florida. Occurs in non-breeding season from

s. Alaska, extreme s. Canada, and most of U.S. south through n. Mexico. In West Indies, rare and irregular in Bahamas and Cuba; vagrant in Jamaica, DR, and Puerto Rico.
Local names: DR: Zorzal Migratorio; **Haiti:** Kwèt Kwèt Etranje.

LA SELLE THRUSH—*Turdus swalesi* **Plate 41**
Breeding Resident—*Endangered Endemic*

Description: 26 cm long; 97 g. A large but secretive thrush, with a slaty black head and upperparts, slaty gray upper breast, rich rufous lower breast and sides, and white belly and undertail coverts. Auriculars have a silvery cast, and throat is streaked white. Eye-ring is reddish orange; bill is yellow. Legs are dark.
Similar species: Red-legged Thrush is paler above, lacks red below, has bright red legs and bill, and white on tail. American Robin has reddish underparts that extend to undertail coverts, and prefers more open habitats. Chat-tanagers are browner above and have white underparts.
Voice: Series of deliberate *tu-re-oo* and *cho-ho-cho* calls, often given at dawn and dusk. Also a loud *wheury-wheury-wheury* alarm call. Song is a quiet, mellow mix of gurgling notes.
Hispaniola: First discovered on Massif de la Selle, Haiti, in April 1927 and known only from Haiti until May 1971 when observed in Sierra de Bahoruco on Loma Pie de Palo. First discovered north of the Neiba Valley/Cul de Sac Plain in February 1975 in Sierra de Neiba, and the following May was collected in Cordillera Central near Constanza, DR. Confined to dense undergrowth in moist montane forest, broadleaf forest, and in pine forest with a dense broadleaf understory; often found where there are thick stands of climbing bamboo. Known only within the elevation range of 1,400–2,100 m.
Status: Uncommon and very local breeding resident in the Massif de la Selle of Haiti, where it is restricted to very small montane broadleaf forest fragments, and absent from Massif de la Hotte. Rare and local in the DR where it occurs in Sierra de Bahoruco (Loma de Toro, Zapotén, Pueblo Viejo, and Pie Pol), Sierra de Neiba, and Cordillera Central. Because of this species' retiring habits and preference for dense understory vegetatation, its abundance is difficult to gauge, although in prime habitat several may be encountered in a day. Steady encroachment and loss of the high-elevation forests inhabited by La Selle Thrush have increasingly endangered the species throughout its range. Two subspecies have been described: *T. s. swalesi* of southern Haiti and the Sierra de Bahoruco, and *T. s. dodae* with a more olivaceous brown (rather than black) back in Sierra de Neiba and Cordillera Central, DR.
Comments: Largely terrestrial. Forages for earthworms, insects, and fruits, mostly on the ground where it runs in spurts and then abruptly pauses. Sings from exposed perches at dawn and dusk, but quiet and inconspicuous during midday.
Nesting: Breeding biology is poorly known, but 2–3 greenish blue eggs with spots are laid in a bulky nest of mostly moss and a few twigs that is sparsely lined with a thin layer of mud and dried grass stems. Nests are placed up to 10 m above ground in a shrub or tree. Breeds May to July.
Range: Endemic to Hispaniola.
Local names: DR: Zorzal de La Selle, Cho-chó; **Haiti:** Kwèt-kwèt Lasèl, Kouèt-kouèt Nwa, Merle.

RED-LEGGED THRUSH—*Turdus plumbeus* **Plate 42**
Breeding Resident

Description: 25–28 cm long; 74 g. A large thrush, distinguished by its slaty gray upperparts, reddish legs and bill, red eye-ring, and conspicuous white tail tips. Breast is gray, throat is white with black stripes, and belly and undertail coverts are whitish. Juvenile is duller overall with a gray throat spotted with black.

Similar species: American Robin and chat-tanagers lack the red bill and legs. La Selle Thrush is darker above, has red breast and sides, and lacks red legs and white tail tips.
Voice: Call notes include a low *wéecha* and a rapid, high-pitched *chu-wéek, chu-wéek, chu-wéek*. Alarm call is a loud *chuá-chuá*, from which species takes its local name. Song is a melodious but monotonous series of 1- to 3-syllable phrases, often uttered in pairs, with a distinct pause between each note, *chirruit*-(pause)-*chirruit* or *pert*-(pause)-*squeer*. Song is similar to that of Pearly-eyed Thrasher but more musical, with shorter pauses between phrases.
Hispaniola: Occurs in a wide variety of habitats from xeric and second-growth woodland to humid forest, from sea level to at least 2,440 m elevation. Most numerous in moist broadleaf forest, where it can occasionally be abundant. Occasional to common in pine forest, dry forest, and shade coffee plantations, and rare in semidesert scrub. Also recorded on Île de la Gonâve, Île de la Tortue, and Isla Saona.
Status: Fairly common resident throughout most of its range. Its general abundance does not seem to have changed markedly since the 1920s and 1930s, though the overall reduction of available habitat has probably caused some decrease of numbers.
Comments: Often seen at dawn as it darts from roadsides or forest openings. Forages mainly on the ground for invertebrates among leaf litter. Also eats fruits and lizards. A conspicuous and aggressive bird during the breeding season, but secretive at other times.
Nesting: Eggs are pale greenish blue, heavily speckled with brown. Lays 2–4 eggs in a nest which is a bulky mass of leaves lined with grass or banana fibers, and often some mud, usually 3–9 m above ground in a tree, palm, or stump. Breeds January to September with a peak from April to June.
Range: Caribbean species. Breeds in Bahamas, Cuba, Cayman Islands, Hispaniola, Puerto Rico, Dominica, and formerly Swan Islands.
Local names: DR: Chua-chuá, Flautero, Calandria; **Haiti:** Kwèt-kwèt, Ouèt-ouèt.

MOCKINGBIRDS, THRASHERS, AND ALLIES (MIMIC THRUSHES): FAMILY MIMIDAE

Mimids are noted for their loud, varied songs and ability to mimic other species. Appearing thrushlike, they are typically more slender with a longer, graduated tail and longer, decurved bill. Mostly plumaged in browns or grays, mimids are typically solitary birds, often skulking and difficult to observe, although most sing from exposed perches. Songs feature repeated phrases, sometimes with accomplished mimicry, whereas calls include whistles, mews, and "chucks." Flight is generally low and rapid. Food is mostly insects and fruit. Nests are bulky cups built at low to middle levels in bushes.

GRAY CATBIRD—*Dumetella carolinensis* **Plate 42**
Non-breeding Visitor

Description: 21–24 cm long; 37 g. A small mimid, almost entirely dark gray with a black cap, reddish brown undertail coverts, and a long tail that is often cocked slightly upward.
Similar species: Rufous-throated Solitaire has reddish brown throat and white outer tail feathers.
Voice: Distinctive, hoarse, catlike *meew*; also a soft,

low-pitched *quirt* or *turrr*. Alarm note is a loud, harsh chatter, *chek-chek-chek*. Song is a rambling series of disconnected phrases, including mews, squeaks, gurgles, imitations, and pauses.
Hispaniola: Found in thickets and dense undergrowth to a maximum elevation of 1,370 m. All known records are from Terrier Rouge and Fermate, Haiti; Del Este National Park, Puerto Plata, Sosúa, Manabao, Mencia, and Las Mercedes, DR. Also recorded from Île de la Tortue.
Status: Rare non-breeding visitor, occurring from late October to late May.
Comments: Moderately shy, more often heard than seen. Generally forages near or on the ground for fruits and insects.
Range: Breeds from s. British Columbia east to s. Manitoba, s. Quebec, and Nova Scotia, south through e. Washington, nc. Utah, n. New Mexico, Gulf states, and n. Florida. Occurs in non-breeding season from east coast of U.S. through Florida and Gulf Coast, along Atlantic slope of Mexico south to c. Panama. In West Indies, found in non-breeding period in Bahamas, Cuba, Jamaica, and Cayman Islands; casual on Hispaniola and east to Virgin Islands.
Local names: DR: Zorzal Gato; **Haiti:** Zwazo Chat, Moqueur Chat.

NORTHERN MOCKINGBIRD—*Mimus polyglottos* **Plate 42**
Breeding Resident

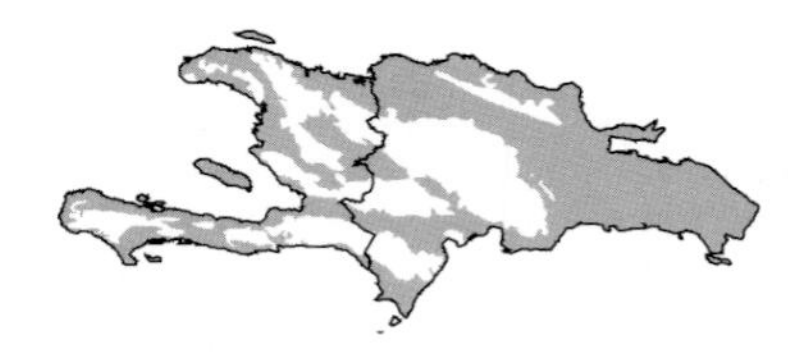

Description: 21–26 cm long; 48 g. An aggressive, conspicuous bird, pale gray or gray-brown above and grayish white below, distinguished by wings and tail conspicuously marked with white, showing clearly in flight. Long tail is often cocked upward. Thin dark line extends through yellowish eye. Juvenile has brownish gray upperparts and buffy white underparts with faint brownish breast spots.
Similar species: Gray Kingbird lacks white wing and tail patches and does not cock its much shorter tail.
Voice: Clear, melodious series of varied phrases, each phrase repeated 2–6 times. Alarm call is a loud, explosive *tchak* or *chat*. Aggressive call is a high, raspy *hew* or *skeeh*. An expert mimic, it often incorporates calls and song phrases of other species in its diverse repertoire. Sings at night as well as during the day.
Hispaniola: Found at low elevations islandwide in nearly every habitat type except dense forest, including open country, semiarid scrub, mangrove edges, gardens, and parks. Seldom found above 500 m elevation. Recorded from most major satellite islands.
Status: Common and conspicuous resident. Appears to have been common at least since the arrival of European settlers in the 16th century, and subsequent reduction of the original forests may have allowed its overall numbers to increase.
Nesting: Lays 2–4 pale bluish green eggs with heavy splotches at the larger end. Nest is a coarse open cup of often spiny twigs lined with finer plant materials and built in a bush, cactus, or tree. Of 46 nests monitored in lowland desert scrub at Cabo Rojo, DR, mean clutch size was 2.9, and mean height above ground was 1.6 m (range 0.5–2.5). Of 43 nests with known outcomes, 40 failed because of depredation. Breeds January to July, with a peak from April to July.
Range: Nonmigratory resident from extreme s. Canada through U.S., south to s. Mexico and Greater Antilles, where it occurs in Bahamas, Cuba, Cayman Islands, Jamaica, Hispaniola, and Puerto Rico east to Anegada.
Local names: DR: Ruiseñor; **Haiti:** Rosinyòl, Rossignol, Moqueur Polyglotte.

PEARLY-EYED THRASHER—*Margarops fuscatus* **Plate 42**
Breeding Resident

Description: 28–30 cm long; 75 g. A large mimid with brown upperparts and dull white underparts that are streaked with brown. Iris is distinctly white, bill is large and yellowish, and tail has large, white tips.

Similar species: Eastern and Western chat-tanagers are uniformly white below with a dark eye, and lacking white tips on tail. Ranges do not overlap.
Voice: Series of 1- to 3-syllabled phrases, such as *pío-tareeu-tsee*, with fairly lengthy pauses separating each. Often sings well into the day and during clear nights. Also many raucous call notes, including a guttural *craw-craw* and a harsh *chook-chook*.
Hispaniola: On Hispaniola and satellite islands, occurs in thickets, woodlands, and coastal forests at low elevations. First discovered to be common on Isla Beata in 1931; later also found on nearby Isla Alto Velo. Occasional visits to Isla Beata in the 1960s and 1970s documented the species' continued abundance there. The first definitive record from the main island was of an adult mist netted in thorn scrub habitat near Guaraguao, Del Este National Park, DR, in February 1984. This bird may have strayed from Mona Island, only 60 km east of mainland DR, where the species is common. In 1999 a resident population was discovered at nearby Punta Cana, DR, and in secondary scrub forest for 2–3 km in all directions from there. This appears to be a recent colonization, as thrashers were not found in the area when it was explored by ornithologists from 1974 to 1976; colonization of the eastern mainland appears to be underway.
Status: Locally common resident on satellite islands adjacent to Hispaniola and in a small portion of the extreme eastern mainland.
Comments: Arboreal. Normally feeds on fruits, invertebrates, and small vertebrates, but this extraordinarily aggressive species also eats the eggs and young of other birds. It competes with other birds for cavity nest sites.
Nesting: Lays 2–3 glossy, deep blue eggs in a bulky twig nest in a cavity, or rarely in a tree or bush. Nesting period rangewide is December to September, but no specific nesting data are available for Isla Beata, Isla Alto Velo, or Hispaniola.
Range: Resident in s. Bahamas, a small portion of mainland Hispaniola and satellite islands, Puerto Rico, Virgin Islands, Lesser Antilles south to St. Lucia, and Bonaire and Horquilla islands off n. Venezuela; vagrant on Barbados and Jamaica.
Local names: DR: Zorzal Pardo; **Haiti:** Zwazo-kowosòl, Moqueur Corossol.

PIPITS: FAMILY MOTACILLIDAE

Motacillids are characterized by a slender, pointed bill; short neck; slender body; and long legs, feet, and tail. They walk or run rather than hop, and they spend most of their time on the ground. All have a strong, undulating flight. Most are rather gregarious except in the breeding season. Motacillids occur worldwide except in the extreme north and some oceanic islands. Only one species reaches the West Indies.

AMERICAN PIPIT—*Anthus rubescens* **Plate 42**
Vagrant

Description: 16.5 cm long; 21 g. A thin-billed, long-tailed terrestrial bird with a habit of regularly bobbing its tail while walking. Upon taking flight, it displays conspicuous white outer tail feathers. In non-breeding adult, note buffy supercilium, two faint wing bars, and underparts that vary from pinkish buff to pale gray, with blackish stripes concentrated on breast.
Similar species: Distinguished from sparrows by more slender bill and body, longer legs, and longer tail.
Voice: Flight call is a high, distinctive *sip-it* or *sip*.
Hispaniola: Expected to occur in open fields and sandy areas. One record of a single bird seen at Trou Caïman, Haiti, in December 2002.
Status: Vagrant.
Range: Breeds from Alaska east to Baffin Island and Quebec, south in mountains from British

Columbia to California and Arizona, and from w. Montana east to n. Ontario and n. Maine. Winters south to Guatemala. In West Indies, a vagrant in Bahamas, Hispaniola, Jamaica, Swan, Providencia, and San Andres islands.
Local names: DR: Pipit Americano; **Haiti:** Pipit, Pipit d'Amerique.

WAXWINGS: FAMILY BOMBYCILLIDAE

This small family of gregarious, northern forest birds consists of only three species. All have sleek plumage and a conspicuous crest and possess bright red, waxy structures on the tips of their secondaries. The bill is short and stubby, and the legs are short. Wings are fairly long and pointed, and the tail is squared. Waxwings are arboreal and highly frugivorous, feeding primarily on berries, although they sometimes capture flying insects near streams or ponds by sallying from exposed perches. Only one species reaches the West Indies.

CEDAR WAXWING—*Bombycilla cedrorum* **Plate 42**
Non-breeding Visitor

Description: 15–18 cm long; 32 g. A sleek, round-bodied bird with overall grayish brown plumage. Crown and back are warm cinnamon brown; wings, rump, and uppertail coverts are plain gray. Underparts are tan, fading to pale yellowish on belly; undertail coverts are whitish. Note the sharp black facial mask edged with white, black chin patch, conspicuous back-pointed crest, and yellow-tipped (occasionally orange) tail. Secondaries may have variable numbers of red, waxlike tips. Juvenile is grayer overall, with broad streaking on underparts.
Similar species: Unmistakable.
Voice: Thin, high-pitched, slightly-trilled *sreee*.
Hispaniola: All records are from the highlands, including 3 collected at Piedra Blanca near Bonao, DR; a small, compact flock seen on several occasions on Morne La Visite; recorded at Macaya Biosphere Reserve and on Plateau Rochelrois above Miragoâne, Haiti; a flock of about 14 seen on Loma de Toro, Sierra de Bahoruco, DR, in December 1982; and a flock of 13 seen repeatedly in February 2005 near Mencia in the Sierra de Bahoruco.
Status: Rare non-breeding visitor, only known from December to late February. May be more regular than has been documented, particularly in seldom-visited higher elevations and especially in the western part of the island.
Comments: Typically occurs in flocks, often visiting trees or shrubs with ripe berries.
Range: Breeds from se. Alaska east to n. Saskatchewan, n. Ontario, c. Quebec, and Newfoundland, south to n. California, s. Colorado, nw. Arkansas, n. Alabama, and nw. South Carolina. Non-breeding range variable from year to year but includes extreme s. Canada south through U.S. to Costa Rica, and rarely Panama. In West Indies, occurs irregularly in non-breeding season in Bahamas, Greater Antilles, and Cayman Islands; vagrant to Guadeloupe and Dominica.
Local names: DR: Cigua Alas de Cera; **Haiti:** Zwazo-maske, Jaseur des Cèdres.

PALMCHAT: FAMILY DULIDAE

The Dulidae is a monotypic family, consisting only of the Palmchat of Hispaniola. Its taxonomy is disputed, but the family appears to be more closely related to the waxwings (Bombycillidae) than to other passerines. Palmchats are arboreal, noisy, and gregarious, often building large, communal nests.

PALMCHAT—*Dulus dominicus* **Plate 43**
Breeding Resident—*Endemic*

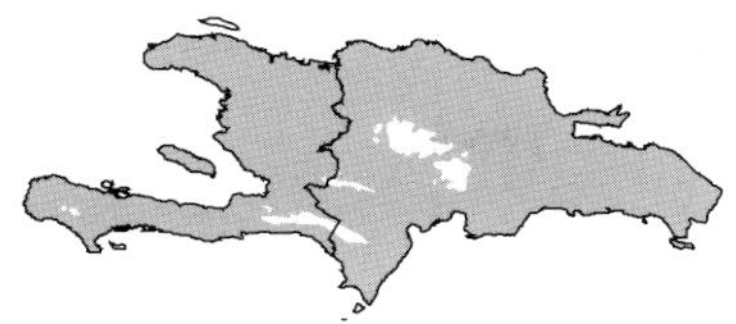

Description: 20 cm long; 42 g. A conspicuous, flocking bird, most often seen flying to and from treetops in semiopen country. Upperparts are dark brown; underparts are creamy white and heavily streaked with brown. Tail is medium long and squared. Bill is yellow, and eye is reddish.
Similar species: Superficially similar to Pearly-eyed Thrasher, but the thrasher has a very restricted distribution, is larger with a much longer, white-tipped tail, and has a white eye.
Voice: Quite noisy, particularly around its nest, producing an array of strange, slurring, whistled call notes.
Hispaniola: Resident islandwide in nearly all habitat types in the lowlands and midelevations, including deciduous and lowland evergreen forest and secondary woodland, especially where there are royal palms. Highest elevation reported is 1,825 m, though rarely found above 1,500 m. Also found on Isla Saona and Île de la Gonâve.
Status: Common and conspicuous breeding resident, but there is some evidence of fluctuations in numbers over time. Many fewer nest colonies were found in appropriate habitat in DR in March–April 1996 than were found in the same areas in 1974–1976. The form *D. d. dominicus* occurs on the main island and on Isla Saona, where it was first discovered in 1964. The form *D. d. oviedo* is confined to Île de la Gonâve, although its validity as a distinct subspecies has been questioned. The precise taxonomic position of *Dulus* has been the subject of much debate.
Comments: The national bird of the DR, the Palmchat is the sole member of its genus and family, making it unique from an evolutionary and taxonomic standpoint. It feeds primarily on fruits and berries. The Dulidae and the Todidae (todies) are the only bird families endemic to the West Indies.
Nesting: Eggs are white with many dark purplish gray spots, particularly at the large end. Lays 2–4 eggs in a large nest structure up to 2 m wide made of twigs. These large nests are built and used communally by many pairs, with each nest chamber having a separate entrance. Nests are most often built in royal palms, but have been recorded in a variety of other trees including yagrumo (*Cecropia schreberiana*) and Hispaniolan pine (*Pinus occidentalis*), as well as utility poles, and exceptionally in a crevice on a rock at sea. Reportedly an occasional host of the brood parasite Shiny Cowbird. Breeds March to June.
Range: Endemic to Hispaniola, including Île de la Gonâve and Isla Saona.
Local names: DR: Cigua Palmera; **Haiti:** Zwazo-palmis, Esclave Palmiste.

WOOD-WARBLERS: FAMILY PARULIDAE

Wood-warblers are small, active, insectivorous birds with a thin, pointed bill. Plumage is varied: in general, canopy-dwelling species are brightly colored, whereas ground-dwelling species tend to be drab. Non-breeding plumages of many species are considerably drabber than the breeding plumages that are acquired in late winter, so field identification on Hispaniola can be challenging. Most North American species are sexually dimporphic in all plumages. Many warblers join mixed-species flocks in the non-breeding season, whereas others remain solitary. Many defend territories throughout the non-breeding period, and they return to these same territories year after year. Most warblers forage actively in foliage, gleaning prey from leaf surfaces or the ground or sallying after flying insects. But many warblers are opportunistic foragers during the non-breeding period and frequently consume fruit or nectar. Songs of most North American migrants are rarely if ever heard during the non-breeding season. Calls include varied *chip* notes often used in territorial interactions, and many of these can be readily identified. High-pitched contact or *tsit* notes often given in flight may also be heard, but these are more difficult to identify to species.

BLUE-WINGED WARBLER—*Vermivora pinus* **Plate 49**
Non-breeding Visitor

Description: 11–12 cm long; 8 g. In non-breeding plumage, bright yellow on crown and underparts, with a bold black eye-line, greenish yellow back and nape, bluish gray wings and tail, and two distinct white wing bars. Female is slightly duller overall than male, with a more olive crown and grayer eye-line; wing bars are often less distinct. Breeding plumage is similar but brighter.
Similar species: Prothonotary Warbler lacks white wing bars and black eye-line. Pine Warbler is larger and less boldly marked, lacks eye-line, and has duller wing bars and underparts. Yellow Warbler lacks dark eye-line and has olivaceous wings and yellow tail spots.
Voice: Call notes include a sharp, dry *chik* or *jeet*; also a short, high, and slightly buzzy *tzip* or *dzit*. Primary song is a somewhat harsh, buzzy *beeee-BZZZZ*.
Hispaniola: Nearly always found in highland regions. All known records are from late October to early May and include single birds in the DR from Hatillo, Hato Mayor, Sosúa, Jarabacoa, Constanza, Medina, Pueblo Viejo, Las Mercedes, and Zapotén. The only Haitian record is from La Visite National Park headquarters.
Status: Rare non-breeding visitor.
Comments: "Brewster's Warbler," the hybrid offspring of Golden-winged and Blue-winged warblers, has been recorded twice on Hispaniola, including once at Manabao and once in the eastern Sierra de Bahoruco.
Range: Breeding range fragmented, from s. Minnesota east to s. Ontario, c. Vermont, and s. Maine, south to extreme ne. Oklahoma, n. Alabama, n. Georgia, and w. Virginia. Occurs in non-breeding season along Atlantic slope of Central America from c. Mexico south to c. Panama. In West Indies, a rare non-breeding visitor in Bahamas and Greater Antilles east to Virgin Islands; vagrant to St. Barthélemy and Guadeloupe.
Local names: DR: Ciguíta Ala Azul; **Haiti:** Ti Tchit Zèl Ble, Paruline à Ailes Bleues.

GOLDEN-WINGED WARBLER—*Vermivora chrysoptera* **Plate 46**
Non-breeding Visitor

Description: 12–13 cm long; 9 g. A small, boldly plumaged warbler with yellow crown, grayish nape and back, and a distinctive, large yellow wing patch. Head is boldly patterned with broad black auriculars bordered above and below by white, and by a black throat. Rest of underparts are grayish white, but undertail coverts are white. Female and immature birds are duller; throat and auriculars are pale gray, and yellow wing patch is typically smaller.
Similar species: Blue-winged Warbler is yellow below, with a black eye-line and white wing bars. Interbreeding of these two species has been recorded frequently and results in hybrids with variable, intermediate characteristics.
Voice: Call note, similar to that of Blue-winged Warbler, is a sharp, loud *chip* or *jeet*; also

gives thin *tzip* calls. Primary song is a fine, high buzz, *zeee bee-bee-bee*, with first note slightly higher in pitch.
Hispaniola: Usually occurs in semiopen forests and forest borders or gaps. Known records, all from DR, include: 1 collected near Pedro Sánchez, 1 in Santo Domingo, 1 near Puerto Escondido, 1 in Sierra de Bahoruco, and 2 at Cabo Rojo.
Status: Rare non-breeding visitor from mid-October to early February.
Comments: Frequents mid- to upper-tree-canopy levels, where it usually probes dead leaves for insects. Generally not gregarious, but may join mixed-species flocks in non-breeding season.
Range: Breeds from s. Manitoba east to n. Minnesota, s. Ontario, w. Vermont, and Massachusetts, south to e. Tennessee and n. Georgia. Occurs in non-breeding season from extreme s. Mexico south to n. and w. Venezuela and c. Colombia. In West Indies, rare or vagrant in Bahamas and Greater Antilles east to Virgin Islands.
Local names: DR: Ciguíta Ala de Oro; **Haiti:** Ti Tchit Zèl Dore, Paruline à Ailes Dorées.

TENNESSEE WARBLER—*Vermivora peregrina* **Plate 44**
Non-breeding Visitor

Description: 10–13 cm long; 10 g. A small to medium-sized warbler, lacking distinctive markings. Adults in non-breeding plumage are grayish green above, dull grayish white below, with white undertail coverts and rest of underparts often tinged with yellow. Note a distinct dusky eye-line and narrow whitish supercilium. Male tends to have grayer crown, brighter olive green upperparts, and less yellowish wash on underparts than females and immatures, especially in breeding season.
Similar species: Similar in plumage to vireos, but vireos are larger, with a heavier, hooked bill.
Voice: Sharp, high *tsip* or *tseet*; also short, fine *tsit* calls given frequently while foraging.
Hispaniola: Most often reported from montane forests between mid-October and early May. All known records are from DR localities and include Jarabacoa, Manabao, Jumunucú, Mencia, and Las Cruces in Sierra de Bahoruco.
Status: Rare non-breeding visitor. May occur throughout the non-breeding season more regularly than documented, because small numbers were found annually in intensively studied shade coffee plantations of Cordilllera Central, DR, during 1990s.
Comments: Frequents the canopies of tall trees, but regularly forages at all heights. Commonly feeds on nectar in non-breeding season and may defend feeding territories in flowering trees.
Range: Breeds from se. Yukon, n. Saskatchewan, n. Quebec, and Newfoundland south to s. British Columbia, n. Minnesota, n. New York, and Nova Scotia. Occurs in non-breeding season from s. Mexico through Central America to w. Colombia, n. Venezuela, and n. Ecuador. In West Indies, occurs in Bahamas and Greater Antilles east to Hispaniola; vagrant in Virgin Islands and Barbados.
Local names: DR: Ciguíta de Tenesí; **Haiti:** Ti Tchit Gri, Paruline Obscure.

NASHVILLE WARBLER—*Vermivora ruficapilla* **Plate 48**
Vagrant

Description: 11–12 cm long; 9 g. A small warbler with unmarked olive green upperparts contrasting with a grayish head that is washed brownish gray in the non-breeding season. Underparts, including throat and undertail coverts, are yellow, and belly is white. Distinguished in all plumages by a conspicuous white eye-ring. Adults may have a chestnut crown patch, but it is usually concealed. Breeding birds are similar, but head is more distinctly gray, especially in male.
Similar species: Female or immature Mourning Warbler and Connecticut Warbler are larger, with entirely gray hood, including chin and throat, and solid yellow belly. Female and immature Northern Parula have wing bars and white undertail coverts.

Hispaniola: Two November-December reports, from La Visite National Park, Haiti and from Cabo Rojo, DR.
Status: Vagrant.
Range: Two disjunct breeding populations. Eastern birds breed from c. Saskatchewan, c. Ontario, s. Quebec, and Nova Scotia south to c. Minnesota, c. Michigan, ne. West Virginia, and n. New Jersey. Western population breeds from s. British Columbia south and east to Oregon, extreme w. Nevada, and s. California. Populations apparently mix in non-breeding range from extreme s. Texas and nc. Mexico south to s. Guatemala, and locally to n. El Salvador. Some western birds spend non-breeding season in coastal lowlands of California. In West Indies, rare or vagrant in Bahamas and Greater Antilles east to Puerto Rico.
Local names: DR: Ciguíta de Nashville; **Haiti:** Ti Tchit Tèt Gri, Paruline à Joues Grises.

NORTHERN PARULA—*Parula americana* **Plate 49**
Non-breeding Visitor

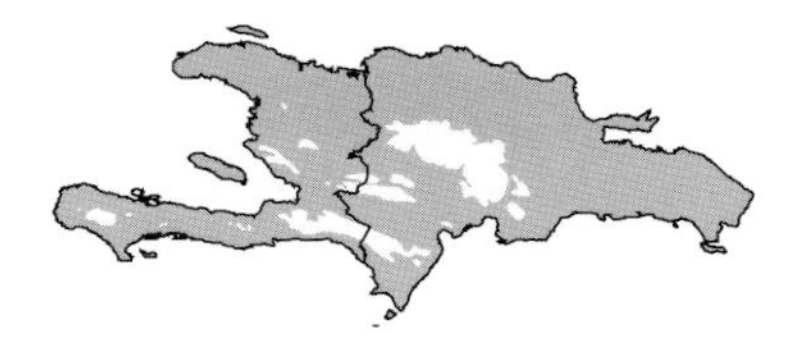

Description: 11–12 cm long; 9 g. A small warbler. In non-breeding birds, upperparts are bluish gray washed green, with a distinct greenish yellow patch on mantle. Chin, throat, and breast are yellow, but belly and undertail coverts are white. Note two white wing bars, white tail spots, and a prominent broken white eye-ring. Male is marked variably across chest with a narrow black and chestnut band. Female and immatures are duller overall with breast band indistinct or lacking. Breeding birds are similar, but upperparts are a brighter blue-gray, and breast bands of male are more distinct. Breeding male also has black lores.
Similar species: Nashville Warbler has unmarked wings, complete eye-ring, and more extensively yellow underparts. Yellow-throated Warbler has head and flanks boldly marked with black and white.
Voice: Call note is a distinct *tschip*, frequently repeated, similar to that of American Redstart but slightly softer and a bit more clipped. Flight call is a high, clear, descending *tseep*. Song is an ascending buzz with a sharp final note, *zeeeeeeee-tsup*, often heard late in the non-breeding season.
Hispaniola: Most numerous in dry broadleaf forest, scrub forest, and second growth from sea level to about 350 m, but reported as high as 1,700 m. This was the second most abundant non-breeding warbler recorded in secondary sclerophyll scrub at 320 m; seen regularly in small numbers in shade coffee plantations in Cordillera Central, DR, at 450–800 m; but rare in broadleaf forest in Sierra de Bahoruco at 745 m. Recorded from several major satellite islands, including Île à Vache, Île de la Gonâve, Île de la Tortue, and Islas Beata and Saona.
Status: Common non-breeding visitor islandwide from late July to late April. Usual fall arrival and transient period is September to October, and peak spring departure period is mid-March to early April.
Comments: This is one of six migratory warblers that occur primarily in the West Indies in the non-breeding season. Most often occurs singly.
Range: Disjunct breeding range. Breeds from se. Manitoba, s. Quebec, and Nova Scotia south to ne. Minnesota, s. Ontario, and s. Maine; also from e. Nebraska, s. Ohio, and s. New York south to s. Texas, Gulf Coast, and s. Florida. Occurs in non-breeding season from extreme s. Florida through Caribbean and Atlantic slope of Mexico south to Belize and Honduras, rarely to Nicaragua and Panama. Found virtually throughout West Indies, but uncommon to rare in southern Lesser Antilles.
Local names: DR: Ciguíta Parula; **Haiti:** Ti Tchit Ble Pal, Paruline à Collier.

YELLOW WARBLER—*Dendroica petechia* **Plate 49**
Breeding Resident; Passage Migrant; Non-breeding Visitor?

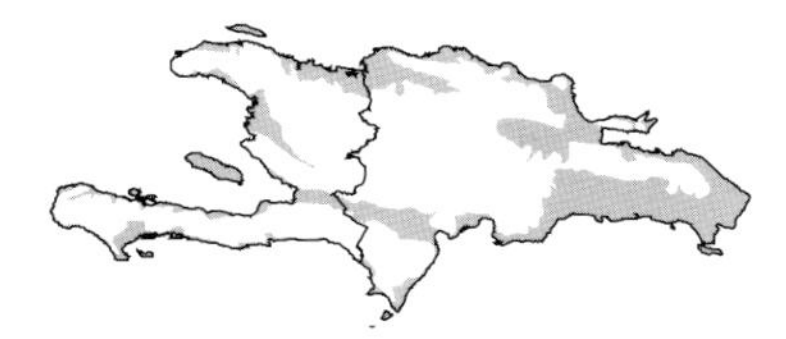

Description: 12–13 cm long; 10 g. A medium-sized, strikingly yellow warbler. Non-breeding birds have olive-yellow upperparts, yellow head, and bright yellow underparts. Wing bars, inner webs of tail, and tail spots are yellow. Male usually shows light chestnut red streaking on breast. Female is similar to male but is less bright, with little or no streaking below. Females may also show variable amounts of gray on the nape. Breeding birds are similar to non-breeding birds but are brighter. Males often show more pronounced chestnut streaking on breast and may have a pale chestnut patch on crown.
Similar species: Female and immature Wilson's Warbler have a more greenish cast to upperparts and lack wing edgings, tail spots, and markings on underparts. Female and immature Hooded Warbler have white tail spots and unmarked underparts and wings. Prothonotary Warbler has blue-gray wings and tail, unmarked underparts, and white undertail coverts. Blue-winged Warbler has blue-gray wings and tail, white tail spots, white wing bars, and black or dusky eye-line.
Voice: Song is variable, but typically a loud, clear, and rapid *sweet-sweet-sweet-ti-ti-ti-weet.* Call notes include a loud, clear *chip* and a high, thin *zeet.*
Hispaniola: Occurs along virtually the entire coastline of the main island and most of its major satellites wherever substantial stands of mangroves are found. Sometimes moves into adjacent dry scrub, acacia scrub, and desert thorn scrub habitat for much of the non-breeding season.
Status: Locally common to abundant breeding resident. A 1994 taxonomic review concluded that the endemic subspecies *D. p. albicollis* is resident on the main island, Île à Vache, and Île de la Tortue, Haiti (and presumably on Grande Cayemite, Haiti, and Islas Beata and Saona, DR, where it also breeds); the endemic subspecies *D. p. chlora* is resident on the Cayos Siete Hermanos, DR; and the endemic subspecies *D. p. solaris* is resident on Île de la Gonâve and Petite Gonâve, Haiti. Although some North American breeding birds certainly pass through regularly as migrants, and some may occasionally remain through the non-breeding season, this is not yet documented by specimens or mist-netting studies. Transients or non-breeding visitors from the north are expected to be of the subspecies *D. p. amnicola* or *D. p. aestiva.*
Comments: Several West Indian populations have declined because of intense brood parasitism by Shiny Cowbirds, but effects on Hispaniolan populations are not known. Gleans insects from foliage, and may occasionally take fruit.
Nesting: Surprisingly few specific Hispaniolan breeding data are available. General West Indies data indicate that Yellow Warblers lay a clutch of 2–4 eggs that are bluish white spotted with brown in a wreath around the large end. The nest is compact and cup shaped, built of grass and fibers, and usually placed in a bush, tree, or mangrove near water, rarely more than 3 m above ground. Breeds March to July.
Range: North American populations breed from n. Alaska east to n. Manitoba, c. Quebec, and Newfoundland south to n. Baja California, s. Mexico, c. Oklahoma, c. Alabama, and c. North Carolina. Occurs in non-breeding season from n. Mexico south to n. Bolivia, c. Peru, and Amazonian Brazil. Other subspecies are permanent residents throughout West Indies.
Local names: DR: Canario de Manglar; **Haiti:** Ti Jòn, Sicrye Mang, Paruline Jaune.

CHESTNUT-SIDED WARBLER—*Dendroica pensylvanica* **Plate 46**
Vagrant

Description: 12–13 cm long; 9 g. A distinctive, medium-sized warbler. Non-breeding plumage is bright yellowish green above with two yellowish white wing bars and variable black spotting; pale gray below. Males may show some chestnut on sides. Note the conspicuous white eye-ring on a gray face. Adult male in breeding plumage is characterized by a yellow

forecrown, black eye-line and malar stripe, chestnut band along sides, and white underparts. Female is duller with less extensive chestnut on sides.
Similar species: Bay-breasted and Blackpoll warblers in non-breeding plumage have more olive-gray and streaked upperparts, more yellowish underparts, and lack eye-ring.
Hispaniola: Four reports of birds in October and April, from Port-au-Prince and Saint-Marc, Haiti, and Cabo Rojo and Pedernales, DR. These birds were probably passage migrants.
Status: Vagrant.
Comments: A specialist of second-growth habitats during breeding season, but inhabits a variety of forest types in non-breeding season. Gleans insects from foliage, and occasionally takes fruits.
Range: Breeds from sc. Saskatchewan east to c. Manitoba, c. Quebec, and Nova Scotia, south to e. Iowa, se. Pennsylvania, and in Appalachians to n. Georgia. Occurs in non-breeding season on Atlantic slope of s. Mexico south to e. Panama. In West Indies, rare in Bahamas and Greater Antilles east to Virgin Islands, casual in Lesser Antilles south to Barbados.
Local names: DR: Ciguíta de Costados Castaños; **Haiti:** Ti Tchit Kòt Mawon.

MAGNOLIA WARBLER—*Dendroica magnolia* **Plate 47**
Non-breeding Visitor

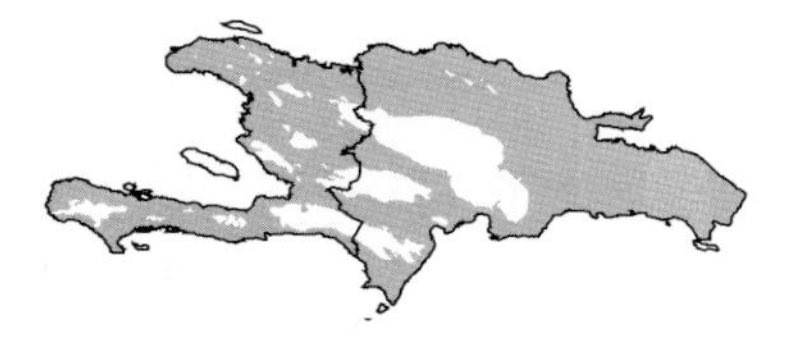

Description: 11–13 cm long; 8–11 g. A medium-sized warbler with a distinctive tail pattern. In non-breeding plumage, upperparts are grayish olive streaked with black, rump is conspicuously yellow, and there are two white wing bars. Head and nape are gray. A pale gray neck band is often present. Note a white eye-ring and thin, pale white supercilium. Underparts are bright yellow with variable black streaking on sides (heavier in males). Lower belly and undertail coverts are white. Tail pattern is diagnostic with large white spots in the middle of all but the two central tail feathers. Breeding male has a black auricular, prominent white supercilium, bold white wing patch, and underparts heavily striped with black forming a "necklace" across breast. Breeding female is much paler than male, with gray auriculars, and underparts only moderately streaked.
Similar species: White median tail band is unique among warblers. Male Cape May Warbler in non-breeding plumage is typically more heavily streaked below with a noticeable yellowish facial pattern. Yellow-rumped Warbler has white or pale grayish throat, with yellow on underparts limited to sides. Prairie Warbler shows variable black streaking on face and habitually wags its tail.
Voice: Call note is a hard, nasal *enk* or *clenk;* also a dry, high-pitched *chip* or *tit,* often repeated. Flight call is a high, buzzy *zee*.
Hispaniola: Reported from sea level to a maximum known elevation of 750 m in Sierra de Bahoruco, DR, generally between late September and late March. Has been reported in non-breeding season in mangroves, dry forest, shade coffee plantations, thorn forest, and transitional broadleaf forest. Most records are of single birds, but 12 observed in mid-March near Laguna Saladilla, DR, may have represented a premigratory concentration of birds.
Status: Regular but uncommon non-breeding visitor in small numbers.
Comments: Insectivorous; not known to feed on fruit in non-breeding season.
Range: Breeds from ne. British Columbia east to nc. Manitoba, n. Ontario, and s. Newfoundland, south to sc. Alberta, ne. Minnesota, n. Ohio, se. West Virginia, and Connecticut. Occurs in non-breeding season from s. Mexico and Greater Antilles south to c. Panama. In West Indies, occurs in Bahamas and Greater Antilles east to Virgin Islands; vagrant in Lesser Antilles south to Barbados.
Local names: DR: Ciguíta Magnolia; **Haiti:** Ti Tchit Ke Blanch, Paruline à Tête Cendrée.

CAPE MAY WARBLER—*Dendroica tigrina* **Plate 47**
Non-breeding Visitor

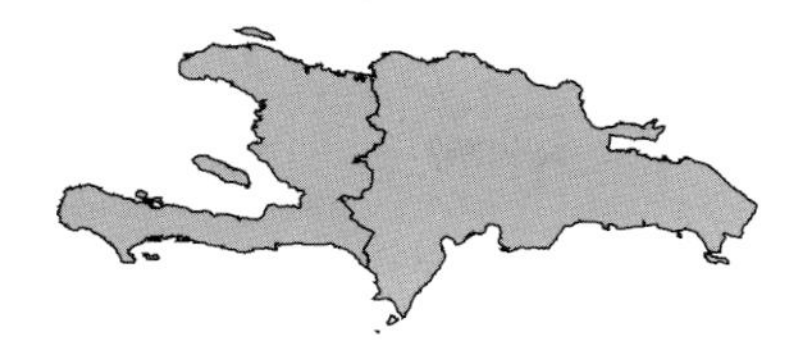

Description: 12–13 cm long; 11 g. A medium-sized warbler with a short tail, thin, decurved bill, and yellowish rump. In non-breeding plumage, adult male has grayish green back with dark streaks, chestnut-tinged auriculars bordered by yellow, including a distinctive yellow patch on sides of neck, and conspicuous white wing patch. Underparts are yellow and heavily streaked with black. Adult female and immatures are similar but duller, with gray auriculars, blurred yellow spot on side of neck, greenish rump, and two whitish wing bars. Adult male in breeding plumage is distinguished by conspicuous chestnut auriculars and large white wing patch.
Similar species: Yellow-rumped Warbler is browner above and has a brighter and more sharply defined rump patch and small yellow patches on sides. Magnolia Warbler has a prominent white band on all but the central two tail feathers and lacks the yellow neck patch. Prairie Warbler has streaks restricted to sides, a dark semicircle under eye, and wags its tail frequently. Palm Warbler has yellow undertail coverts, a distinct pale supercilium, and also pumps its tail.
Voice: Calls include a musical, high-pitched *chip* or *tsit*, similar to Hispaniolan Spindalis; also a high, thin *tsee* in flight or while feeding. Rarely sings a thin, buzzy, ascending song, *tseet-tseet-tseet-tseet*, before migrating north.
Hispaniola: Occurs from late August to early May, but typical fall arrival is mid-October to early November and spring departure from early March to early April. The species appears commonly in many habitats from sea level in thorn scrub and dry forest, to shade coffee and transitional broadleaf forest at midelevations, to open pine forest as high as 2,600 m. Recorded from several major satellite islands, including the Cayemites, Île à Vache, Île de la Gonâve, Île de la Tortue, and Isla Saona.
Status: Common to abundant non-breeding visitor islandwide.
Comments: This is one of six migrant warbler species that occur in the non-breeding season principally in the West Indies. It is highly opportunistic in its feeding habits, and its unique semitubular tongue is specialized for feeding on nectar. Where nectar is not available, insects, small fruit, and homopteran honeydew are taken. May be found in mixed-species flocks but is often solitary and aggressively territorial around favored food sources. Often found in residential areas.
Range: Breeds from se. Yukon east to n. Alberta, c. Ontario, se. Quebec, and sw. Newfoundland south to c. Saskatchewan, n. Wisconsin, s. Ontario, and c. Maine. Occurs in non-breeding season primarily in West Indies, but also extreme s. Florida and along Caribbean coast of Central America from Yucatán Peninsula south to c. Panama. In West Indies, common in Greater Antilles and U.S. Virgin Islands, rare to vagrant elsewhere in Lesser Antilles.
Local names: DR: Ciguíta Tigrina; **Haiti:** Ti Tchit Kou Jòn, Paruline Tigrée.

BLACK-THROATED BLUE WARBLER—*Dendroica caerulescens* **Plate 44**
Non-breeding Visitor

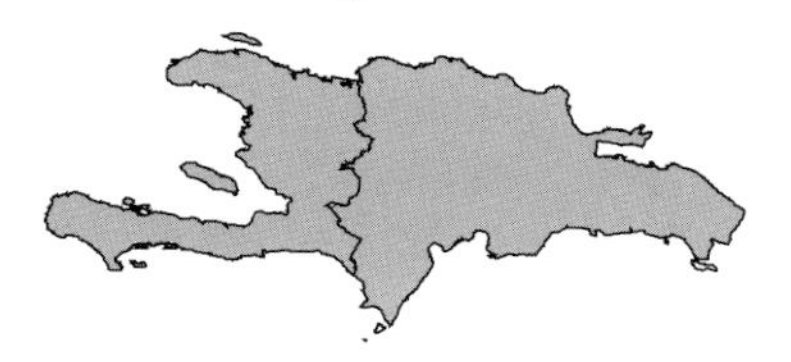

Description: 12–14 cm long; 10 g. A medium-sized, fairly stocky warbler with sexes distinguishable in all plumages. Male is uniformly dark blue above with a small but distinct white patch on wing at base of primaries, and a black face and throat. Rest of underparts are white with a black band along flanks. Female is drab olivaceous gray above with a small white spot at base of primaries, narrow whitish supercilium, and slightly grayer auriculars. Underparts are dull buffy white. Immature male is like adult but with upperparts washed

greenish; immature female is similar to adult, but white wing spot is much reduced or absent.
Similar species: Male is unmistakable. Female is superficially similar to Tennessee Warbler but larger and has a whitish supercilium and wing spot.
Voice: Fairly sharp, dry *tsik* or *ctuk*, reminiscent of two coins being struck together, given frequently by both sexes.
Hispaniola: Occurs in a wide range of habitats from sea level to an extreme high known elevation of 2,450 m in pine forest of the Cordillera Central, DR. Most numerous, however, in shade coffee plantations and midelevation broadleaf forest with a substantial canopy, especially humid montane broadleaf and transitional forest from 900 to 2,100 m. Becomes increasingly scarce at progressively higher or lower elevations. Males prefer more heavily forested sites with a closed canopy, whereas females are found more often in early successional and scrub habitats with a broadleaf component. Typically arrives in late September or early October and departs northward by late April. Recorded from several major satellite islands, including the Cayemites, Île à Vache, Île de la Gonâve, and Île de la Tortue.
Status: Common non-breeding visitor in appropriate habitat islandwide. The northern subspecies, *D. c. caerulescens*, is by far the most numerous form present on Hispaniola, but the southern subspecies, *D. c. cairnsi*, is an uncommon visitor.
Comments: This is one of six warbler species whose non-breeding-season range is concentrated in the West Indies. Usually forages in low vegetation, infrequently on the ground or high in trees. Feeds on insects and spiders, but also takes small fruit and nectar when available.
Range: Breeds from sw. Ontario and s. Quebec east to Nova Scotia, south to ne. Minnesota, n. Michigan, and in Appalachians south to s. Connecticut, n. New Jersey, and ne. Georgia. Occurs in non-breeding season from extreme s. Florida through Greater Antilles, and along Caribbean coast of Yucatán Peninsula, Belize, and Honduras, casually south to n. Colombia and Venezuela. In West Indies, common in Bahamas and Greater Antilles east to Puerto Rico; rare to vagrant south of there to St. Vincent.
Local names: DR: Ciguíta Azul, Garganta Negra; **Haiti:** Ti Tchit Ble Kou Nwa.

YELLOW-RUMPED WARBLER—*Dendroica coronata* **Plate 46**
Non-breeding Visitor

Description: 14 cm long; 12 g. A large warbler, grayish brown above in non-breeding plumage, with indistinct black streaking on back, a small yellow crown patch, and a bright yellow rump. Underparts are dull whitish with blurry streaks on breast and small yellow patches on sides. Note two narrow white wing bars, white tail spots, pale supercilium, and broken white eye-ring. Females and immatures are duller and browner than male, with smaller patches of yellow on sides, and often lacking the yellow crown patch. Male in breeding plumage is bluish gray above, whitish below, with a distinct black breast band and auriculars, and a bright yellow crown and side patches. Breeding female is duller overall, with smaller yellow patches.
Similar species: Magnolia Warbler is mostly yellow below and has a tail marked with a bold white central band. Immature Cape May Warbler has a duller yellow rump and paler area on side of neck. Immature Pine Warbler has little or no streaking below and lacks yellow rump. Palm Warbler has a dull yellow rump but distinctly yellow undertail coverts, and exhibits tail-bobbing behavior.
Voice: Call note is a flat, emphatic *chek* or *chup*, given frequently while foraging. Flight call is a soft, clear, rising *ssit*. Song is a clear, warbling trill, somewhat disorganized and variable.
Hispaniola: Occupies a wide range of elevations from sea level to 2,450 m, in habitats ranging from coastal scrub to pine forest. Most common in pine forest, less frequent in low-elevation scrub. Often occurs in small flocks of 5–10 birds, sometimes up to 30–50, and has been

found to show non-breeding-season site fidelity in pine forests of Sierra de Bahoruco. Typically arrives later and departs earlier than most migrants. May arrive late September but generally not until early November. Usual northbound departure is in March, though some birds may linger to late April or early May. Recorded from several major satellite islands, including the Cayemites, Île à Vache, Île de la Tortue, Cayos Siete Hermanos, and Isla Saona.
Status: Regular and locally common non-breeding visitor, subject to periodic irruptions when it becomes more abundant. For unknown reasons, the species seems to occur much more regularly and in greater numbers on Hispaniola than on Cuba.
Comments: In the southeastern U.S., winter abundance of this species in pine stands is associated with the amount of fruit found in the broadleaf understory; this seems to be true in the Caribbean as well. Also feeds on insects, primarily as a foliage gleaner, but may also sally for flying insects.
Range: Breeds from w. Alaska east to nw. Mackenzie, n. Ontario, n. Quebec, and Newfoundland, south to n. British Columbia, ne. Minnesota, s. Ontario, and s. Maine, south in Appalachians to ne. West Virginia. Occurs in non-breeding season along both Atlantic and Pacific coasts from Massachusetts and s. Washington south throughout Gulf States and Mexico to Costa Rica, irregularly to Panama. In West Indies, common to uncommon in Bahamas and Greater Antilles east to Virgin Islands; rare to vagrant south through Lesser Antilles to St. Vincent.
Local names: DR: Ciguíta Mirta; **Haiti:** Ti Tchit Dèyè Jòn, Paruline à Croupion Jaune.

BLACK-THROATED GREEN WARBLER—*Dendroica virens* **Plate 47**
Non-breeding Visitor

Description: 11–12 cm long; 9 g. A fairly small, foliage-gleaning warbler. In non-breeding plumage, male is bright olive green above with a bright yellow face and dull olive gray auriculars. Note the distinctive black chin and throat patch extending onto upper breast; two prominent white wing bars; and white tail spots. Underparts are whitish, sides are streaked with black, and there is a yellowish wash across undertail coverts. Adult female and immature male are duller, with yellowish chin and throat. Immature female is duller still, and lacks black on underparts except for faint streaking on sides. Breeding plumage is similar but bolder.
Similar species: Female and immature Blackburnian Warblers lack black markings on a more yellowish orange throat and breast, have whitish streaks on a darker back, and have darker auriculars.
Voice: Short, sharp *tsip* or *tek*, somewhat hoarse. Flight call is a clear, high, rising *sweet* or *see*. Male song has two distinct patterns of short, level buzzes: a fast *zee-zee-zee-zee-zo-zeet* and a more relaxed *zooo-zee-zo-zo-zeet*.
Hispaniola: Occurs from sea level to a maximum known elevation of 2,060 m, though most often found in moist broadleaf forest, pine forest, and shade coffee plantations in the range of 500–1,500 m. Usually solitary, but rarely 2–3 are found together. Usually shows strong site persistence throughout the non-breeding season. Generally occurs from mid-October to early May. Also reported from Île à Vache and Isla Beata.
Status: Uncommon but regular non-breeding visitor in small numbers.
Comments: Arboreal. Forages actively for insects, occasionally takes fruits. Primarily a foliage gleaner, but may sally for flying insects.
Range: Breeds from e. British Columbia east to c. Manitoba, c. Quebec, and Newfoundland, south to c. Alberta, n. Wisconsin, and s. New England, and south in Appalachians to n. Georgia. Also a disjunct breeding population confined to coastal plain of Virginia and Carolinas. Occurs in non-breeding season from extreme s. Florida and s. Texas south to n. Colombia and w. Venezuela. In West Indies, uncommon in Bahamas and Greater Antilles east to Hispaniola, vagrant east of there and south through Lesser Antilles to Barbados.
Local names: DR: Ciguíta Pechinegro; **Haiti:** Ti Tchit Fal Nwa, Paruline à Gorge Noire.

BLACKBURNIAN WARBLER—*Dendroica fusca* **Plate 47**
Vagrant

Description: 11–12 cm long; 10 g. A fairly small, canopy-dwelling warbler. Male non-breeding plumage is dusky olive to blackish above with pale buffy streaks on back; black auriculars bordered by yellowish orange; and two bold white wing bars. Chin, throat, and breast are yellowish orange; belly is buffy white; sides are streaked black; and there are white spots in outer web of the two to three outermost tail feathers. Female and immature plumages are similar but duller, with yellow replacing orange, and less prominent streaking on back and sides. Breeding male is blackish above with black auriculars and brilliant orange throat and facial markings. The two white wing bars form a prominent patch. Breeding female is dusky olive above with dusky auriculars and yellowish orange facial markings.
Similar species: Immature Black-throated Green Warbler has a buff-colored throat and breast, unstreaked back, and lacks dark auriculars. Yellow-throated Warbler has a white supercilium and unstreaked gray back.
Voice: Thin but sharp *tchip* or *tsip*, often 2-syllabled. Flight call is a high, thin buzz. Song is dry, thin, high pitched, and ascends at the end: *tsi tsi tsi tsi tsi titititi tseeeee.*
Hispaniola: All records from DR in October and March or April, suggesting that this species migrates through Hispaniola in small numbers. Individuals recorded near Puerto Escondido, and twice at Santo Domingo. Also recorded at Isla Muertos, Cayos Siete Hermanos.
Status: Vagrant.
Range: Breeds from c. Alberta east to c. Ontario and sw. Newfoundland, south to s. Manitoba, c. Wisconsin, and s. New Hampshire, and south in Appalachians to w. Pennsylvania and n. Georgia. Occurs in non-breeding season from Costa Rica south to Venezuela and c. Peru. In West Indies, rare in Bahamas and most of Greater Antilles; vagrant from Hispaniola east and south through Lesser Antilles to Grenada.
Local names: DR: Ciguíta del Frío; **Haiti:** Ti Tchit Flanbwayan.

YELLOW-THROATED WARBLER—*Dendroica dominica* **Plate 46**
Non-breeding Visitor

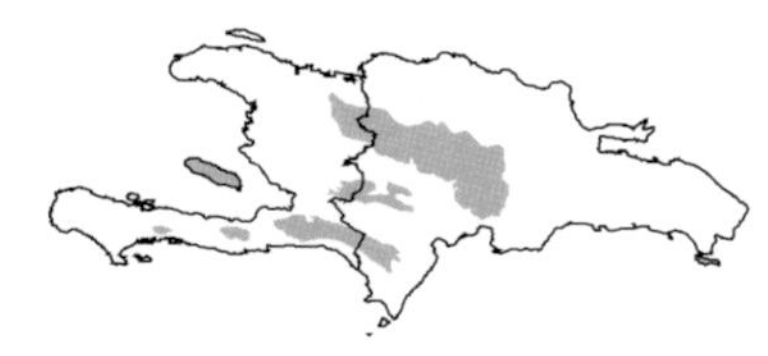

Description: 13–14 cm long; 9 g. A medium-sized, strikingly patterned warbler with a long bill. Upperparts are gray, throat and upper breast are yellow, and belly is white with black streaks on sides. Note triangular black auriculars broken by a white crescent below eye and bordered by a white supercilium and a white patch on side of neck. There are two white wing bars. Female and immature are similar but duller, with black markings slightly reduced.
Similar species: Blackburnian Warbler has triangular auriculars bordered by orange or yellow, lacks white neck patch, and has whitish streaks on back.
Voice: Call note is a clear and high-pitched, slightly descending *chip* or *clip*.
Hispaniola: With the exception of migration periods when it has been recorded at sea level and in lowlands, this species is nearly always found in pine habitat and thus most often at or above 700 m elevation, with a maximum reported elevation of 2,060 m. It is one of the earliest warblers to arrive and depart from Hispaniola. Some birds arrive in mid-August, but general fall arrival is late September. In spring, northbound departure is in March. Reported from several major satellite islands, including Île à Vache, Île de la Gonâve, and Isla Saona.
Status: Moderately common non-breeding visitor.
Comments: Generally a canopy bird occupying the higher tree branches, and often occuring in mixed-species flocks. Entirely insectivorous; forages deliberately by creeping along pine tree limbs and trunks, probing bark crevices and clumps of pine needles with its long bill. In the absence of pines, often forages in palm trees.

Range: Breeds from s. Wisconsin east to n. Ohio and c. New Jersey, south to e. Kansas, s. Texas, Gulf Coast, and c. Florida. Occurs in non-breeding season from coastal South Carolina south to s. Florida, throughout West Indies, and coastal Texas south to Costa Rica, casually to Panama. In West Indies, common in Bahamas and Greater Antilles; vagrant in Lesser Antilles south to Barbados.
Local names: DR: Ciguíta Garganta Amarilla; **Haiti:** Ti Tchit Fal Jòn, Paruline à Gorge Jaune.

PINE WARBLER—*Dendroica pinus* **Plate 49**
Breeding Resident

Description: 13–14 cm long; 12 g. A large warbler with a fairly stout bill, rarely found outside pine forests. Adult male is distinguished by unstreaked greenish olive upperparts and a bright yellow throat and breast with variable amounts of diffuse dark streaking. Greenish olive auriculars are bordered sharply below and toward the rear by yellow. Note indistinct yellow eye-ring and supercilium, two white wing bars, white spots on outermost two to three tail feathers, and white lower belly and undertail coverts. Adult female is similar to male but duller overall with more grayish or brownish wash above, and less streaked below. Immatures are browner and duller than adults.
Similar species: White-eyed and Yellow-throated vireos have heavier bills, distinct yellow spectacles, and lack streaking below. In the non-breeding season, Bay-breasted and Blackpoll warblers have a streaked back and appear short tailed because of longer undertail coverts.
Voice: Song is a simple, rapid, musical trill, usually at the same frequency. Call note is a strong, sharp *tzip* or *tchik*.
Hispaniola: Resident of pine forest above 700 m in the interior hills and mountains to a maximum known elevation of at least 2,600 m. This species is essentially confined to the mountain ranges of the Cordillera Central, Sierra de Neiba, and Sierra de Bahoruco, where it is common to occasionally abundant. Also occurs in the westward continuations of these mountain ranges into Haiti.
Status: The resident endemic subspecies on Hispaniola is *D. p. chrysoleuca*, and it is now considered threatened in much of Haiti because of pine forest destruction. Overall, it has probably declined islandwide since the 1930s. A reported increase in numbers of this species in the non-breeding season, attributed to migrants from North America, is not supported by specimens of the North American migratory subspecies *D. p. pinus*. Furthermore, plumage differences that distinguish the two subspecies are not seen in Hispaniolan birds, and Pine Warblers are most often found in pairs in the non-breeding season, suggesting that they are all permanent residents.
Comments: Forages from ground level to canopy for insects, spiders, pine seeds, and berries.
Nesting: Lays 2–4 white eggs with brown splotches in a cup-shaped nest of pine needles and grass, often in the fork of a narrow branch high in a pine tree; also documented to nest on the ground. In Sierra de Bahoruco, DR, the nesting density of this species is high, site persistence within and between years is high, home ranges average 3.1 ha, and 5 arboreal nests averaged 8.0 m above ground and 1.0 m from the trunk of pine trees. Breeds February to June.
Range: Breeding range in e. North American fragmented, from se. Manitoba east to n. Wisconsin, s. Quebec, and se. New Brunswick, south to e. Texas, Gulf Coast, and s. Florida. Occurs in non-breeding season in southern part of breeding range, rarely to s. Texas and ne. Mexico. In West Indies, resident populations occur in Bahamas and on Hispaniola.
Local names: DR: Ciguíta del Pinar; **Haiti:** Ti Tchit Bwa Pen, Paruline des Pins.

PRAIRIE WARBLER—*Dendroica discolor* **Plate 47**
Non-breeding Visitor

Description: 11–12 cm long; 8 g. A small warbler; male in non-breeding plumage is olivaceous above with chestnut streaks on back and bright yellow below with pronounced black streaks on sides. The distinctive facial pattern features a yellow supercilium, blackish eye-line, broad yellow crescent below eye, and blackish malar crescent. Also note pale yellow wing bars and white tail spots. Non-breeding female is duller and less distinctly marked than male. Immature is similar to female but duller still, and lacks chestnut streaking on back; facial contrasts are reduced, and auriculars are grayish. Breeding plumage is similar but brighter and more contrasting. Prairie Warblers characteristically pump their tail.
Similar species: Palm Warbler is grayish brown above and has dully streaked, whitish underparts with contrasting yellow undertail coverts, and a pale whitish supercilium. Pine Warbler is larger, has a white belly, and lacks black facial marks and chestnut streaking on back. Magnolia Warbler in non-breeding plumage has a grayish olive back streaked with black, yellow rump, and tail has a distinct white central band.
Voice: Call note is a dry, husky *chip* or *chek*. Song is a thin, ascending buzzy *zee-zee-zee-zee-zee-zee-zee-ZEEET*.
Hispaniola: Preferred habitat is dry thorn scrub forest and the margins of mangrove areas, but generally absent from midelevation dry broadleaf forests. Most birds are found at elevations of 300 m or below, but where the species occurs at higher elevations it is often found in pine forest where there is a scrubby broadleaf understory. Highest known elevation is 1,140 m. General arrival date is mid-September through October. In spring, general departure is March, but may persist until mid-April. Reported from most major satellite islands.
Status: Common to abundant and widespread non-breeding visitor islandwide.
Comments: One of six migrant warblers whose entire population is virtually restricted to the West Indies in the non-breeding season. Like many other warblers, generally defends exclusive territories on Hispaniola, to which individuals return each year. Occasionally roosts communally in mixed-species groups in mangroves. Feeds on insects, spiders, and fruits.
Range: Somewhat fragmented breeding range from e. Oklahoma north and east to nc. Illinois, nw. Michigan, s. Ontario, and s. Maine, south to e. Texas, s. Alabama, and s. Florida. Occurs in non-breeding season from Florida through Greater Antilles, and Caribbean coast of Yucatán Peninsula, Belize, and Honduras, casually along Pacific slope from s. Mexico to El Salvador. In West Indies, common in Bahamas and Greater Antilles, rare to vagrant south through Lesser Antilles to Grenada.
Local names: DR: Ciguíta de los Prados; **Haiti:** Ti Tchit Zèl Jòn, Paruline des Prés.

PALM WARBLER—*Dendroica palmarum* **Plate 45**
Non-breeding Visitor

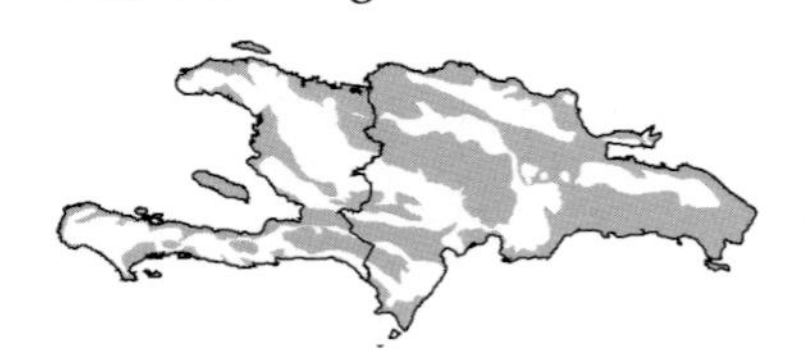

Description: 12–14 cm long; 10 g. Non-breeding plumage is grayish brown to cinnamon brown above, dull whitish to pale buff below, with faint streaking on breast and flanks. Note brick red crown patch, pale supercilium, dark eye-line, yellow chin and throat, and two indistinct buffy wing bars. Rump is yellow-olive; undertail coverts are yellow. Breeding plumage has a more pronounced reddish brown crown, and yellow chin and throat. Sexes are similar in all plumages. Frequently pumps its tail, and often seen on or near the ground.
Similar species: Prairie Warbler is entirely yellow below, with heavily streaked sides and blackish facial markings. Female and immature Cape May and Yellow-rumped warblers have white undertail coverts, brighter yellow rump, and do not pump their tail.

Voice: Call is a distinctive, sharp, slightly metallic *chip* or *chik*. Flight call is a thin *tsip*.
Hispaniola: Occurs primarily in low vegetation or on the ground. Tends to prefer drier, open areas without thick ground cover, such as thorn scrub, desert, and semiarid woodlands. Occurs only rarely in moister transitional broadleaf habitat. At higher elevations, found most often in open pine forest and in disturbed areas. Reported from sea level to a maximum known elevation of 2,400 m, but is more numerous at elevations of 500 m and below. In fall, arrival is often not until early November; in spring, northbound departure is usually in March. Reported from most major satellite islands.
Status: Common to abundant non-breeding visitor, known from early October to early May. Collected specimens identified to subspecies have all been of the form *D. p. palmarum*, but about 2% of mist-net captures in the Sierra de Bahoruco, DR, have been of *D. p. hypochrysea*.
Comments: This is one of six migratory warblers that occur primarily in the West Indies in the non-breeding season.
Range: Breeds from se. Yukon east to n. Saskatchewan, n. Ontario, e. Labrador, and Newfoundland, south to ne. British Columbia, s. Manitoba, n. Michigan, c. Maine, and Nova Scotia. Occurs in non-breeding season in North America, on East Coast from Delaware south through Florida to s. Texas, and on West Coast from Oregon to s. California. Also occurs throughout West Indies, and along Caribbean coast of Central America from Yucatán Peninsula to Honduras and Nicaragua. In West Indies, common in Bahamas and Greater Antilles, rare to vagrant south through Lesser Antilles to Barbados.
Local names: DR: Ciguíta Palmar; **Haiti:** Ti Tchit Palmis, Bilbelé, Ti Bon Ami.

BAY-BREASTED WARBLER—*Dendroica castanea* **Plate 46**
Passage Migrant

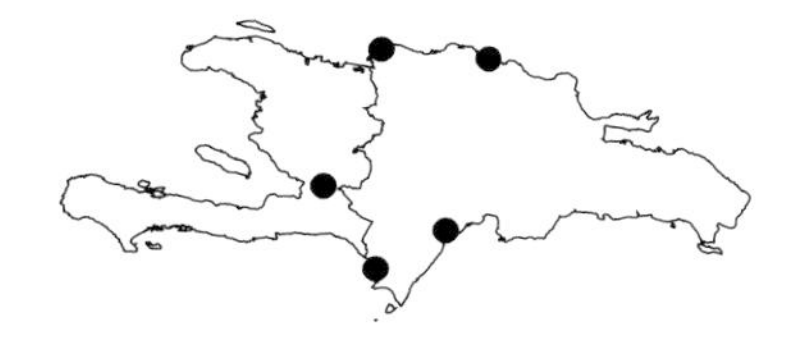

Description: 13–15 cm long; 13 g. A fairly large, nondescript warbler in non-breeding plumage, with greenish gray upperparts faintly streaked blackish, and whitish, unstreaked underparts. Undertail coverts are buffy, and two wing bars are white. Adult males have variable amounts of chestnut on sides, but females and immatures have a yellowish wash to sides. Legs and feet are grayish. Breeding male has a dark chestnut cap and band on the chin, throat, and sides; black mask; and buffy patch on sides of neck. Breeding female is duller; crown, breast, and sides are only washed with chestnut.
Similar species: Non-breeding adult and immature Blackpoll Warblers are finely streaked below and have pale legs, yellow feet and white undertail coverts. Pine Warbler has an unstreaked back, is yellow only from the chin to upper belly, and has black legs and feet. Cape May Warbler has a yellowish rump, often a yellow patch on side of neck, and is usually heavily streaked below.
Voice: Call notes include a clear, metallic *chip*; a high, loud, buzzy *tsip* or *tsee* given in flight or while foraging; and a thin *tsee-tsee-tsee* or *tititi*. Song is series of high, thin, squeaking notes that vary little in pitch.
Hispaniola: Eleven DR records, all from coastal areas in October, suggesting that this species is a rare fall passage migrant. Records are from Puerto Plata, Monte Cristi, Puerto Alejandro, and Cabo Rojo. A single record from Trou Caïman, Haiti, is also from October.
Status: Rare passage migrant.
Comments: A relatively slow-moving warbler, foraging methodically in the canopy. Primarily insectivorous, but may take fruits, especially in the dry season.
Range: Breeds from se. Yukon east to n. Alberta, c. Ontario, and w. Newfoundland, south to e. British Columbia, s. Manitoba, n. Michigan, n. New York, and s. Maine. Occurs in non-breeding season from Costa Rica south to n. Colombia and nw. Venezuela. In West Indies, rare in Bahamas, Greater Antilles, and south through Lesser Antilles to St. Vincent.
Local names: DR: Ciguíta Castaña; **Haiti:** Ti Tchit Fal Mawon, Paruline à Poitrine Baie.

BLACKPOLL WARBLER—*Dendroica striata* **Plate 44**
Passage Migrant

Description: 13–15 cm long; 13 g. In non-breeding plumage, upperparts are grayish olive variably streaked with black; underparts are whitish variably washed yellow on throat, breast, and sides, with diffuse olive streaking on sides. Lower belly and undertail coverts are pure white, and two wing bars are also white. Legs and feet are distinctively yellowish. Female and immatures are similar but have reduced streaking. Breeding male has black cap and malar stripe and broad white auriculars.
Similar species: Non-breeding adult and immature Bay-breasted Warblers lack streaking below, have buffy undertail coverts and black legs. Pine Warbler is unstreaked above and has black legs and feet.
Voice: Call is a sharp, clear *chip*. Song is a thin, high *tsit-tsit-tsit-tsit-tsit-tsit* on same pitch, usually louder in the middle and softer at the end.
Hispaniola: Recorded in fall from dry scrub on the southern coast. Spring records tend to be from similar habitat on the northern coast. There is a surprising absence of records from the interior. General fall arrival is late September, with most birds having passed through by early November. Latest known fall records are from late December. There are no midwinter reports. A few spring transients begin to arrive in late March, with most reports concentrated in late April and early May. Reported from several satellite islands, including Île à Vache, Île de la Gonâve, and Islas Alto Velo and Beata.
Status: Regular and common southbound passage migrant in fall, much less common as a northbound migrant in spring. Blackpoll Warbler is the most abundant migrant warbler in desert thorn scrub forest at Cabo Rojo, DR, in mid-October, but virtually all birds are gone by mid-November.
Comments: Primarily forages by gleaning insects, spiders, and small fruits from twigs and branches. At Cabo Rojo, most birds foraged on abundant lepidopteran caterpillars during migration stopover. Birds tended to arrive in waves with high-pressure systems from North America, but fewer than 15% remained more than 1 day before continuing southward.
Range: Breeds from w. Alaska east to n. Mackenzie, n. Ontario, n. Labrador, and Newfoundland, south to s. British Columbia, c. Saskatchewan, s. Quebec, and Nova Scotia; local in mountains of New England and New York. Occurs in non-breeding season in n. South America east of Andes, from nw. Colombia and nw. Brazil south to ne. Peru and c. Brazil. In West Indies, common to uncommon in Bahamas and Greater Antilles east to Puerto Rico; rare but regular south through Lesser Antilles to Grenada.
Local names: DR: Ciguíta Casco Prieto, Ciguíta Cabeza Negra; **Haiti:** Petit Chit, Ti Tchit Sèjan, Paruline Rayée.

BLACK-AND-WHITE WARBLER—*Mniotilta varia* **Plate 44**
Non-breeding Visitor

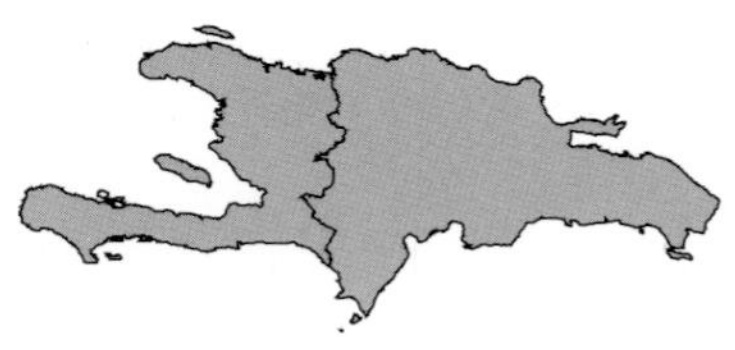

Description: 11–13 cm long; 11 g. A medium-sized, distinctively patterned warbler. Male in non-breeding plumage has upperparts and sides conspicuously streaked black and white, underparts white, crown with bold black and white stripes, and blackish auriculars. Female has buffy white underparts, less distinct streaking on sides, and grayer auriculars. Breeding plumage is similar but bolder with more contrast. Bill is long and slightly decurved.
Similar species: Unmistakable. Black-and-white-striped crown and tree-creeping habit distinguish this species from all other warblers on Hispaniola.

Voice: Call is a sharp, buzzy *chit* or *spik*; also a thin, weak *tsit* or *tseep*. Song, occasionally heard late in the non-breeding season, is a high-pitched, thin series of 2-syllable phrases with 5–10 repetitions, *weesee weesee weesee weesee weesee*.
Hispaniola: Reported from sea level to a maximum known elevation of 2,600 m. Most numerous at mid- and upper elevations in dry forest, moist broadleaf forest, shade coffee, and pine, but is more numerous in lowlands during migration periods. Recorded from late August to early May, with general arrival in October, and usual northbound departure in March. Reported from most major satellite islands.
Status: Moderately common non-breeding visitor.
Comments: Forages distinctively by creeping across branches and up and down tree trunks, probing bark for spiders and insects. May also foliage-glean. Regularly joins mixed-species foraging flocks. Tends to occupy larger home ranges than do most other non-breeding visitors.
Range: Breeds from se. Yukon east to c. Saskatchewan, n. Ontario, and Newfoundland, south to c. Alberta, e. Montana, sc. Texas, s. Alabama, and se. North Carolina. Extensive non-breeding-season range from s. North Carolina south through Florida and West Indies, se. Texas throughout all but northernmost Mexico, south to w. Ecuador and w. Venezuela. In West Indies, common in Bahamas, Greater Antilles, and Lesser Antilles south to Martinique, vagrant south to St. Vincent.
Local names: DR: Pegapalo; **Haiti:** Ti Tchit Demidèy, Demi-deuil, Paruline Noir et Blanc.

AMERICAN REDSTART—*Setophaga ruticilla* **Plate 44**
Non-breeding Visitor

Description: 11–13 cm long; 8 g. A small, active warbler, easily identified in all plumages. Adult male has black upperparts, throat, and breast and large orange patches on wings, tail, and sides. Belly and undertail coverts are white. Adult female is olive gray above, grayer on head, dull white below, with large, pale yellow patches on wings, tail, and sides. Immature male is similar to female, but yellow patches are more orange, and there is often some irregular black flecking on head, throat, and breast.
Similar species: Unmistakable.
Voice: Call is a clear, high *tschip*, somewhat liquid, and quite similar to that of Northern Parula; also a high, weak, rising *tsip* or *tsweet*. Two distinct male song types, both variable and of high-pitched, repeated notes. One has an emphatic, buzzy, down-slurred ending, *tsee tsee tsee tsee tseeo*; the second is softer and lower with an unaccented ending, *tseets tseeta tseeta tseet*.
Hispaniola: Recorded from sea level to a maximum known elevation of 1,500 m. Found in a wide variety of habitats from mangroves, coastal scrub, and lowland dry forest, rarely up to the pine zone, but most abundant in DR in moist broadleaf forest and shade coffee plantations in Cordillera Central and Sierra de Bahoruco. Fall arrival is typically mid-September to mid-October, with adult males tending to arrive later. Usual northbound departure occurs in March. The species was previously reported to be present in the DR during every month of the year, but fieldwork since the 1980s has not confirmed its presence in June or July. No documentation of breeding, though nesting has been confirmed in Cuba. Recorded from most major satellite islands.
Status: Common to locally abundant non-breeding visitor.
Comments: Redstarts display habitat segregation, with older males predominating in tall, relatively moist, mature forests, whereas females and young males tend to occupy scrubby second growth. They are strictly insectivorous, typically foraging at midcanopy level, often alone or in pairs, occasionally in mixed-species flocks. Opportunistic feeder, but most often forages by aerial sallies and flush-chases. This characteristic behavior includes acrobatic flycatching, with birds sometimes appearing to fall through the vegetation with much flapping and flashing of yellow or orange wing and tail markings.

Range: Patchily distributed when breeding from se. Alaska and nw. Yukon east to s. Mackenzie, n. Ontario, s. Labrador, and Newfoundland, south to nw. California, ec. Arizona, e. Texas, and nw. Florida. Occurs in non-breeding season from extreme s. Florida throughout West Indies, coastal s. California through Pacific slope of Mexico and Atlantic slope of c. Mexico south to Peru and nw. Brazil. Common throughout Bahamas and Greater Antilles, uncommon to rare in Lesser Antilles.
Local names: DR: Bijirita, Candelita; **Haiti:** Ti Tchit Dife, Paruline Flamboyante.

PROTHONOTARY WARBLER—*Protonotaria citrea* **Plate 49**
Passage Migrant; Non-breeding Visitor?

Description: 13–15 cm long; 15 g. A fairly large, strikingly yellow warbler with a large bill and prominent black eye. In non-breeding plumage, male has a bright golden yellow head, breast, and belly, contrasting with greenish yellow back and unmarked bluish gray wings and tail. Lower belly and undertail coverts are white; tail has conspicuous white outer webs. Females and immatures are duller and washed greenish, especially on crown and nape. Breeding birds are similar, but male is a rich golden yellow. Females are strongly washed green above, with yellow on head and underparts less intense than that of males.
Similar species: Blue-winged Warbler has white wing bars and a black eye-line. Yellow Warbler has yellow tail spots and undertail coverts, and yellowish wing bars on olive-brownish wings in all plumages.
Voice: Call is a clear, loud, metallic *tink* or *tchip*; also a quieter, thin *tsip*. Song is a repeated series of high-pitched, metallic, upslurred notes, *tsweet tsweet tsweet tsweet.*
Hispaniola: Usually found in coastal lowlands, most often in mangroves, but 1 unusual record from montane pine understory at 1,100 m. All records are from the DR; most are during the migratory period and include individual birds at Santo Domingo, Valle Nuevo, near Baní, Cabral, near Barahona, Cabo Rojo, and the mouth of Río Culebra. Six birds were mist netted in Cabo Rojo in 2 days during October 1997, suggesting that this species may occasionally move through in numbers.
Status: Uncommon passage migrant and possible non-breeding visitor.
Comments: This warbler may be easily overlooked because of its similarity to the common Yellow Warbler which also frequents mangroves. Reported to be nonterritorial in the non-breeding season. Feeds primarily on insects and spiders, but may take seeds, fruit, and even nectar.
Range: Breeds from ec. Minnesota east to s. Ontario, w. New York, and s. Connecticut, south to ec. Texas, Gulf Coast, and s. Florida. Occurs in non-breeding season from Caribbean coast of s. Mexico south to Panama, n. Colombia, and n. Venezuela. In West Indies, uncommon in Bahamas and Greater Antilles; rare to vagrant throughout Lesser Antilles.
Local names: DR: Ciguíta Cabeza Amarilla; **Haiti:** Ti Tchit Tèt Jòn, Paruline Orangée.

WORM-EATING WARBLER—*Helmitheros vermivorum* **Plate 45**
Non-breeding Visitor

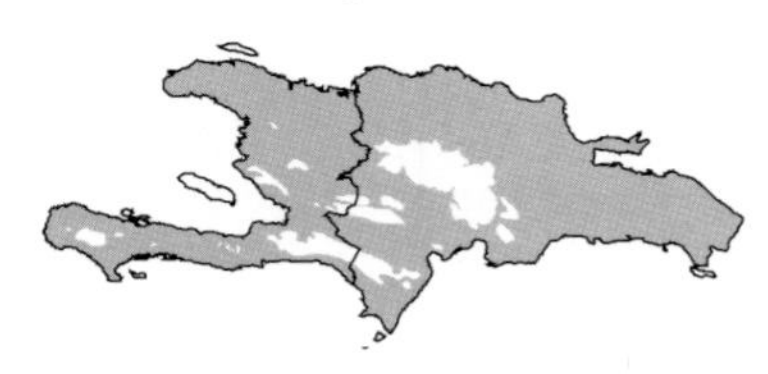

Description: 11–13 cm long; 13 g. A medium-sized, stocky warbler with a large bill and flat head. Upperparts are olive-brown, and head is buffy with bold black lateral crown stripes and eye stripes. Underparts are buffy but slightly whiter on throat and belly. There are no wing bars or tail spots. Tail is fairly short and squared.

Similar species: Swainson's Warbler has chestnut brown crown; long, creamy buff supercilium; and large bill.
Voice: Call is a sharp, loud *chip* or *tchik*; flight call is a short, high buzz, *dzt*. Song, rarely heard in non-breeding season, is a rapid, flat, buzzy trill.
Hispaniola: Recorded from sea level to a maximum known elevation of at least 2,100 m at Pueblo Viejo, Sierra de Bahoruco, though most records are below 1,000 m. Typically found in dry forest, wet limestone forest, sun coffee plantations, or transitional broadleaf forest. Occurs between late September and late March.
Status: Uncommon non-breeding visitor.
Comments: A secretive species that is typically distributed sparsely through its habitat, more often captured in mist nets than observed. Like many North American migrants that are territorial in the non-breeding season, tends to show strong year-to-year fidelity to specific sites. Strictly insectivorous. Often forages at suspended clusters of dead, curled leaves at all levels in the forest.
Range: Tends to breed locally within range, which extends from ne. Kansas and se. Iowa east to sw. Michigan, se. New York, and se. Massachusetts, south to e. Texas, s. Mississippi, and nw. Florida. Occurs in non-breeding season from s. Mexico south along Atlantic and Pacific slopes of Central America to c. Panama. In West Indies, common to uncommon in Bahamas and Greater Antilles; rare to vagrant from Puerto Rico south to Guadeloupe.
Local names: DR: Ciguíta Cabeza Rayada; **Haiti:** Ti Tchit Tèt Plat, Paruline Vermivore.

SWAINSON'S WARBLER—*Limnothlypis swainsonii* — Plate 45
Non-breeding Visitor

Description: 13–14 cm long; 19 g. A medium-sized warbler with a long, heavy, sharply pointed bill. Upperparts are unmarked brown or olive-brown. Crown is chestnut brown with a contrasting pale white or creamy buff supercilium, and a dusky line through eye. Underparts are yellowish white to white, but grayer on sides.
Similar species: Worm-eating Warbler has bold black stripes on head.
Voice: Call note is a strong, liquid, descending *chip* or *sship*; also a short, metallic *ziip* and a high, thin *zeep* or *sreee* flight call. Song, unlikely to be heard in non-breeding season, is a loud and ringing series of clear, slurred notes, *whee whee whee whip-poor-will*.
Hispaniola: Only 13 reports known, all of mist-netted individuals from four sites since 1997: 2 from western Sierra de Martín García, 5 from montane broadleaf forest at about 1,400 m elevation in Sierra de Bahoruco, 5 at wet broadleaf forest sites in Massif de la Hotte at elevations of 1,200 m and 1,850 m, and 1 in a broadleaf forest fragment of Massif de la Selle at 2,025 m elevation. At all sites, birds occupied dense understory of vines and shrubs beneath a solid canopy of broadleaf trees 8–15 m high. One individual in the Bahorucos was later recaptured in the same season, and 2 in following years. These reports suggest that the species is probably a regular, if rare or locally uncommon, non-breeding-season resident in montane forest. Intensive mist netting or focused surveys in appropriate habitat of Sierra de Neiba and Cordillera Central seem likely to reveal the species' presence there also.
Status: Rare but regular non-breeding visitor.
Comments: Strictly insectivorous, foraging primarily on the ground for insects and spiders among the leaf litter. Retiring and extremely difficult to locate. Likely occurs more frequently than records suggest.
Range: Breeds from e. Oklahoma east to s. Illinois, s. West Virginia, and s. Delaware, south to e. Texas, Gulf Coast, and n. Florida. Occurs in non-breeding season in Bahamas, Greater Antilles, Yucatán Peninsula, n. Guatemala, and Belize. In West Indies, regular in n. Bahamas, Cuba, and Jamaica, rare to casual in s. Bahamas and Hispaniola, vagrant to Puerto Rico and Virgin Islands.
Local names: DR: Ciguíta de Swainson; **Haiti:** Ti Tchit Bèk Pwenti, Paruline de Swainson.

OVENBIRD—*Seiurus aurocapilla* **Plate 45**
Non-breeding Visitor

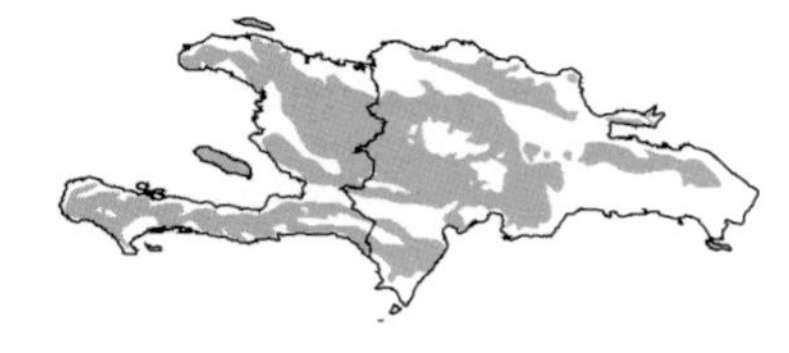

Description: 14–16 cm long; 19 g. A large, ground-dwelling warbler. Upperparts uniformly grayish olive, with orangish crown bordered by blackish stripes and a bold white eye-ring. Underparts white boldly streaked with blackish on throat, breast, and sides. Legs are pinkish. Sexes identical. Immature is similar to adults but with rusty-fringed tertials which are difficult to see.
Similar species: Migratory thrushes are larger and lack the blackish crown stripes. Northern and Louisiana waterthrushes have a conspicuous white supercilium, unmarked crown, no eye-ring, and often wag their tail.
Voice: Fairly vocal, giving a loud, sharp *chup* or *chek* which is frequently repeated in territorial interactions. Song, rarely heard in late non-breeding season, is an explosive, ringing *teacher teacher teacher teacher*, increasing in volume.
Hispaniola: Reported from sea level to a known maximum elevation of 2,100 m, though most numerous at midelevations from 200 to 1,500 m. Often abundant in shade coffee plantations and moist broadleaf forest in Cordillera Central, DR. Abundant in dry forest and moist broadleaf habitats in Sierra de Bahoruco, less numerous but still common in the pine zone there at higher elevations. Rare in lower elevation desert thorn scrub. Reported from most major satellite islands.
Status: Common non-breeding visitor islandwide. The form *S. a. aurocapilla* is by far the most common subspecies present. The westcentral U.S. subspecies *S. a. cinereus* is a rare non-breeding-season resident. The subspecies *S. a. furvior* from Newfoundland may occur but has not yet been documented on Hispaniola.
Comments: Territorial. In Sierra de Bahoruco, highly faithful to individual territories from year to year. Ovenbirds glean prey from the leaf litter, primarily taking ants, beetles, and seeds, also snails and spiders.
Range: Breeds from nw. British Columbia east to s. Mackenzie, c. Manitoba, c. Quebec, and Newfoundland, south to s. Alberta, ne. Wyoming, se. Oklahoma, n. Alabama, and e. North Carolina. Occurs in non-breeding season from c. Florida through West Indies, and n. Mexico south to Panama, occasionally to n. Venezuela and n. Colombia. In West Indies, common thoughout Bahamas and Greater Antilles, rare to vagrant in Lesser Antilles.
Local names: DR: Patico, Ciguíta Saltarina; **Haiti:** Ti Tchit Dore, Paruline Couronnée.

NORTHERN WATERTHRUSH—*Seiurus noveboracensis* **Plate 45**
Passage Migrant; Non-breeding Visitor

Description: 13–15 cm long; 18 g. A large, terrestrial warbler with dark olive-brown, unmarked upperparts and buffy white underparts with well-defined dark brown streaks that are finer on throat. A prominent buffy white supercilium narrows behind eye. Sexes are identical. Waterthrushes have the habit of constantly bobbing and teetering as they walk and of rhythmically pumping their tail.
Similar species: Louisiana Waterthrush is distinguished by its unstreaked whitish throat and conspicuous white supercilium, which is broader and extends well behind eye. Ovenbird has an orange crown bordered with black stripes and a prominent white eye-ring.
Voice: Call note is a recognizable loud, and quite metallic, *chink*. Song, rarely heard on the wintering grounds, is a loud, ringing series of notes, generally falling in pitch and accelerating, *sweet sweet sweet swee-wee-wee chew chew chew chew*.

Hispaniola: Reported from sea level to a maximum known elevation of about 1,640 m, though the great majority of reports are in lowlands at or below 150 m elevation. Virtually always found near water, whether streams, rivers, coastal lagoons, or borders of lakes and ponds, but most often close to mangroves. General arrival in fall is from September to mid-October; typical northbound spring departure is from mid-March to mid-April. Reported from most major satellite islands.
Status: Moderately common to abundant passage migrant and non-breeding visitor. Large numbers pass through on migration, and many remain for the non-breeding season. Traditionally, at least two subspecies are expected in the West Indies: eastern *S. n. noveboracensis* is the most numerous, whereas *S. n. notabilis* from the northwest portion of the species' range appears to be less common but perhaps more likely to be found at interior freshwater localities. Relative abundance and distribution of these two subspecies on Hispaniola are not known.
Comments: Territorial and solitary in the non-breeding season, rarely encountered far from water. Terrestrial and fairly tame, it walks deliberately along the water's edge, bobbing and teetering, stopping to pick insects or larvae from the substrate. Feeds primarily on aquatic insects, but may take snails, small clams, crabs, and occasionally seeds.
Range: Breeds from w. Alaska east to nw. Mackenzie, n. Manitoba, n. Quebec, and Newfoundland, south to s. Alaska, sw. Oregon, s. Michigan, nw. Virginia, and Rhode Island. Occurs in non-breeding season from s. Florida, Baja California, and e. Mexico south through West Indies and Central America to Venezuela, Colombia, and ne. Peru. In West Indies, common in Bahamas and Greater Antilles, uncommon to rare in Lesser Antilles.
Local names: DR: Ciguíta del Agua; **Haiti:** Ti Tchit Mang Lanmè, Paruline des Ruisseaux.

LOUISIANA WATERTHRUSH—*Seiurus motacilla* **Plate 45**
Passage Migrant; Non-breeding Visitor

Description: 13–15 cm long; 20 g. A large, terrestrial warbler, uniformly dark olive-brown above and white below, with dark brown streaks on breast and sides, and flanks washed buffy. Distinctive field marks include a white supercilium which flares and broadens behind eye, and an unspotted white throat. Sexes are identical. Like Northern Waterthrush, this species has the habit of constantly bobbing and teetering, and rhythmically pumping its tail.
Similar species: Northern Waterthrush is washed with yellow or buffy below, has fine streaks on throat, and has a buffier supercilium which narrows behind eye. Ovenbird has an orange crown bordered with black stripes and a prominent white eye-ring.
Voice: Call note is a loud, metallic *chink* or *chip*, similar to that of Northern Waterthrush but slightly softer and higher pitched. Song is a ringing, musical, clear series of slurred whistles, *seeup seeup seeup*, followed by a complex jumble of shorter, rapid chips.
Hispaniola: Reported from sea level to a maximum known elevation of 1,800 m, though few records exist above 1,000 m. During migration periods, often found in mangroves, along streams and marsh borders at sea level, but otherwise prefers streams with flowing freshwater in hills and mountains of the interior or the edges of freshwater lakes such as Étang Miragoâne, Haiti. Earliest arrivals frequently reported in mid-August, always at coastal localities, though 1 was seen in late August 2003 at a stream near Sabanita de Pallavo, Don Juan de Monte Plata; these are believed to be transients. General fall arrival at inland sites is late August and September. Northbound spring departure occurs in early to mid-March. Recorded from several major satellite islands, including Navassa Island, Île de la Gonâve, and Islas Beata and Saona.
Status: Uncommon non-breeding visitor and passage migrant.
Comments: Establishes non-breeding-season territories along streams. Terrestrial foraging behavior similar to that of Northern Waterthrush. Feeds primarily on aquatic insects and other invertebrates; reported to occasionally take small frogs.

Range: Breeds from e. Minnesota and Nebraska east to c. Michigan, s. Ontario, ne. New York, and s. Maine, south to e. Texas, s. Louisiana, and nw. Florida. Occurs in non-breeding season from n. Mexico south to c. Panama, rarely to ne. Colombia and n. Venezuela. In West Indies, uncommon in Greater Antilles, and rare to vagrant south through Lesser Antilles to St. Vincent.
Local names: DR: Ciguíta del Río; **Haiti:** Ti Tchit Rivyè, Paruline Hochequeue.

KENTUCKY WARBLER—*Oporornis formosus* **Plate 48**
Passage Migrant; Non-breeding Visitor?

Description: 12–14 cm long; 14 g. A medium-sized, ground-dwelling warbler. Adult male has plain olive green upperparts and solid bright yellow underparts. Crown, forecrown, and side of head black, forming a distinctive mask; yellow supercilium and partial eye-ring form bold spectacles. Legs are pinkish. Adult female and immature male are similar to adult male but with less extensive black on face and crown. Black on face of immature female is replaced by grayish olive. Breeding plumage is similar, but black on face and crown is more pronounced.
Similar species: Hooded Warbler has variable black hood, entirely yellow face, and white tail spots. Canada Warbler has bluish gray upperparts, streaked breast, and white undertail coverts. Immature male Common Yellowthroat lacks yellow spectacles and is not uniformly yellow below.
Voice: Calls vary from a low, hollow *chok* or *chup* to a higher, sharper *chip* or *chek* when agitated. Song is a rolling series of 2-syllable phrases, *prr-reet prr-reet prr-reet prr-reet prr-reet*.
Hispaniola: Five DR records, from Pepillo Salcedo; Puerto Escondido and Aceitillar, Sierra de Bahoruco; Loma Guaconejo, Cordillera Septentrional; and Los Haitises National Park. Another 5 birds were recorded in mid-October 2003 in the interior of Los Haitises National Park, suggesting that this species may occur in larger numbers than currently suspected. Normally occupies dense undergrowth and thickets in moist forest understory.
Status: Rare but regular passage migrant and possible non-breeding visitor on Hispaniola.
Comments: Typically inhabits understory of moist forests in non-breeding season. Solitary and territorial. Little information on diet, but presumed to forage primarily on ground for insects and spiders.
Range: Breeds from sw. Wisconsin and e. Nebraska east to n. Indiana, Pennsylvania, and se. New York, south to e. Texas, Gulf Coast, and nw. Florida. Occurs in non-breeding season from e. Mexico south through Panama to n. Colombia and n. Venezuela. In West Indies, rare in Bahamas, Greater Antilles, and U.S. Virgin Islands; vagrant on Antigua and Guadeloupe.
Local names: DR: Ciguíta de Kentukí; **Haiti:** Ti Tchit Tè, Paruline du Kentucky.

CONNECTICUT WARBLER—*Oporornis agilis* **Plate 48**
Passage Migrant

Description: 13–15 cm; 15 g. A large, terrestrial warbler with a stocky build. Olive to olive-brown above, with a pale gray to brownish hood extending to lower throat, and a complete white eye-ring in all plumages. Remaining underparts are pale yellow with long undertail coverts extending nearly to tip of tail. Adult male has a bluish gray hood; adult female and immature have a grayish brown hood, and throat is whitish or pale buffy.
Similar species: Mourning Warbler is similar but has a narrow, broken white eye-ring and shorter undertail coverts.
Voice: Usually silent in non-breeding season, but call note is a soft *pwik* or *poit*.
Hispaniola: Seldom observed because of its extremely retiring habits, but mist netting in the 1990s showed it to occur much more often than previously known. Records are all from DR and include individual birds observed at Santo Domingo, Laguna Villa Isabella, and between

Baní and Azua. Several birds were banded and others observed at Cabo Rojo in mid-October 1997, suggesting the species may regularly move through the island during migration. Most records are from October and early November, but once recorded in spring.
Status: Regular passage migrant in small numbers.
Comments: Typically stays on or near the ground in dense understory of moist woodlands, usually near water. Non-breeding-season diet not well known but likely consists of insects and spiders, and may include some fruits. Migratory patterns also poorly known overall. There is some evidence for a nonstop transoceanic migration by Connecticut Warblers in fall, but most are believed to migrate northward from South America through the West Indies in spring. The non-breeding ecology of this secretive species needs further study.
Range: Breeds from ec. British Columbia east to c. Manitoba and sc. Quebec, south to s. Manitoba, n. Wisconsin, and sc. Ontario. Non-breeding-season range poorly known but primarily in n. South America, from Colombia and Venezuela south to c. Brazil and Peru. In West Indies, regular passage migrant through Hispaniola, otherwise a vagrant in Bahamas and rest of Greater Antilles; also St. Martin, and St. Barthélemy.
Local names: DR: Ciguíta de Lentes; **Haiti:** Ti Tchit Fal Gri, Paruline à Gorge Grise.

MOURNING WARBLER—*Oporornis philadelphia* **Plate 48**
Vagrant

Description: 13–15 cm long; 13 g. A fairly large, sturdy warbler with olive green upperparts and bright yellow underparts. In non-breeding plumage, a gray to brownish gray hood extends to lower throat, but upper throat is often yellowish. Legs and feet are pinkish. Both sexes may have thin, white arcs above and below eye. In breeding plumage, male has a bluish gray hood, including a broad black bib on lower edge, and lacks an eye-ring. Female is similar to male but has a pale gray or brownish gray hood without black on lower edge, and a broken eye-ring.
Similar species: Connecticut Warbler has a complete, bold white eye-ring and longer undertail coverts.
Voice: Distinctive harsh, metallic *chik* or *jink*.
Hispaniola: Two DR reports: 1 individual collected near Zapotén, Sierra de Bahoruco in October 1974; another passed the non-breeding period of 1973–1974 near Santiago.
Status: Vagrant.
Comments: Typically found on or near the ground. Prefers wet habitats, including dense thickets, second growth, abandoned fields, and swamp edges. Primarly insectivorous, but may take seeds and fruit in the non-breeding season.
Range: Breeds from ne. British Columbia east to ne. Alberta, c. Ontario, and Newfoundland, south to Manitoba, c. Minnesota, n. Illinois, w. Massachusetts, and in Appalachians south to West Virginia. Occurs in non-breeding season from s. Nicaragua south to w. Colombia and s. Venezuela. A vagrant in West Indies in Bahamas and Greater Antilles.
Local names: DR: Ciguíta Triste; **Haiti:** Ti Tchit Tris, Paruline Triste.

COMMON YELLOWTHROAT—*Geothlypis trichas* **Plate 46**
Non-breeding Visitor

Description: 11–13 cm long; 10 g. A medium-sized, compactly built warbler. Adult male has plain olive green upperparts, with a conspicuous black facial mask edged above by white or gray. Throat, breast, and undertail coverts are bright yellow; belly and sides are dusky or brownish white. Adult female is like male but lacks a facial mask and has paler buffy yellow underparts. Immature is duller and browner than adult female; male may have traces of black mask.

Similar species: Kentucky Warbler is entirely bright yellow below, with yellow spectacles. Female and immature Connecticut and Mourning warblers have underparts entirely yellow, and a grayish or pale brownish hood.
Voice: Distinctive call note is a dry, husky *tchep* or *tchat*. Clear song, *witchety, witchety, witchety, witch*, may rarely be given late in the non-breeding season before northward migration.
Hispaniola: Reported from sea level to a maximum known elevation of at least 2,300 m. Common to abundant in dry scrubby growth, overgrown fields, regenerating thickets in forest openings, herbaceous scrub understory of pine forest, and sun coffee plantations. Also inhabits the borders of lowland marshes and swampy meadows. At higher elevations frequents patches of weeds and bracken. Appears to be common in suitable habitat at almost any elevation. Recorded from late September to late April or even early May. Reported from several major satellite islands, including Navassa Island, Île à Vache, Île de la Gonâve, and Île de la Tortue.
Status: Common non-breeding visitor. Most birds are of the subspecies *G. t. trichas* from eastern North America, but *G. t. typhicola* from southeastern U.S. is also present in small numbers.
Comments: Highly territorial on the non-breeding grounds. May show habitat segregation by sex, with males more common in second growth and scrub and females in open pasture and regenerating fields. Forages for insects and spiders on or near the ground.
Range: Breeds from se. Alaska east to n. Alberta, c. Manitoba, c. Quebec, and Newfoundland south to n. Baja California, c. Mexico, s. Texas, Gulf states, and s. Florida. Occurs in non-breeding season from c. California, s. Texas, and North Carolina south to c. Panama, casually to n. Ecuador and w. Venezuela. In West Indies, common in Bahamas and Greater Antilles, vagrant in Lesser Antilles south to Dominica.
Local names: DR: Ciguíta Enmascarada; **Haiti:** Ti Tchit Figi Nwa, Paruline Masquée.

HOODED WARBLER—*Wilsonia citrina* **Plate 48**
Non-breeding Visitor

Description: 12–14 cm long; 10 g. Distinctive adult male plumage, with olive green upperparts and a black hood and throat contrasting with a bright yellow forecrown and auriculars; rest of underparts are bright yellow. Adult female is similar to male, but extent of black hood varies, ranging from almost complete to having black markings only on crown. Immature male similar to adult but with olive-yellow tipping on throat and crown feathers. Immature female lacks black hood and has a yellow face contrasting with olive green crown and nape. Note the large, dark eye and characteristic tail-fanning behavior, which exposes large white spots in outer tail feathers.
Similar species: Kentucky Warbler has variable black on face in all plumages, yellow spectacles, and lacks white tail spots. Adult female and immature Wilson's Warblers are smaller, have a yellow supercilium with no black bordering face, and lack white in tail feathers.
Voice: Call is a loud, metallic *chink* or *chip*. Song is a loud, clear, musical series of slurred notes, *ta-wit ta-wit ta-wit teeoo*.
Hispaniola: All known reports are from DR and include individuals at Santo Domingo; San Pedro de Macorís; Puerto Plata; Loma Guaconejo in the Cordillera Septentrional; El Limón; La Vega; Manabao; near Laguna de Oviedo; and Mencia, Pedernales, and Aceitillar in the Sierra de Bahoruco. Also reported from Isla Saona. Reported from late September to early April.
Status: Uncommon but regular non-breeding visitor.
Comments: Strongly territorial in non-breeding season. Segregates by sex, with males predominating in mature forest and females occupying scrub, secondary forest, and disturbed habitats. Insectivorous, foraging on or close to ground.

Range: Breeds from ne. Iowa east to c. Wisconsin, s. Ontario, se. New York, and Rhode Island, south to c. Texas, Gulf Coast, and n. Florida. Occurs in non-breeding season from e. Mexico south to Panama, rarely to n. Colombia and Venezuela. Bulk of non-breeding population found in Yucatán Peninsula, Mexico. In West Indies, uncommon to rare in Bahamas and Greater Antilles, vagrant south in Lesser Antilles to St. Vincent.
Local names: DR: Ciguíta Gorra Negra; **Haiti:** Ti Tchit Kagoul Nwa.

WILSON'S WARBLER—*Wilsonia pusilla* **Plate 48**
Vagrant

Description: 10–12 cm long; 8 g. A small, spry warbler. Adult male has upperparts uniformly yellowish olive-green, a glossy black crown, and bright yellow forecrown and supercilium. Underparts are entirely yellow. Adult female and immature male are similar to adult male, but duller with little or no black in crown. Immature female invariably lacks black cap and has a more olive forecrown. Characteristically flicks its tail upward and hops restlessly from perch to perch.
Similar species: Adult female and immature Hooded Warblers are larger and have white tail patches and a yellow face patch, rather than a supercilium. Immature Yellow Warbler lacks yellow supercilium and has yellow primary edges and inner webs of tail feathers.
Voice: Call is a husky, nasal *chip* or *jip*. Song, rarely heard in non-breeding season, is a rapid series of 10–15 short, whistled notes with sharp, chattery quality, slowing and dropping in pitch toward the end, *chchchchchchchch*.
Hispaniola: Two DR sight records by experienced field observers: 1 at La Ciénaga near Santo Domingo in early March 1978, and 1 at Aguacate, Sierra de Bahoruco, in early November 1987.
Status: Vagrant.
Range: Breeds from nw. Alaska, nw. Mackenzie, n. Manitoba, n. Quebec, and Newfoundland south to c. California, n. New Mexico, ne. Minnesota, s. Quebec, ne. Vermont, and Nova Scotia. Occurs in non-breeding season from se. Texas, s. Louisiana, and n. Mexico south to w. Panama. Vagrant in West Indies on Bahamas, Cuba, Jamaica, DR, and Puerto Rico.
Local names: DR: Ciguíta de Wilson; **Haiti:** Ti Tchit Kepi Nwa, Paruline à Calotte Noire.

CANADA WARBLER—*Wilsonia canadensis* **Plate 47**
Vagrant

Description: 12–15 cm long; 10 g. Adult non-breeding male has bluish gray upperparts, black forecrown contrasting with complete white eye-ring, and yellow lores that form conspicuous spectacles. Underparts are yellow with bold black breast streaks forming a necklace; undertail coverts are white. Adult female is similar to male but duller overall, with paler, less distinct markings on head and breast. Immatures are like adults but duller still, with less distinct breast streaking and facial markings. Breeding birds are similar to non-breeding male but with bolder black markings.
Similar species: Kentucky Warbler has olive upperparts, yellow undertail coverts, and lacks streaking on breast. Female Hooded Warbler has a yellow mask, yellow undertail coverts, and conspicuous white tail spots.
Voice: Calls include a low, subdued *chip* or *tchup* and a loud, sharp *check*. Song is a series of rich, jumbled, liquid notes, often preceded by a loud *chip* note.
Hispaniola: Two late-fall DR reports: 1 male carefully identified near Constanza, and 1 male at La Placa, Sierra de Bahoruco.
Status: Vagrant.
Range: Breeds from se. Yukon, n. Alberta, n. Ontario, sc. Quebec, and Nova Scotia south to c. Alberta, s. Wisconsin, c. Pennsylvania, s. Rhode Island, south in Appalachians to w. Maryland and nw. Georgia. Occurs in non-breeding season in South America, mainly east of

Andes, and from Venezuela and Colombia south to e. Ecuador and c. Peru. Vagrant in West Indies to Bahamas, Greater Antilles, Guadeloupe, St. Lucia, and Barbados.
Local names: DR: Ciguíta del Canadá; **Haiti:** Ti Tchit Kanada, Paruline du Canada.

YELLOW-BREASTED CHAT—*Icteria virens* **Plate 49**
Vagrant

Description: 18–19 cm long; 25 g. A large, aberrant warbler with a robust build, heavy bill, and long tail. Upperparts, wings, and long tail are olive green. Throat, breast, and upper belly are yellow; lower belly and undertail coverts are white. Note the white spectacles. Lores are black in males, gray in females, and are bordered above and below by white.
Similar species: Unmistakable.
Voice: Calls are variable; include a harsh, nasal *cheow* or *chak*, a sharp *cuk* or *kook*, and a low, soft *tuk*. Song is extremely varied, a collection of whistles, cackles, mews, rattles, squawks, and gurgles.
Hispaniola: One October 2000 record of a bird mist netted in Del Este National Park near Guaraguao at Bayahibe, DR.
Status: Vagrant.
Comments: Although genetic studies suggest that this species is not closely related to other warblers, it is also not yet clear with what other species it is allied. The chat prefers dense scrub-shrub habitat or thick forest undergrowth and feeds on insects, spiders, and fruit during migration and in the non-breeding season.
Range: Breeds from s. British Columbia, s. Saskatchewan, s. Wisconsin, s. Ontario, and c. New York south to n. Baja California, sw. Texas, Gulf Coast, and n. Florida, locally to sc. Mexico. Occurs in non-breeding season along Pacific and Atlantic slopes of Mexico south to w. Panama. In West Indies, rare in Bahamas and Cuba, vagrant in Cayman Islands and DR.
Local names: DR: Ciguíta Grande; **Haiti:** Gwo Tchit Fal Jòn.

BANANAQUIT: FAMILY COEREBIDAE

The Bananaquit, the sole representative of its family, is widely distributed throughout the Caribbean and Central and South America. Formerly classified in a subfamily with honeycreepers, Bananaquits are small arboreal birds with a slender, curved bill used to pierce the base of flowers to obtain nectar and insects. They have a specially adapted brushlike tongue for extracting nectar. Soft fruits also form part of their diet.

BANANAQUIT—*Coereba flaveola* **Plate 53**
Breeding Resident

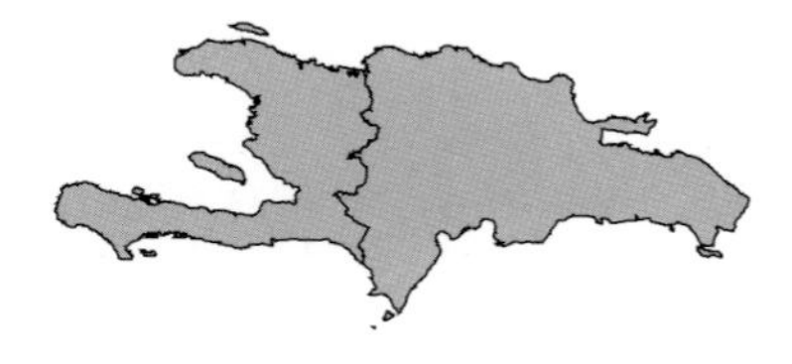

Description: 10–11 cm long; 9 g. A small, colorful bird with a short tail and slender decurved bill. Adult is grayish black above with yellow rump, bold white supercilium, and small white spot on wing. Underparts, including breast and belly, are yellow, but throat is gray. Bill has a reddish pink spot at base. Immature is paler and duller than adult and has a yellowish white supercilium.
Similar species: Unmistakable.
Voice: Song is a thin, wheezy, high-pitched series of rapid notes, *tzi-tzi-tzi-tzi-tziit-tzi*. Call note is a short, high, unmusical *tsip*.
Hispaniola: Found in nearly all habitat types, including desert, but most numerous in mesic scrub and forest, particularly moist broadleaf forest; less often found in numbers in pine habi-

tat, but usual in disturbed areas and openings in pine forest even at higher elevations. Also common in gardens, suburban areas, town parks, and plantations. Reported from sea level to at least 2,500 m elevation. Reported from most major satellite islands.

Status: Locally common, often abundant, breeding resident. Appears to have been common to abundant since at least 1900. Two endemic subspecies are recognized: the subspecies on the main island, and presumably on all the satellite islands except Île de la Tortue, is *C. f. bananivora*; the subspecies confined to Île de la Tortue is *C. f. nectarea*.

Comments: Many nests are built to serve only as roost sites. While foraging, the Bananaquit commonly pierces the base of flowers to steal nectar, its primary food.

Nesting: Eggs are dull whitish, marked with warm brown and reddish spots, especially at the larger end. Lays 2–4 eggs in a globular nest of grass and plant fibers with an opening in the side or partially underneath. The nest is usually built at the tip of a branch 1–6 m above ground. Of 19 nests monitored in desert thorn scrub of Sierra de Bahoruco, 15 were built in cacti at a mean height of 1.5 m above ground, with a mean clutch size of 2.7 eggs and fledging dates that ranged from April 30 to July 11. Believed to breed year-round.

Range: Breeds in Bahamas and throughout rest of West Indies except Cuba, and from s. Mexico south to nw. Peru and n. Argentina.

Local names: DR: Ciguíta Común; **Haiti:** Sucrier, Kit.

TANAGERS: FAMILY THRAUPIDAE

The tanagers are a large Western Hemisphere assemblage. The males are often beautifully plumaged but are endowed with limited vocal ability. They are typically arboreal forest-dwelling species that feed almost exclusively on fruits, but here we include both *Microligea*, which is primarily insectivorous, and *Xenoligea* and *Phaenicophilus*, which both consume considerable quantities of insects as well.

GREEN-TAILED GROUND-TANAGER—*Microligea palustris* **Plate 50**
Breeding Resident—*Endemic*

Description: 12–14 cm long; 13 g. A small, slender, active bird with a long tail. Upperparts are unmarked olive greenish, except head, face, and nape are grayish. Underparts are uniformly dull white. Note the incomplete white eye-ring. Adults have a distinctive red iris; immatures have a brown iris and a greenish wash on head and nape.

Similar species: Hispaniolan Highland-Tanager has a small white patch on outer primaries and in front of eye, has white tail spots, and lacks red iris.

Voice: Calls are a varied series of rasping and squeaking notes, similar to typical alarm calls of small song birds. Song is a high *sip sip-sip*.

Hispaniola: Found in lowland dry scrub habitats, desert thorn scrub, and dry broadleaf forest, and at higher elevations in moist broadleaf forest, especially with dense thickets of climbing bamboo. Also found in dense understory of pine habitat. Recorded from sea level to a maximum known elevation of 2,925 m in Cordillera Central, DR. In DR, occurs in the western third of the country from near Monte Cristi southeasterly in the foothills and mountains of the Cordillera Central to about Nizao, in the lowlands from Azua to Barahona, in Sierra de Neiba, Sierra de Bahoruco, the Neiba Valley, and western Pedernales Province, but absent from the coastal strip from Barahona to Cabo Beata. Also an isolated population inhabits coastal lowlands of DR from near San Pedro de Macorís to Isla Saona. Reported at San Lorenzo, Samaná Bay, and Los Haitises National Park, but not regular there. In Haiti, known only from two areas: higher elevations in Massif de la Selle, but to date not recorded west of Jacmel on the Tiburón

Peninsula; and an isolated population in the coastal lowlands up to 330 m elevation in extreme northwestern Haiti on Morne Chien and between Bombardopolis and Môle Saint-Nicolas, an area where the species apparently did not occur in the early 1900s.
Status: Endemic, locally common resident. Formerly known as Green-tailed Warbler or Green-tailed Ground-Warbler. Genetic data published in 2001 suggest that the Green-tailed Ground-Tanager (*Microligea*) and Hispaniolan Highland-Tanager (*Xenoligea*) are closely related to the endemic palm-tanagers (*Phaenicophilus*) and are not warblers as previously suspected. Although its numbers have been described as "common" to "abundant" at all reported elevations, the Green-tailed Ground-Tanager's population density appears to be greatest at middle and higher elevations. On Isla Beata the local race is considered threatened, but the cause of the decline is unknown. The endemic subspecies *M. p. palustris* occupies most of the species' range on the main island. The endemic subspecies *M. p. vasta* was described as possibly confined to Isla Beata, but birds resident in the xeric lowland habitat from southern Pedernales Province north and east to southwestern Peravia Province may pertain to this form. The highland *palustris* subspecies is larger and darker than the smaller, paler *vasta* form of the lowlands. Subspecific determination of the isolated populations in northwestern Haiti and southeastern DR is lacking.
Comments: Often feeds alone or in pairs, but regularly joins mixed-species foraging flocks in pine habitat in the non-breeding season. Banded individuals in Sierra de Bahoruco have been sighted or mist netted repeatedly at same sites over as many as 7 years. Primarily insectivorous, the ground-tanager feeds low to the ground and in thick tangles of brush and vines. A fairly inquisitive species, it often readily responds to *pishing* noises.
Nesting: Lays 2–4 pale green, spotted eggs in a cup-shaped nest of plant materials, typically built low to the ground. Among 18 nests monitored in desert thorn scrub of Sierra de Bahoruco, 15 were built in shrubs and 3 in cacti at an average height of 1.2 m above ground (range 0.5–2.5). Average clutch size was 3.2 eggs (range 2–4), and average hatching date was May 11. In montane broadleaf forests of Sierra de Bahoruco, average height above ground of 14 tree and understory shrub nests was 2.2 m (range 1.0–10), average clutch size was 2.2 eggs (range 2–3), and average hatch date was June 12. Fledging rates were low in both habitats: 30% in desert, 38% in montane forests. Most nests failed because of predation. Most nesting attempts occur April to June, but it is suspected to sometimes nest year-round.
Range: Endemic to Hispaniola, including Islas Beata and Saona.
Local names: DR: Ciguíta Cola Verde, Tangara Cola Verde; **Haiti:** Ti Tchit Lasèl.

HISPANIOLAN HIGHLAND-TANAGER—*Xenoligea montana* **Plate 50**
Breeding Resident—*Endangered Endemic*

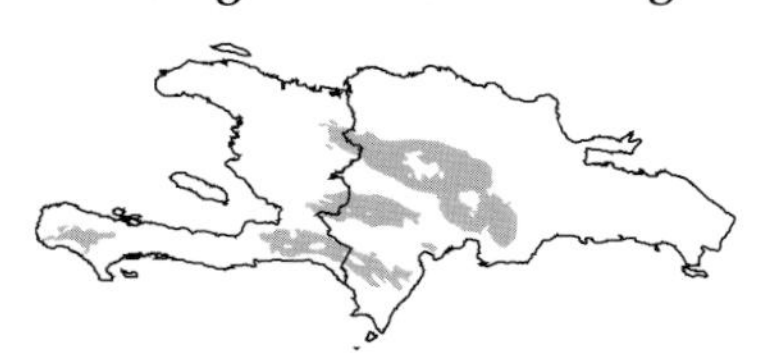

Description: 13–14 cm long; 12 g. A small but robust tanager with a fairly long tail and heavy bill. Back and rump are bright olive greenish. Underparts are white, washed gray on sides of breast and flanks. Head and nape are gray, with a blackish spot between eye and bill bordered above by a white line, and narrow white crescents above and below eye. Wings and tail are blackish, and outer primaries are edged white, forming a prominent white patch in the closed wing. There are also conspicuous white spots on the outer tail feathers.
Similar species: Green-tailed Ground-Tanager has a red iris and lacks white in wing, tail, and in front of eye.
Voice: A low chattering *suit . . suit . . suit . . chir . . suit . . suit . . suit . . suit . . chir . . chi*. Also a thin *tseep*. Song is a series of high-pitched, squeaky notes, often accelerating at the end.
Hispaniola: Locally common within a known extreme elevation range of 875 to 2,000 m, but in general found only in high-elevation moist montane forest of broadleaf or mixed broadleaf and pines above 1,300 m elevation. In DR, found in parts of Cordillera Central, Sierra de Bahoruco, and on the south side of the crest of Sierra de Neiba. In Haiti, formerly common in parts of Massif de la Selle, but very rare and possibly extirpated from La Visite National Park.

Still fairly common in wet karst limestone forests of Macaya Biosphere Reserve, but considered endangered overall in Massif de la Hotte. Also once common in the Morne Malanga, Morne Tranchant, and Crête-à-Piquant areas above Port-au-Prince, but all of these areas have been severely deforested, and the Hispaniolan Highland-Tanager may be nearly or completely extirpated. Not recorded from any major satellite island.
Status: Breeding resident. Originally placed in the genus *Microligea* with *palustris, Xenoligea* was later raised to full generic rank and both genera were placed with the warblers. Formerly known as White-winged Warbler, but genetic data published in 2001 suggest that the Hispaniolan Highland-Tanager (*Xenoligea*) and Green-tailed Ground-Tanager (*Microligea*) are closely related and are not wood-warblers, but are more closely related to the endemic palm-tanagers (*Phaenicophilus*). The highland-tanager is now locally common, but its numbers have likely declined substantially since the mid-1970s because of severe habitat loss, particularly in Haiti, but also in DR localities such as Sierra de Neiba which is heavily deforested even at high elevations. The species is now considered endangered.
Comments: Usually seen in pairs or in mixed-species flocks, actively foraging at all levels in the forest from the understory to the high canopy. Feeds on insects and seeds. This species is often associated closely with the Florida trema (*Trema micrantha*) which produces one of its favorite fruits.
Nesting: A nest with 1 egg and 1 nearly fledged chick, discovered at Pueblo Viejo, Sierra de Bahoruco, in June 2004, is the only nest to have been documented with absolute certainty. The nest, similar to that of *Microligea,* was an open, cup-shaped structure of twigs, plant fibers, and moss, placed in a vine tangle 2.5 m above ground. Eggs are pale greenish and faintly marked with reddish brown blotches and scrawls. Believed to breed May to July.
Range: Endemic to Hispaniola.
Local names: DR: Ciguíta Aliblanca, Tangara Aliblanca; **Haiti:** Petit Quatre-yeux, Ti Tchit Kat Je, Tangara des Montagnes.

BLACK-CROWNED PALM-TANAGER—*Phaenicophilus palmarum* Plate 51
Breeding Resident—*Endemic*

Description: 18 cm long; 29 g. A robust bird with yellow-green upperparts, gray nape, and black crown and facial mask. The mask contrasts sharply with three white spots around the eye, including a large white spot above the eye, a narrow crescent below the eye, and a small spot between the eye and bill. Underparts are pale gray, with a diffuse white chin and throat. Immature is duller than adult; crown is variable and may be gray to dark gray or black.
Similar species: Gray-crowned Palm-Tanager has a gray crown and sharp contrast between a reduced white area on the throat and the gray breast. Immature Black-crowned Palm-Tanager may be easily confused with Gray-crowned Palm-Tanager, but look especially for the amount of white in the throat and whether it is diffuse or sharply defined.
Voice: Nasal *pi-au* is most frequently heard, but there is also a pleasant dawn song. Call is a low *chep,* frequently given while foraging.
Hispaniola: Common and widespread islandwide except absent from the Tiburón Peninsula of southern Haiti west of the Jacmel-Fauché depression. Reported from sea level to a known maximum elevation of at least 2,500 m, though decidedly less common above 2,000 m. Found in all habitat types from lowland desert scrub to pine forest. Not reported from any major satellite islands except Isla Saona.
Status: One of the most common, ubiquitous endemics on Hispaniola. Little change in its abundance has been reported since the early 1900s, although its numbers probably have declined somewhat because of the loss of all types of forest, its primary habitat.
Comments: The white spots on the head give the impression of "four eyes," hence its local Spanish name. Feeds on seeds, fruits, and insects. Usually found in heavy cover, where it moves about in a slow and deliberate fashion, occasionally twitching its tail and probing into

crevices and holes. It hops among branches and over huge limbs and hangs on the bark of large tree trunks, often probing bromeliads in moist forest. Serves as the central, or "nuclear," species for mixed-species flocks occurring in pine forests. Roosts at night in dense vegetation; known to return to favorite sites nightly.

Nesting: Eggs vary in color from whitish to pale green with spots, and also vary substantially in size. Lays 2–4 eggs in a loose, deep-cup nest which is almost never lined and is built in a bush or small tree. Of 72 nests monitored in Sierra de Bahoruco in desert thorn scrub (53 nests), dry broadleaf forest (8 nests), and pine forest (11 nests), average clutch size was 2.9 eggs, and average nest height above ground was 2.0 m in desert (range 1.0–5.0), 2.8 m in dry broadleaf (range 2.0–5.0), and 3.6 m in pine (range 0.5–8.0). Only 25% of nests fledged young. Breeds February to July, possibly throughout the year.

Range: Endemic to Hispaniola, including Isla Saona.

Local names: DR: Cuatro Ojos; **Haiti:** Quatre-yeux, Kat-je Tèt Nwa.

GRAY-CROWNED PALM-TANAGER—*Phaenicophilus poliocephalus* Plate 51
Breeding Resident—*Threatened Endemic*

Description: 18 cm long; 27 g. Adult is unmarked yellow-green above, light gray below, with a gray crown and nape, blackish facial mask, and sharply defined white chin and throat. Three white spots contrast sharply with black mask. Immature is duller than adult.

Similar species: Adult Black-crowned Palm-Tanager has a black crown, and white of throat blends into gray of breast. Immature Black-crowned Palm-Tanager may have a gray crown but will not have the sharply defined white chin and throat.

Voice: Call is *peee-u*, shorter than similar note of Black-crowned Palm-Tanager. During courtship, a lovely song is heard and, less often, a canarylike "whisper song."

Hispaniola: This species is effectively confined to the western end of the Tiburón Peninsula in southern Haiti west of the Jacmel-Fauché depression and to three major western satellite islands: the Cayemites, Île à Vache, and Île de la Gonâve. It is the only full species effectively restricted to Haiti. Reported from sea level to 2,400 m on Pic de Macaya. Primarily a forest species like its congener, it has been recorded in both dry and moist broadleaf forest, pine forest in the mountains, and in mangroves, arid lowland scrub, and second growth on satellite islands. Has apparently remained common since the early 1900s, though its absolute numbers have probably declined as forest habitat has been cut. The extent of colonization of Gray-crowned Palm-Tanager east of the Jacmel-Fauché depression is not known in detail now, but a few appear to have found their way to Morne La Visite in Massif de la Selle. This species has apparently occurred just barely over the border into DR, with birds recorded near Aguacate, and in desert scrub and in pine forest along Aceitillar, Sierra de Bahoruco, but these sightings are very rare and have not been confirmed with photographs. A zone of hybridization between Gray-crowned and Black-crowned palm-tanagers has been documented near the Jacmel-Fauché depression where the two species meet.

Status: Endemic to Hispaniola. Common but local on the Tiburón Peninsula of southern Haiti where it is considered threatened because of habitat destruction. Also common on the satellite islands. Three endemic subspecies are recognized: *P. p. poliocephalus* occurs on the main island and on Grande Cayemite; *P. p. coryi* is larger and paler than *P. p. poliocephalus* and is confined to Île de la Gonâve; and *P. p. tetraopes* is confined to Île à Vache.

Nesting: Eggs are pale green to bluish white marked with cinnamon and blackish brown spots and scrawling lines, tending to form a wreath around the larger end. Lays 2–4 eggs in a frail nest of twigs with a deep cup, which is built in a dense bush or tree 1–9 m above ground. Breeds May to July.

Range: Endemic to Hispaniola and several satellite islands.

Local names: DR: Cuatro Ojos Cabeza Gris; **Haiti:** Quatre-yeux, Quatre-yeux de Sud, Kat-je Tèt Gri, Tangara Quatre Yeux.

WESTERN CHAT-TANAGER—*Calyptophilus tertius* **Plate 52**
Breeding Resident—*Critically Endangered Endemic*

Description: 17–20 cm long; 49 g. A shy, skulking, mockingbird-shaped endemic, with a long, rounded tail. Dark chocolate brown above, mostly white below, with a bright yellow spot in front of eye and a yellow fringe on bend of wing. Legs and feet are robust.
Similar species: Larger La Selle Thrush is also ground-dwelling but has a red breast and sides. Red-legged Thrush has a red bill and legs. Western Chat-Tanager is distinguished by being substantially larger than Eastern Chat-Tanager; Eastern Chat-Tanager also has a brighter yellow eye spot, yellow eye-ring, and more extensive yellow-orange patches at bend of wing and on underwing coverts. Although the ranges of the two species are not supposed to overlap, see "Comments" below.
Voice: Melodious song can be heard at any time of the year, especially at dawn: an emphatic, clear whistling *chip-chip-swerp-swerp-swerp* or *chirri-chirri-chirri-chip-chip-chip*, repeated many times. Call note is a sharp *chin-chin-chin*. Also a *tick, tick, tick* contact call, and a 2-note call similar to that of La Selle Thrush, made when adults meet at the nest.
Hispaniola: Resident at higher elevations in Massif de la Hotte and Massif de la Selle, Haiti, and in Sierra de Bahoruco, DR, east through Pueblo Viejo to at least Polo. The population in Massif de la Hotte is separated from that in Massif de la Selle by lowlands and a distance of 150 km. Reported from an elevation as low as 745 m at Aceitillar, Sierra de Bahoruco, to a maximum recorded elevation of 2,200 m. Found in dense, often damp, understory of moist broadleaf and pine forest, frequently in heavily vegetated ravines, and typically spending much of its time on or near the ground. Shy and usually difficult to observe.
Status: Generally uncommon and local permanent resident. Never a numerous or abundant species anywhere, though sometimes described as locally common in areas such as La Visite National Park and Macaya Biosphere Reserve. Although quantitative data are lacking, this species must have declined since 1930 because of habitat loss, particularly in Massif de la Selle. Considered critically endangered because of declining habitat and extremely restricted distribution.
Comments: The distribution of Western and Eastern chat-tanagers is still very much in question, and the taxonomy of these species is confused and requires further study. Two forms of the Western Chat-Tanager have been described: *C. t. tertius* in Massif de la Hotte and *C. t. selleanus* in Massif de la Selle and presumably, based on discoveries in the early 2000s, in Sierra de Bahoruco, DR. Although some consider *selleanus* as synonymous with *tertius*, others have argued that the birds from la Hotte area are noticeably larger and darker in coloration with more dramatic yellow lores than the chat-tanagers on La Visite, and that their vocalizations are more elaborate. To confuse the situation further, birds in the most eastern extension of the Sierra de Bahoruco, between Polo and Barahona, are said to be similar in appearance to the Eastern Chat-Tanager and to sing their song, whereas birds in the Sierra de Neiba have been reported singing the Western-type song.
Nesting: The recently discovered (2002) nest is an unusual bulky cup shape with a dome over much of it but a large opening toward the front. The nest is constructed mostly of moss, with some bromeliad leaves, other dried leaves, twigs, and fine vines in the inner cup. Nests are usually placed in extremely dense tangles of vines or broadleaf understory 1–5 m above ground. Two light blue eggs with reddish brownish flecking are laid. Breeds May to July.
Range: Endemic to Hispaniola.
Local names: DR: Chirrí de los Bahorucos, Patico Chirrí; **Haiti:** Cornichon, Gwo Kònichon.

EASTERN CHAT-TANAGER—*Calyptophilus frugivorus* **Plate 52**
Breeding Resident—*Critically Endangered Endemic*

Description: 17–20 cm long; 32 g. A shy, skulking, mockingbird-shaped endemic, with a long, rounded tail. Dark chocolate brown above, mostly white below with a bright yellow

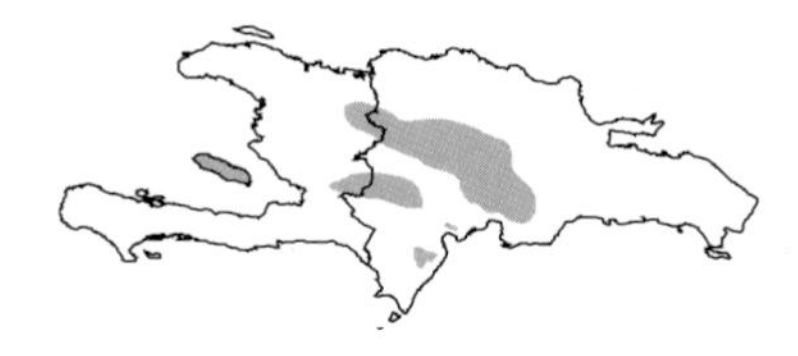

spot in front of eye, yellow fringe on bend of wing, and a distinct yellow eye-ring which is broken anteriorly and posteriorly, and which is not found in Western Chat-Tanager. Legs and feet are robust.
Similar species: Larger La Selle Thrush is also ground-dwelling but has a red breast and sides. Red-legged Thrush has a red bill and legs. See Western Chat-Tanager.
Voice: Song similar to that of Western Chat-Tanager with only subtle differences, but more study needed; an emphatic, clear whistling *chip-chip-swerp-swerp-swerp* or *chirri-chirri-chirri-chip-chip-chip*, repeated many times. Call note is a sharp *chin chin chin*, heard especially at dawn.
Hispaniola: Historically this species has occurred in three separate areas, all north of the Neiba Valley/Cul de Sac Plain: Île de la Gonâve, Haiti; in the lowlands of the DR near Villa Riva and Samaná; and at mid- to upper elevations in the Sierra de Neiba and the Cordillera Central, DR, particularly from areas around San Juan de la Maguana and La Vega (Ebano Verde Reserve), and in the Sierra de Ocoa in the southern Cordillera Central. The species has been reported from sea level on Île de la Gonâve and near Samaná to at least 2,000 m in parts of Cordillera Central. It has also been documented through mist netting in the western Sierra de Martín García, and the most eastern extension of the Sierra de Bahoruco, between Polo and Barahona, including Monteada Nueva. Preferred habitat is thick, moist underbrush of humid broadleaf forest in the mountains and dense, often swampy, undergrowth and jungle at low elevations. Inhabits dense semiarid scrub on Île de la Gonâve, where the species may be less shy than on the main island.
Status: Generally uncommon and increasingly local permanent resident. The distribution of Eastern and Western chat-tanagers is still in question (see "Comments" under Western Chat-Tanager). Three subspecies of Eastern Chat-Tanager are described. A pale and slightly smaller form on Île de la Gonâve is *C. f. abbotti*. Populations from the lowlands near Samaná, from Villa Riva, and from the highlands above La Vega and San Juan de la Maguana in the foothills of Cordillera Central, DR, have been assigned to the race *C. f. frugivorus*. A population in the Sierra de Neiba has been described as *C. f. neibei*. The race from Île de la Gonâve, formerly fairly common, has not been reported in recent years. Concerted efforts to relocate this species in its historical range in the lowlands at or anywhere near Samaná and Villa Riva have not been successful since 1960, and it is believed that the species is extirpated from this area. Rapid forest destruction in the Sierra de Neiba since the mid-1980s has greatly reduced one of the last refuges of this species as well. It may now be principally restricted to the most remote areas of the Cordillera Central. It is seriously at risk and considered critically endangered because of declining habitat and extremely restricted distribution.
Comments: Mainly ground dwelling. The large, strong feet are adapted for a terrestrial lifestyle and for searching the leaf litter for food. Feeds primarily on invertebrates, including centipedes and hairy spiders, and on some seeds. Very secretive.
Nesting: Nest and eggs unknown, but expected to be similar to those of Western Chat-Tanager.
Range: Endemic to Hispaniola, including Île de la Gonâve.
Local names: DR: Chirrí de Cordillera Central, Patico Chirrí; **Haiti:** Cornichon, Kònichon.

SUMMER TANAGER—*Piranga rubra* **Plate 53**
Vagrant

Description: 18–20 cm long; 29 g. A large tanager with a relatively long, heavy bill; head may appear slightly crested. Non-breeding male is yellowish olive green above and orangish yellow below, but often with scattered red feathers, becoming more mottled red and yellow by spring. Non-breeding female is similar to breeding female and is yellowish olive green above and orangish yellow below, with a faint dusky eye-line. Adult male is entirely rose red, brighter below, with wings and tail subtly darker.

Similar species: Western Tanager has pale wing bars in all plumages.
Voice: Call is a distinctive staccato, descending *pit-i-tuck*. Song, unlikely to be heard in the non-breeding season, is a series of musical, rolling, 3-syllable phrases.
Hispaniola: One DR record of 3 seen together at about 760 m elevation in the canopy of a guama (*Inga vera*) tree in a shade coffee plantation in Manabao. Typical habitats include woodlands, forest edges, thickets, and gardens, primarily at midelevations.
Status: Vagrant.
Comments: Expected singly or in pairs. When encountered during migration, this species is frequently in the company of thrushes, warblers, and vireos. Feeds on large insects, including wasps and bees, fruit, and seeds, including those of royal palm (*Roystonesa hispaniolana*).
Range: Breeds from se. California east to sw. Utah, c. Texas, s. Iowa, c. Indiana, West Virginia, and s. New Jersey, south to n. Baja California, c. Mexico, Gulf Coast, and s. Florida. Occurs in non-breeding season from c. Mexico south to n. Bolivia and c. Brazil. In West Indies, uncommon in Bahamas and Greater Antilles east to Jamaica; vagrant to DR, Saba, and Guadeloupe.
Local names: DR: Tangara del Verano; **Haiti:** Tangara Wouj, Tangara Vermillon.

WESTERN TANAGER—*Piranga ludoviciana* — Plate 53
Vagrant

Description: 17–19 cm long; 28 g. A medium-sized tanager with a relatively small bill, short tail, and pronounced wing bars. Adult non-breeding male has a greenish yellow head, often washed reddish on face; blackish back, wings, and tail; and yellow nape, rump, uppertail coverts, and underparts. Note the two distinctive wing bars with the upper bar yellow and lower bar white. Breeding male is similar but brighter, with most or all of head reddish. Adult female has olive green upperparts, which are grayer on back and more yellow on rump and uppertail coverts. Underparts are variable, from bright yellow to grayish white, with undertail coverts invariably yellow. Wings and tail are grayish brown with two yellowish white wing bars. Immature birds are similar to adult female, but males are brighter overall.
Similar species: Summer Tanager lacks wing bars.
Voice: Call is a quick, soft chatter of 2–3 syllables, *pit-ick*, *pit-er-ick*, or *pri-di-dit*. Song is a series of fairly rapid, slurred, hoarse notes.
Hispaniola: Two DR records: 1 male seen in early spring 1996 near Puerto Escondido, and a male and a female seen in mid-December 1999 at Valdesia near San Cristóbal.
Status: Vagrant.
Comments: Typically occurs singly or in small flocks, frequenting a wide variety of forest, scrub and semiopen habitats, also parks and gardens. Fairly sluggish foraging behavior is reminiscent of vireos; feeds on insects and fruits.
Range: Breeds from se. Alaska and sc. Mackenzie east to e. Saskatchewan, south to n. Baja California, c. Colorado, and s. New Mexico, east to w. Texas. Occurs in non-breeding season from s. coastal California, s. Baja California, and c. Mexico south to Costa Rica, casually on Gulf Coast east to Florida. In West Indies, vagrant to Bahamas, Cuba, and DR.
Local names: DR: Tangara del Oeste; **Haiti:** Tangara Tèt Wouj, Tangara à Tête Rouge.

HISPANIOLAN SPINDALIS—*Spindalis dominicensis* — Plate 51
Breeding Resident—*Endemic*

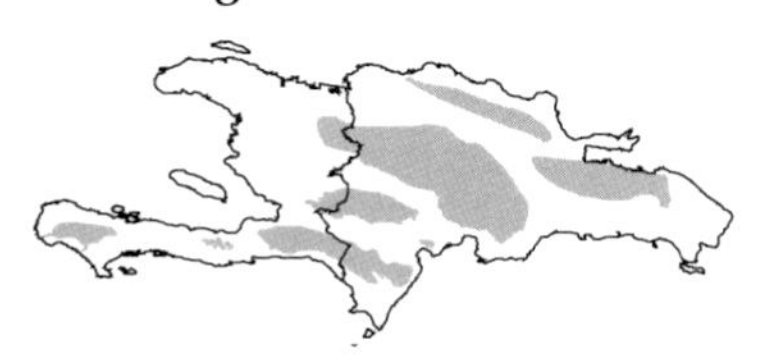

Description: 16 cm long; 31 g. A distinctive, small-billed tanager with a striking difference between the sexes. Male is distinguished by its black head boldly striped with white. Nape is brilliant yellow with an orange tinge; belly is yellow, washed orange; and primarily yellow back is broken by a dark yellow-green mantle. Underparts are yellow with a reddish brown wash on breast. Tail and wings are black, edged whitish, with a bold chestnut patch at bend of wing. Female has olive-brown upperparts with a distinctly yellowish rump. Underparts are

dull whitish with fine dusky streaks. Immatures are similar to adult females, but males are more heavily streaked, including indistinct black streaks on dark green back. Rump of immature male is yellowish tinged chestnut.
Similar species: Unmistakable.
Voice: Weak, very high pitched *seep*, feeble and insectlike. Dawn song is a thin, high-pitched whistle.
Hispaniola: Locally common resident in montane areas islandwide where there is adequate forest habitat. Reported from sea level to at least 2,450 m, but is typically uncommon in any habitat below 700 m, especially xeric habitats. On the main island, most often found in midelevation evergreen and montane evergreen forest, both broadleaf and pine. In DR, regularly found in Sierra de Bahoruco, Sierra de Neiba, and Cordilleras Central, Septentrional, and Oriental. In Haiti, known historically from Massif des Montaignes Noires and Massif du Nord, though present status there is unknown, and reported as common in moist broadleaf forest habitat in La Visite National Park and very common in Macaya Biosphere Reserve in Massif de la Hotte. Tends to move frequently in search of large fruit crops.
Status: Common permanent resident locally in mountains. Has certainly declined to some degree since 1930 because of habitat loss.
Comments: Arboreal. Feeds primarily on fruits and flower buds, but may also take some insects and young leaves. Until recently it was considered part of the Stripe-headed Tanager species complex.
Nesting: Eggs are cream colored to pale bluish green marked with dark brown scrawls and blotches, particularly at the large end. Lays 2–3 eggs in a small, cup-shaped nest built of grass, usually at a moderate height in a tree or bush. Breeds primarily April through June. Some pairs may breed more than once annually.
Range: Endemic to Hispaniola, including Île de la Gonâve.
Local names: DR: Cigua Amarilla; **Haiti:** Moundélé, Bannann-mi-mòn.

EMBERIZINE SPARROWS AND ALLIES: FAMILY EMBERIZIDAE

The Emberizid finches are primarily ground-dwelling birds that are mostly brownish and often steaked. They occur in grasslands and shrubby habitats and are usually secretive. Emberizids have a short, conical bill, and they mostly feed on seeds, some fruit, and, especially during the breeding season, insects. Emberizids on Hispaniola include both permanent resident and migratory species.

YELLOW-FACED GRASSQUIT—*Tiaris olivaceus* **Plate 53**
Breeding Resident

Description: 12 cm long; 9 g. A small finch of scrubby habitats, greenish olive above, slightly grayer below. Male recognized by distinctive golden yellow throat, supercilium, and crescent below eye, and by black upper breast. Female and immature usually show a faint yellowish supercilium, crescent, and chin, and they lack black on breast.
Similar species: Female and immature Black-faced Grassquit are generally more olive in coloration, and facial markings are faint at best. Female Antillean Siskin has pale yellow bill and yellow wing bars.
Voice: Call note is usually a soft *tek*, imitated by removing one's tongue from against the upper palate. The song is a distinctive thin trill, sometimes uttered sequentially on different pitches.
Hispaniola: Resident islandwide, primarily of open grassy areas from the lowlands to moder-

ate elevations, but occurs even into the high mountains where habitat is suitable. Recorded from sea level to at least 2,500 m, but is more numerous at lower elevations than at high ones. Typical habitats are thickets, the bracken of the pinelands, weeds of abandoned fields and pastures, edges of clearings, and agricultural areas; usually absent from moist broadleaf forest. Common in dry forest and desert scrub, in pines with a broadleaf scrub understory, and in sun coffee plantations. Recorded from several major satellite islands, including Navassa Island, the Cayemites, Île à Vache, Île de la Gonâve, and Île de la Tortue.
Status: Common to abundant resident. In Haiti and elsewhere, this species may be an unfortunate indication of ruinate conditions. This species has probably increased at all elevations since 1930 as forest has been cleared.
Comments: Solitary or in small groups. Forages on seeds, usually from grassheads, but sometimes on the ground. Often seen singing from a conspicuous spike of grass or cane.
Nesting: Lays 2–3 white to bluish white eggs, heavily blotched with brown on the large end and faintly spotted with brown elsewhere. Nest is a domed structure of grass which has an entrance on the side and is typically built low off the ground in a clump of grass or in the branches of a low tree. Breeds year-round.
Range: Resident from c. Nuevo León and e. San Luis Potosí south to nw. Ecuador and nw. Venezuela. In West Indies, common in Cuba, Cayman Islands, Jamaica, Hispaniola, and Puerto Rico.
Local names: DR: Ciguíta de Hierba; **Haiti:** Tizèb, Zèbable.

BLACK-FACED GRASSQUIT—*Tiaris bicolor* **Plate 53**

Breeding Resident

Description: 12 cm long; 10 g. A small, dark olive finch. Male identified by black head and underparts. Female and immature are uniformly drab brownish olive. Juveniles up to 3 months of age can be distinguished from females by having a lighter bill; young males show progressive development of the black hood and bib.
Similar species: Female and immature are drabber than their Yellow-faced Grassquit counterparts and lack the faint yellowish facial markings. Male is much smaller than immature Greater Antillean Bullfinch which can also combine black and olive-brown plumages.
Voice: Emphatic buzz that is often followed by a second, louder effort. Call note is a soft musical *tsip*.
Hispaniola: Locally common resident islandwide reported from sea level to at least 1,800 m, but seems to prefer grassy patches and open areas at midelevations and well into the mountains. Found in semiarid as well as moist habitats, but virtually always in open situations, including pastures, cane fields, clearings, agricultural areas, streamsides, and open broadleaf woodland. Recorded from several major satellite islands, including Navassa Island, Île à Vache, Île de la Gonâve, and Île de la Tortue.
Status: Breeding resident, generally less numerous overall than Yellow-faced Grassquit, but has surely increased in abundance since 1930 as forest has been cleared. Reported from several major satellite islands.
Comments: Often coexists with Yellow-faced Grassquit, but usually in smaller numbers. Occurs in high-elevation moist broadleaf forest which Yellow-faced Grassquit avoids. Feeds on seeds of grasses and other plants. The male displays for females with an arching flight, landing on a perch near the female, where he spreads his vibrating wings laterally and slightly downward, while giving a buzzy, slurred vocalization.
Nesting: Lays 3 whitish eggs, heavily flecked with reddish brown at the large end, in a domed nest with a side or bottom entrance. The nest is built of stems and leaves of grasses and weeds without a lining of finer material, and placed in bunches of grass or low shrubs. Breeds year-round.
Range: Resident throughout West Indies (except Cuba) and in n. Colombia and n. Venezuela.
Local names: DR: Juana Maruca; **Haiti:** Sisi Zèb.

GREATER ANTILLEAN BULLFINCH—*Loxigilla violacea* **Plate 54**
Breeding Resident

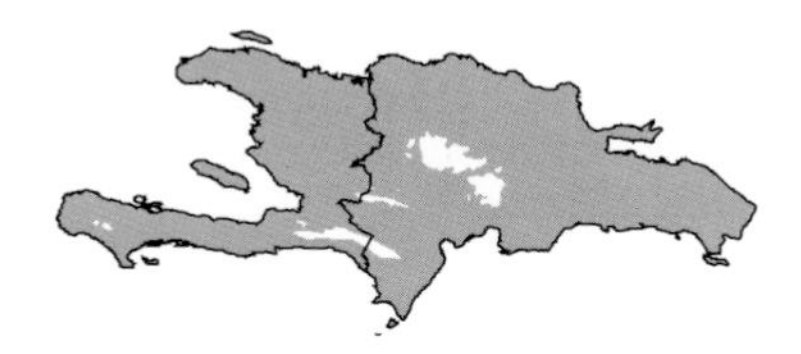

Description: 15–18 cm long; 20 g. A chunky, heavy-billed bird with diagnostic orange-red supercilium, throat, and undertail coverts. Adult male is deep black overall; adult female is slate gray. Immatures are generally olive-brown with orange-red markings similar to those of adults, but begin to show the gray or black feathers of adults, resulting in a mottled appearance.
Similar species: Much larger than grassquits, which in some plumages are olive-brown and black.
Voice: Song is a repetition of the shrill, insectlike call note, *t'zeet, t'seet, t'seet, tseet, seet, seet, seet, seet, seet*. Alarm note is a thin *spit*.
Hispaniola: Widely distributed in drier habitats islandwide. Reported from sea level to at least 2,070 m, but rare above 1,500 m. At all elevations where it occurs, typically found in relatively dense growth such as thickets or understory in dry broadleaf forest. May also occur in humid forest and pine forest, but is more common in drier sites below 800 m, especially where gumbo limbo (*Bursera simaruba*) fruits are found in season. Reported from several major satellite islands.
Status: Common year-round resident, but population has probably declined since the 1930s because of forest clearing. In general, two endemic subspecies are recognized in the Hispaniolan region: *L. v. maurella* is confined to Île de la Tortue, and *L. v. affinis* occurs on the main island and other satellite islands. A third form, *L. v. parishi*, described from Île à Vache and Islas Beata and Catalina, has not been supported by recent taxonomies. *L. v. maurella* is larger (at least in wing length) than those taken elsewhere in the Hispaniolan region.
Comments: Feeds primarily on fruits and seeds, but also on flower parts and snails. Despite being relatively common, it is difficult to detect because of keeping to dense vegetation, and because it has a high-pitched song that can be difficult to discern.
Nesting: Lays 3–4 bluish white to white eggs with brown spots principally at the larger end, in a deep, domed nest with a side entrance. The nest is built in a clump of grass, among cacti, or in a low bush or tree branches, typically 1–2 m above ground. Breeds through much of the year, but especially March to June. Individuals will reuse their own nests.
Range: Resident in Bahamas, Hispaniola, and Jamaica.
Local names: DR: Gallito Prieto, Pico Cotorra; **Haiti:** Ti Kòk, Tchitchi Gwo Bèk, Gros-bec, Père-noir, Sporophile Petit-coq.

SAVANNAH SPARROW—*Passerculus sandwichensis* **Plate 54**
Vagrant

Description: 15–19 cm long; 20 g. A slender sparrow with upperparts and underparts heavily streaked with brown. Supercilium and lores are usually yellowish and conspicuous, though sometimes buff colored. Also has a pale central crown stripe, dark malar stripe, and pink legs. Tail is short and slightly notched. Typically seen on the ground.
Similar species: Grasshopper Sparrow has a more golden supercilium and lacks a malar stripe. Only immature Grasshopper Sparrow has streaks below, and these are much finer and paler than those of Savannah Sparrow. When landing, Savannah Sparrow does not turn slightly to one side as does Grasshopper Sparrow.
Voice: Pleasant, high-pitched, melodious call of 3 *chips* followed by 2 wispy notes, the last one shorter and lower, *chip-chip-chip-tisisiiii-tisi*.
Hispaniola: One adult female collected at Saint-Michel de L'Atalaye, Haiti. To be looked for in open fields, pastures, bushy savannas, and sparse thickets near the coast.
Status: Vagrant.

Range: Breeds from nw. Alaska east to n. Ontario and Newfoundland, south to wc. California, s. Utah, n. New Mexico, e. Tennessee, and n. New Jersey, locally in Baja California and c. Mexico. Occurs in non-breeding season from s. U.S. south to n. Honduras. In West Indies, uncommon to rare non-breeding visitor to Bahamas, Cuba, and Cayman Islands; vagrant to Haiti.
Local names: DR: Gorrión de Sabana; **Haiti:** Zwazo Savann, Bruant des Prés.

GRASSHOPPER SPARROW—*Ammodramus savannarum* **Plate 54**
Breeding Resident

Description: 12 cm long; 17 g. A small, flat-headed sparrow, intricately patterned with rufous, buff, and gray above; a golden or orangish mark forward of eye; a white eye-ring; and a whitish central crown stripe. Plain buffy below. Juveniles possess a paler mark by bill and have fine streaks on breast and flanks.
Similar species: Savannah Sparrow is more heavily streaked than immature Grasshopper Sparrow. Also, Grasshopper Sparrow turns slightly to one side upon landing, Savannah Sparrow does not. Rufous-collared Sparrow is quite different and has a black band on foreneck, reddish brown nape, and gray crown with black stripes. It is found only in the mountains on Hispaniola.
Voice: Two distinct songs. A long, thin, insectlike buzz, followed by what sounds like a hiccup, *zzzzzzz-hic*. Also a thin, high-pitched twitter or tinkling song, like fairy bells. Call note is a high-pitched, gritty, insectlike *kr-r-it*.
Hispaniola: Usually occurs in small colonies in lowlands islandwide. Preferred habitat is open fields, cattle ranches, savannas, rice fields, and areas of tall grass from sea level to only 400 m. Not recorded from any major satellite island.
Status: Locally common but declining resident. Formerly considered common on the northern and central plains of Haiti, but its current status in that country is unknown. Seasonal burning of fields, both in the past and currently, planting of large expanses of sugarcane, introduction of the mongoose, and more mechanized agricultural practices reduced the population substantially in the 20th century. The subspecies *A. s. intricatus* is endemic to Hispaniola.
Comments: Feeds primarily on grass seeds. Very secretive, it is much more often heard than seen. Although not yet documented by a specimen, the migratory form *A. s. pratensis* from eastern mainland North America is to be expected as an occasional non-breeding visitor since it occurs regularly in the Bahamas and Cuba.
Nesting: Lays 3 white eggs with reddish spots in a domed nest which is built virtually on the ground in a grass clump. Breeds primarily May to August.
Range: Breeds from e. Washington east to s. Alberta, Ontario, and s Maine, south to s. California, n. Sonora and Chihuahua, c. Georgia, and e. Virginia. Occurs in non-breeding season from s. U.S. south to n. Costa Rica. In West Indies, resident on Jamaica, Hispaniola, and Puerto Rico; non-breeding visitor to Bahamas and Cuba.
Local names: DR: Tumbarrocío; **Haiti:** Zwazo-kann, Bruant Sauterelle.

SONG SPARROW—*Melospiza melodia* **Plate 54**
Vagrant

Description: 15–16 cm long; 21 g. A medium-sized, stocky sparrow with a fairly long tail. Upperparts are grayish with coarse brown steaks; underparts are whitish with coarse brown streaks converging into a central breast spot. Wings and tail are brownish. Crown has broad brown lateral stripes bordering a paler central stripe. Also note broad gray supercilium, and conspicuous brown malar stripes bordering unmarked whitish throat. Often feeds on the ground with the tail cocked.

Similar species: Lincoln's Sparrow is smaller, less heavily streaked above and below, and has a buffy wash on breast.
Voice: Call notes include a sharp, husky *tchunk* or *tchip*; also a high-pitched contact note, *tseep* or *seeet*.
Hispaniola: One mid-November 1997 DR record of a single bird in the western Sierra de Neiba on the border of Bahoruco and San Juan provinces at about 1,900 m elevation.
Status: Vagrant.
Range: Breeds from s. Alaska east to s. Mackenzie, n. Ontario, and sw. Newfoundland, south to c. Baja California and c. Mexico, ne. Kansas, ne. Alabama, and South Carolina. Occurs in non-breeding season in southern part of breeding range and to Gulf Coast and s. Florida. In West Indies, vagrant in Bahamas and DR.
Local names: DR: Gorrión; **Haiti:** Zwazo-kann Chantè, Bruant Chanteur.

LINCOLN'S SPARROW—*Melospiza lincolnii* **Plate 54**
Vagrant

Description: 14–15 cm long; 18 g. A small sparrow with retiring behavior. Upperparts are grayish brown finely streaked with black; breast and sides are washed buffy and finely streaked with black. Belly is whitish, and wings and tail are brown. Crown is brown streaked black, with a narrow grayish median stripe, broad gray supercilium, and buffy malar stripes.
Similar species: Song Sparrow is larger, more boldly streaked above and below, and lacks the buffy wash on breast.
Voice: Calls include a sharp, low-pitched *chip* and a soft, high-pitched *zeet*.
Hispaniola: Non-breeding-season vagrant known only from Haiti, with 3 records in early 1980s: 1 bird collected on Morne La Visite, and sight records from La Visite National Park in Massif de la Selle, and Macaya Biosphere Reserve in Massif de la Hotte. All birds found in understory of moist broadleaf forest, including bracken fern tangles at the edge of cutover forest at about 1,800 m elevation. Because of its skulking habits, this species may occur more frequently than recorded.
Status: Vagrant.
Comments: Very shy, usually solitary, but during migration it is sometimes observed in association with other sparrows.
Range: Breeds from w. Alaska east to nw. Mackenzie, n. Ontario, n Quebec, and Newfoundland, south to se. Alaska, in mountains to s. California and s. New Mexico, ne. Minnesota, n. New Hampshire, and Nova Scotia. Occurs in non-breeding season from extreme coastal sw. British Columbia to s. Baja California; also from sw. Utah, e. Kansas, and s. Alabama south to El Salvador and Honduras. In West Indies, rare in Bahamas, Cuba, and Jamaica; vagrant on Hispaniola and Puerto Rico.
Local names: DR: Gorrión de Lincoln; **Haiti:** Zwazo-kann Sousi Gri.

RUFOUS-COLLARED SPARROW—*Zonotrichia capensis* **Plate 54**
Breeding Resident

Description: 15–16 cm long; 20 g. A handsome, medium-sized, and stocky sparrow, with brown upperparts with coarse dark streaks and grayish white underparts. Identified by its black neck band, reddish brown nape, and gray crown with black stripes. Often displays a slight crest. Juvenile is duller, somewhat spotted below, and lacks black or reddish brown markings.
Similar species: Grasshopper Sparrow lacks black band on foreneck, reddish brown nape, and gray on crown, and does not occur in mountain forests. Song and Lincoln's sparrows are streaked above and below.

Voice: Accelerating trill, *whis-whis-whis-whis-whiswhisu-whiswhis*.
Hispaniola: Common resident confined to higher elevations of Sierra de Neiba and Cordillera Central, DR. Lowest elevation reported is 900 m and highest is at least 2,450 m. Typical habitat is scrub and understory in the pine zone. Never observed in Haiti, but known from areas literally on the DR-Haiti border and expected to inhabit the Montagnes de Trou-d'Eau west of the Sierra de Neiba. A single straggler was found in 1996 at Zapotén, Sierra de Bahoruco, DR, the only known report south of the Cul de Sac Plain.
Status: Locally common year-round resident. The subspecies *Z. c. antillarum* is endemic to Hispaniola.
Comments: Shy and retiring in habits. Feeds on seeds, usually in pairs.
Nesting: Lays 2 pale bluish eggs, heavily spotted and scrawled with brownish, in a compact cup nest about 10 cm in diameter on the inside, with walls about 2 cm high. The nest is made of lichens, moss, and pine needles and lined inside with fine grass. It is placed about 2 m above ground in the middle of a small shrub. Timing of breeding is unknown but probably May and June.
Range: Widespread in mainland Middle and South America. Resident in West Indies only on Hispaniola.
Local names: DR: Cigua de Constanza; **Haiti:** Zwazo-kann Chingolo.

CARDINALINE FINCHES AND ALLIES: FAMILY CARDINALIDAE

The Cardinalidae are characterized by a thick, conical bill used for eating seeds. In general appearance they resemble the grassquits, bullfinch, and sparrows of the Emberizidae. They are often found in weedy or brushy habitats, although the Rose-breasted Grosbeak is more of a forest dweller. Some species, such as the Indigo Bunting, may occur in large flocks during the non-breeding season. On Hispaniola, all of the Cardinalidae are neotropical migrants that spend the non-breeding season on the island.

ROSE-BREASTED GROSBEAK—*Pheucticus ludovicianus* **Plate 54**
Non-breeding Visitor

Description: 19–20 cm long; 46 g. A large, robust migratory finch with a heavy, pale bill. Non-breeding male is distinguished by its pinkish breast which may have a considerable buffy wash, black or blackish brown head and back, white spots on upperwing coverts and primaries, and large white patches on outer tail feathers. Lower breast and belly are white; rump is white with gray barring. Female has upperparts brown and streaked, head boldly striped whitish and brown, and underparts whitish, streaked brown. In flight, male displays rosy underwing coverts; female's are yellow. Breeding male is brighter and more contrasting, with black replacing brown, and red replacing pink. Breeding female is similar to non-breeding female.
Similar species: None.
Voice: Call is a sharp, metallic *chink* or *eek*. Song is a rich, melodious warble.
Hispaniola: Occurs at all elevations; frequently seen in pairs or small groups. Most often found in broadleaf forest canopy or subcanopy. Known reports from Haiti include Port-au-Prince and Caracol. In the DR, sightings have been recorded at Los Haitises National Park, Puerto Plata, Punta Rucia, and Oviedo; and in Sierra de Bahoruco at Aceitillar, near Pedernales, Los Arroyos, Mencia, Zapotén, and near Puerto Escondido. Also reported from Île de la Gonâve.

Status: Uncommon but regular non-breeding visitor from mid-November to late April.
Comments: Often migrates in small conspecific flocks, but may be found in mixed flocks. Feeds on fruit and seeds, also insects.
Range: Breeds from ne. British Columbia east to n. Alberta, c. Ontario, and Newfoundland, south to se. Alberta, se. Kansas, s. Illinois, n. Georgia, and Delaware. Occurs in non-breeding season along both Pacific and Gulf slopes of Mexico south to Venezuela and e. Peru. In West Indies, uncommon in Bahamas, Hispaniola, Cuba, and Cayman Islands; rare in Jamaica east to Virgin Islands; vagrant in Lesser Antilles from Antigua to St. Vincent.
Local names: DR: Degollado; **Haiti:** Kadinal Fal Wòz, Cardinal à Poitrine Rose.

BLUE GROSBEAK—*Guiraca caerulea* **Plate 54**
Non-breeding Visitor

Description: 16–18 cm long; 28 g. A large finch with a heavy bill and a long, rounded tail. Immature male and female are similar: plain warm rufous brown overall. Breeding male is deep blue with reddish brown wing bars and dull black flight feathers. A small black patch on face between eye and bill extends to chin. Bill is two toned, with a black upper mandible and silver lower mandible. Female is paler gray-brown throughout, with occasional blue feathers, and a buffy brown patch on wing. Commonly flicks and spreads its tail.
Similar species: Indigo Bunting is smaller, with a smaller bill, and lacks conspicuous wing bars.
Voice: Call is a metallic, hard *tink* or *chink*. Song is a long, rich, husky warble.
Hispaniola: All known reports are from DR and include sightings from Villa Mella, Cabral, Santiago, Jarabacoa, Puerto Plata, Punta Rucia, and near Los Arroyos in the Sierra de Bahoruco. Although occasionally found in moist broadleaf forest, typically occurs in weedy fields, rice plantations, dry forest edges, and lowland scrub.
Status: Rare non-breeding visitor from mid-November to early April.
Comments: Opportunistic feeder; diet includes insects, snails, and wild and cultivated grass seeds. Not documented to feed on fruit. Throughout much of its migratory range, often congregates in flocks in rice and other grain fields.
Range: Breeds from n. California east to s. Idaho, sw. Minnesota, s. Ohio, and s. Pennsylvania, south to n. Baja California, Costa Rica, Gulf Coast, and Florida. Occurs in non-breeding season from c. Mexico south to c. Panama. In West Indies, uncommon to rare in Bahamas and Greater Antilles east to Puerto Rico; vagrant in Virgin Islands.
Local names: DR: Azulejo Real, Azulejón; **Haiti:** Giraka Ble, Guiraca Bleu.

INDIGO BUNTING—*Passerina cyanea* **Plate 54**
Non-breeding Visitor

Description: 14 cm long; 14 g. A small, active finch, often encountered in flocks. Non-breeding male is brown overall but with traces or patches of blue in wings and tail. Female is entirely dull brown with pale breast streaks, pale wing bars, and no conspicuous markings. Breeding male is entirely bright blue but is observed in this plumage only late in the non-breeding season before its northward migration.
Similar species: Female's faint breast stripes and wing bars distinguish it from immature mannikins.
Voice: Call note is an emphatic *twit*. Sometimes gives a thin song, usually of paired phrases such as *sweet-sweet, chew-chew, sweet-sweet*.

Hispaniola: Known records are all from the DR and include single birds or small groups at Villa Elisa, San Francisco de Macorís, Puerto Plata, Reserva Científica Loma Quita Espuela, and Mencia and Aceitillar sector of the Sierra de Bahoruco. A minimum of 8 birds occurred together near Jarabacoa in pine forest with a recently burned understory from at least mid-February to early April 1995. More typical habitat may include rice fields, grassy areas bounded by heavy thickets, rows of trees, pasture edges, and fairly dry scrub.
Status: Uncommon non-breeding visitor, probably overlooked and occurring more frequently than available reports suggest.
Comments: Typically flocks.
Range: Breeds from se. British Columbia east to n. Minnesota, sw. Quebec, and Nova Scotia, south to s. New Mexico, Texas, Gulf Coast, and Florida. Occurs in non-breeding season from n. Mexico south to Panama and nw. Colombia. In West Indies, common to uncommon in Bahamas and Greater Antilles; vagrant at Saba and Antigua.
Local names: DR: Azulejo; **Haiti:** Zwazo-digo, Passerin Indigo.

BLACKBIRDS AND ALLIES: FAMILY ICTERIDAE

Blackbirds and orioles are characterized by a fairly slender, pointed, conical bill. Almost all of the species on Hispaniola have a substantial amount of black in their plumage, and several show substantial yellow or yellowish orange. Icterids are typically found in open areas, and some species frequent urban parks and human habitations. Blackbirds and grackles are often seen in mixed-species flocks, especially during the non-breeding season. Orioles feed on insects, fruit, and nectar, whereas blackbirds feed on insects, fruit, and grain.

BOBOLINK—*Dolichonyx oryzivorus* **Plate 56**
Passage Migrant

Description: 18 cm long; male 47 g, female 37 g. Non-breeding adult is warm buffy brown overall with a streaked back, rump, and sides. Note unmarked, buff-colored throat and buffy central crown stripe. Tail feathers are noticeably pointed. Breeding male is black below with a distinctive yellowish buff hindneck and white patches on wings and rump. Breeding female is similar to non-breeding birds but paler, with a whitish, rather than buff-colored, throat.
Similar species: Non-breeding adult Bobolink differs from the various sparrows by its much larger size and its streaked sides and lower belly.
Voice: Very distinctive *pink*.
Hispaniola: Reported surprisingly seldom given the frequency of its occurrence in Cuba and Jamaica. Four known records, including a flock of about 100 at Île de la Tortue, Haiti; several collected near San Juan de la Maguana, DR; 1 mist netted at Isla Beata, DR; and 1 found dead aboard a ship 65 km south of Tiburón, Haiti. To be looked for in rice fields, pastures, and areas where grass is seeding.
Status: Rare spring and fall passage migrant.
Comments: Typically flocks.
Range: Breeds from c. British Columbia east to s. Manitoba, s. Quebec, and Newfoundland, south to s. Washington, n. Utah, n. Missouri, and c. New Jersey. Occurs in non-breeding season in s. South America. In West Indies, uncommon passage migrant in Bahamas, Cuba, Cayman Islands, and Jamaica; rare to vagrant on Hispaniola, Puerto Rico, and Lesser Antilles south to Grenada.
Local names: DR: Bobolink; **Haiti:** Gwo-bèk Ke Pwenti, Goglu des Prés.

TAWNY-SHOULDERED BLACKBIRD—*Agelaius humeralis* **Plate 56**
Breeding Resident—*Threatened*

Description: 19–22 cm long; male 37 g, female 29 g. A medium-sized bird, black overall with a tawny shoulder patch that is most conspicuous when flying. Shoulder patch of immature birds is much smaller.
Similar species: Shoulder patch of Tawny-shouldered Blackbird is sometimes not visible in perched birds, particularly immatures. Greater Antillean Grackle is larger, has a long V-shaped tail, and lacks a shoulder patch.
Voice: Sometimes emits a harsh call, *wiii-wiiii-wiiii*. Typical call note is a strong, short *chic-chic* that resembles the *chip* of Common Yellowthroat but is stronger.
Hispaniola: To date certainly recorded only from westcentral and northwestern Haiti near Port-de-Paix, the lower Artibonite River delta, and near Saint-Marc. It frequents coastal marshes and scrub areas and can be locally common where it occurs. Also may be found near small woodlands, gardens, rural farms, swamp edges, pastures, and rice fields, but only in the lowlands. Perhaps established in the 1920s or 1930s on the island by individuals coming from Cuba, but the species' range does not seem to have expanded since that time, as would be expected of a new colonist, so it may have just been overlooked earlier. Regarded as hypothetical from DR, although an unverified report exists from southeast of Pepillo Salcedo in extreme northwestern DR. Should be looked for in that area, as well as in extreme western Lago Enriquillo.
Status: Very local resident. Habitat loss in Haiti has resulted in declining populations to the point where this blackbird's survival is threatened. The subspecies *A. h. humeralis* is endemic to Hispaniola.
Comments: Usually in flocks, it forms pairs during the breeding season. Primarily feeds on seeds, but also fruit, pollen, nectar, flowers, small lizards, and domestic animal feed in farms and dairies. It sometimes enters open restaurants for food scraps. On Cuba, its nest is parasitized by the Shiny Cowbird.
Nesting: No specific Hispaniola data. In Cuba, a clutch of 3–5 bluish or greenish eggs with brown spots concentrated on the large end is laid in a rough cup-shaped nest of dried grass, moss, hair, feathers, and twigs. Breeds in April and May.
Range: Resident in Cuba and w. Hispaniola; vagrant to Florida.
Local names: Haiti: Merle, Ti Kawouj, Petit Carouge.

GREATER ANTILLEAN GRACKLE—*Quiscalus niger* **Plate 56**
Breeding Resident

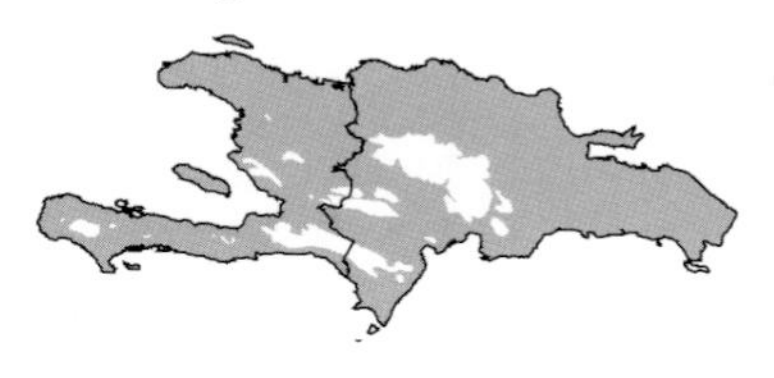

Description: 25–30 cm long; male 86 g, female 62 g. A fairly large blackbird with dark plumage, a long tail, and a long, conical, sharply pointed bill. Adult male has glossy metallic blue to violet-black plumage and a deep V-shaped tail. Adult female is duller than male, and tail has a smaller V. Adults of both sexes have a yellow iris. Among immatures, plumage is dull brownish black, tail is flat, and iris is light brown.
Similar species: All other blackbirds within its range lack a V-shaped tail. Male Shiny Cowbird is smaller, without a yellow iris, and has a thinner bill. Tawny-shouldered Blackbird is smaller, shorter-tailed, with a tan shoulder patch.
Voice: Highly variable repertoire, including a high *cling, cling, cling*. Also clear, musical notes and wheezy gasps. Song is a 4-syllabled phrase. Call note is *chuck*.
Hispaniola: Resident islandwide, recorded from sea level to at least 2,180 m, but most numerous below 1,000 m. Not abundant anywhere, with flocks rarely exceeding 20–40 birds.

Typically found in lowlands in mangroves, marshes, savannas, open fields, and agricultural areas, but rare in lowland desert scrub. When found at higher elevations, it is around clearings in open pine and broadleaf forest. Reported from several major satellite islands, including Île à Vache, Île de la Gonâve, Île de la Tortue, and Isla Beata.
Status: Moderately common breeding resident. Has probably increased in numbers to some degree since 1930 because of forest clearing. The subspecies *Q. n. niger* is endemic to Hispaniola and associated islands.
Comments: Typically flocks. Often forages around livestock, feeding on insects. Also forages on lawns, and commonly scavenges food scraps from humans in parks and around restaurants. Roosts in large flocks.
Nesting: Lays 3–4 pale blue eggs spotted with blackish and darker blue, with some spots fine and others forming large blotches that are often arranged in irregular lines. Nest is a bulky structure of grass and other plant materials, often including Spanish moss, and occasionally lined with mud. Nest is usually built in a tree, shrub, or vine tangle at some distance above ground. Breeds February to September, with peaks from April to August.
Range: Resident on Cuba, Cayman Island, Jamaica, Hispaniola, and Puerto Rico.
Local names: DR: Chinchilín; **Haiti:** Merle Diable, Mèl Diab, Quiscale Noir.

SHINY COWBIRD—*Molothrus bonariensis* **Plate 56**
Breeding Resident

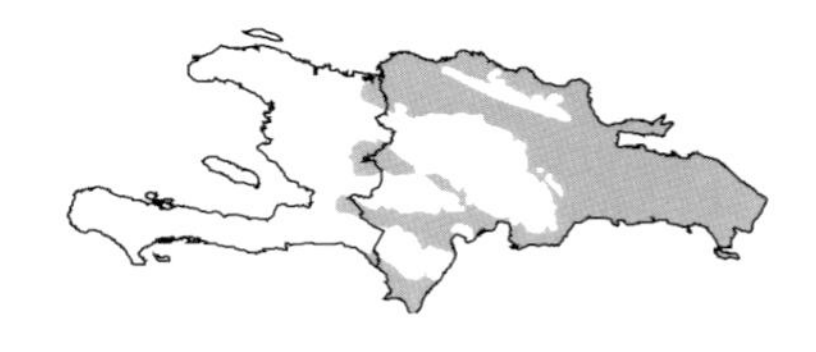

Description: 18–20 cm long; male 39 g, female 32 g. A medium-sized dark bird with a conical bill. Adult male is uniformly glossy black with a purplish sheen. Adult female has drab grayish brown upperparts, lighter brown underparts, and a faint supercilium. Immature resembles adult female, but underparts are streaked with pale gray and supercilium is even more indistinct.
Similar species: Greater Antillean Grackle is much larger with a heavier bill and longer tail. Similar-sized Tawny-shouldered Blackbird has a tan shoulder patch, most visible in flight.
Voice: Song consists of one or more whistles, followed by a melodious trill. Also a variety of short call notes.
Hispaniola: A widespread resident that has been invading steadily northward through the West Indies since first recorded on Grenada in 1891. First record for Hispaniola was at Hato Nuevo, between Herrera and San Cristóbal, DR, in October 1972. By 1982 the species had spread through most of eastern DR west to Nizao and Samaná Bay, along the northern coast to Monte Cristi, and as far west as Neiba. By 1987 it had reached Puerto Viejo, and 1 was found singing on both sides of the Pedernales Province-Haiti border in November 1996, the first known record for Haiti. This species is found primarily in open areas in lowlands, especially near rice fields, livestock, marshes, woodland edges, and second-growth scrub.
Status: Increasingly common and widespread resident. As more and more land is cleared in the mountains, this species may be expected to spread into the interior where many more bird species, some of them already quite rare, may suffer from brood parasitism.
Comments: Shiny Cowbird is a brood parasite, laying its eggs in the nests of other species, and often causing the loss of the host's own eggs or nestlings. The decline of several Puerto Rican birds, including the Puerto Rican Vireo, Yellow-shouldered Blackbird, and Puerto Rican Oriole, has been attributed to Shiny Cowbird. Feeds heavily on rice and grains.
Nesting: An aggressive brood parasite on other species. Eggs vary greatly in color and markings to mimic those of its hosts. In the DR, parasitism has been confirmed in Palmchat, Yellow Warbler, Hispaniolan Oriole, and Village Weaver, but it is unknown what effects this has had on populations of these species. Breeds March through July.
Range: Resident virtually throughout West Indies, from e. Panama through most of South America south to Chile and Argentina, and in Florida.
Local names: DR: Pájaro Vaquero; **Haiti:** Reskiyè, Vacher Luisant.

HISPANIOLAN ORIOLE—*Icterus dominicensis* **Plates 43, 56**
Breeding Resident—*Endemic*

Description: 20–22 cm long; 43 g. A showy oriole. Adult is black overall with distinctive yellow shoulders, rump, and undertail coverts. Immature has mainly olive upperparts and dull olive-yellow underparts. Wings are black, and throat is sometimes black or reddish brown.
Similar species: Tawny-shouldered Blackbird has a tan patch on shoulder. Immatures may be similar to female Shiny Cowbird, but oriole is more olive and its bill is longer. Black and orange Baltimore Oriole is a rare non-breeding visitor. Orchard Oriole is black marked with chestnut and has narrow wing bars.
Voice: Call note is a hard, sharp *keek* or *check*, sometimes sounding as if the bird has a cold. The beautiful but rarely heard song, consisting of exclamatory and querulous high-pitched whistles, is given after dawn.
Hispaniola: Occurs primarily in the lowlands, but reported from sea level to at least 1,100 m wherever palms are found. Typically prefers broadleaf forest edges, though usually uncommon to rare in dry lowland forest and desert scrub, and almost never found in pines. Common in shade coffee plantations in Cordillera Central where palms are present; also common in Los Haitises National Park, DR, and coastal hardwood and palm forest in eastern DR. Also routine in lowland coastal Haiti near Gonaïves and in the Artibonite River delta. Occasionally gathers in flocks of up to 50, but usually seen only in family groups. Found on several offshore islands, including Île à Vache, Île de la Gonâve, Île de la Tortue, and Isla Saona.
Status: Common resident islandwide. Probably has declined somewhat since 1930 because of lowland broadleaf forest loss and parasitism by Shiny Cowbird. The Hispaniolan Oriole was proposed to be split from other Greater Antillean orioles in 2005 based on differences in morphometrics, plumages, behavior, and ecology, but this split has not been formerly recognized by the Committee on Classification and Nomenclature of the AOU.
Comments: Feeds on fruits, insects, flowers, and nectar, often on the undersides of palm fronds.
Nesting: A typical clutch consists of 3–4 bluish white eggs with reddish and brownish spots which fuse into a cap at the large end. The nest is a shallow basket woven of palm fibers and padded with cottonlike plant material, typically attached to a palm or banana frond. Breeds throughout the year but primarily March to June.
Range: Endemic to Hispaniola and associated islands.
Local names: DR: Cigua Canaria; **Haiti:** Banane Mûre, Bannann-mi.

BALTIMORE ORIOLE—*Icterus galbula* **Plate 56**
Non-breeding Visitor

Description: 18–20 cm long; 34 g. Adult male is distinguished by its medium size, orange and black plumage, white wing bar, and orange outer tail patches. Adult female and immature are drab olive brownish to brownish orange above, with two whitish wing bars. Underparts are orange-yellow, brightest on breast, with lower belly and sides often pale grayish white
Similar species: Male Village Weaver is yellow and black, and chunkier with a heavier bill and shorter tail. American Redstart also combines orange and black but is a much smaller and a more active warbler.
Voice: Call is a distinctive flutelike, double-noted whistle, infrequently heard in the West Indies.
Hispaniola: A dozen known reports from mid-November to early May, including individual

birds seen in Gonaïves and Moulin sur Mer, Haiti; and another seen in the foothills of Massif de la Hotte, Haiti. In the DR, reports exist from Santo Domingo, Punta Rucia, Laguna de Don Gregorio, Barahona, Puerto Escondido, Mencia, Pedernales, and Isla Beata.
Status: Rare non-breeding visitor. To be looked for at all elevations in treed gardens, semiarid scrubland, open woodlands, swamps, and forest edges.
Range: Breeds from ne. British Columbia east to s. Manitoba, s. Ontario, and Nova Scotia, south to e. Texas, c. Alabama, and c. North Carolina. Occurs in non-breeding season from n. Mexico and Florida south to n. Colombia and n. Venzuela. In West Indies, uncommon to rare in Greater Antilles; vagrant at St. Barthélemy, St. Kitts, St. Lucia, Barbados, St. Vincent, and Grenada.
Local names: DR: Cigua Canaria Americana; **Haiti:** Ti Mèl Ameriken , Oriole du Nord.

ORCHARD ORIOLE—*Icterus spurius* — Plate 56
Vagrant

Description: 15–18 cm long; 19 g. A small, compact oriole with a short tail and bill. Male has entirely black head, upper breast, back, and tail and chestnut lower breast, belly, undertail coverts, and rump. Wings are mostly black with a narrow white lower wing bar and a broad chestnut upper wing bar. Female is olive green above, greenish yellow below, with grayish brown wings with two narrow white wing bars. Immature is similar to female, but first-year male has variable black on throat and chest, and occasional chestnut feathers on body.
Similar species: Immature Hispaniolan Oriole lacks white wing bars and has a more olive back with contrasting yellowish green rump, shoulder patch, and undertail coverts. Male Baltimore Oriole is larger, with orange body rather than chestnut, and tail mostly orange. Female Baltimore Oriole is more orange-yellow, with lower belly and sides often pale grayish white.
Voice: Call is a soft, low *chut* or *chuck*, also a rapid, scolding chatter. Song is a long, rich warbling, often including harsh notes with a down-slurred ending.
Hispaniola: One record of an immature male observed in mesquite trees in dry subtropical forest at Tierra Nueva, DR, in December 2001; another male observed at Rabo de Gato, DR, in January 2004.
Status: Vagrant.
Range: Breeds from se. Saskatchewan, c. Minnesota, s. Ontario, and se. New Hampshire south to c. and e. Mexico, se. Texas, Gulf Coast, and c. Florida. Occurs in non-breeding season from c. Mexico and n. Yucatán Peninsula south to Colombia and nw. Venezuela. In West Indies, vagrant to Cuba, Bahamas, Jamaica, and DR.
Local names: DR: Cigua Canaria de Huertos; **Haiti:** Ti Mèl, Oriole des Vergers.

FRINGILLINE AND CARDUELINE FINCHES AND ALLIES: FAMILY FRINGILLIDAE

The finches are generally characterized by a conical bill for eating seeds, although the crossbills have specialized bills for extracting seeds from pine cones. In appearance these species resemble the weavers (Ploceidae), estrildid finches (Estrildidae), and grosbeaks of the Emberizidae. The males of most species are colorful, at least during the breeding season, and many species flock following nesting. The nest is usually cup shaped. Flight is undulating. The genus *Euphonia*, traditionally thought to be a tanager, has been shown through genetic studies to be closer to the Fringillidae, where it is now placed.

ANTILLEAN EUPHONIA—*Euphonia musica* — Plate 53
Breeding Resident

Description: 12 cm long; 13 g. A small, compact bird with a distinctive sky blue crown and nape. Males on Hispaniola are primarily blackish above, orangish yellow below and on rump

and forehead. Chin and throat are dark violet. Female is duller overall, mostly greenish above and yellowish green below. Rump and forecrown are yellowish.
Similar species: Unmistakable.
Voice: Variety of distinctive call notes, including a rapid, subdued, almost tinkling *ti-tit* (sometimes 1 or 3 syllables), and a hard, metallic *chi-chink*. Also a plaintive *whee*, like some *Myiarchus* flycatchers but more melodious. Also a jumbled, tinkling song mixed with explosive notes.
Hispaniola: Locally common resident almost islandwide where there is forest with mistletoe, its favorite food. Reported from sea level to a maximum recorded elevation of at least 2,300 m. On the main island, typically found in tropical evergreen and deciduous forest; rare in dry forest or desert scrub below 600 m. Not uncommon in shade coffee plantations, and occasional in pine forest in Cordillera Central, DR. In Sierra de Bahoruco, it is moderately common in moist broadleaf forest at 750 m and less common in pine forest above 900 m elevation. Few data from Haiti since the 1980s, but apparently rare in La Visite National Park, on the northern slope of Morne Cabaïo, and in Macaya Biosphere Reserve from 900 to 2,300 m, where it is encountered in moist broadleaf forest. Its current status in the rest of Haiti is unknown. Among satellite islands, reported only from Île de la Gonâve.
Status: Locally common resident, but has surely declined since the 1930s because of habitat loss. The subspecies *E. m. musica* is endemic to Hispaniola, including Île de la Gonâve.
Comments: Active in the dense vegetation of the canopy where it feeds mainly on mistletoe berries. Occasionally joins mixed flocks of warblers and tanagers. This euphonia is easily overlooked despite its colorful appearance. It is most readily located by its call.
Nesting: Lays 4 white eggs spotted with mauve in a domed nest with a side entrance which is usually built high in a tree among epiphytes. Breeds January to June.
Range: Resident on Hispaniola, Puerto Rico, and nearly all of Lesser Antilles from St. Barthélemy south to Grenada.
Local names: DR: Jilguerillo, Onza de Oro; **Haiti:** Lwidò, Louis d'or.

HISPANIOLAN CROSSBILL—*Loxia megaplaga* **Plate 55**
Breeding Resident—*Endangered Endemic*

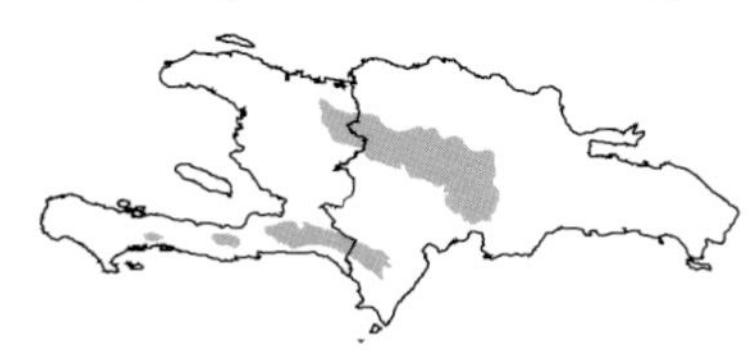

Description: 15–16 cm long; 28 g. A remarkable finch of pine forests that is noted for its uniquely crossed bill tips. Adult male is dusky brown, washed pale red overall, with intensity of red varying among individuals but concentrated on head, upper back, and upper breast. Adult female is generally dusky brown with a finely streaked breast, often with a yellow wash to foreparts and a yellowish rump. Adults of both sexes have two broad white bars on blackish wings. Juvenile is similar to female but is browner and more heavily streaked, especially on breast.
Similar species: Unmistakable.
Voice: High-pitched, emphatic, often repeated *chu-chu-chu-chu*, similar to the keys of an electric typewriter. During breeding season, also a soft, whistling warble.
Hispaniola: Local resident in the higher parts of the Cordillera Central and Sierra de Bahoruco, DR, the latter's continuation into Haiti, the Massif de la Selle, and in the Massif de la Hotte. So far not recorded in Sierra de Neiba. Recorded from as low as 540 m at Jarabacoa and as high as 2,600 m on Pico del Yaque. This species feeds on the seeds of mature Hispaniolan pine (*Pinus occidentalis*) and thus is essentially confined to that habitat type. May wander as food supply varies.
Status: Uncommon and local year-round resident considered endangered on Hispaniola. Total population estimates have ranged from fewer than 1,000 to 3,375 islandwide, with most in the Sierra de Bahoruco. This surely represents a decline since 1930 because so much high-

altitude pine forest has been cut since then. Further, it has been shown that the crossbills nest and feed principally in areas where mature pines are densest and bear many cones and that the current rate of uncontrolled burning will eliminate the type of habitat on which the species depends in 100–150 years if measures are not taken to protect unfragmented pine areas of adequate size.

Comments: Often quiet and secretive, but sometimes in noisy flocks feeding on pine seeds, which it extracts from cones with its specialized bill. Breeds during colder periods of the year, as do North American populations. The residency of this crossbill, a species typical of northern coniferous forests, in the West Indies supports the argument that the region was more temperate during the Pleistocene. As the West Indies warmed up over the past few thousand years, the crossbill was "trapped" in the last cool refugium in the region, the high peaks of Hispaniola. Until 2003 this species was considered part of the White-winged Crossbill species complex.

Nesting: Eggs are bone white with brownish blotches at the larger end. A compact cup-shaped nest of twigs, pine needles, lichens, and Spanish moss is built about 14 m above ground in a tall Hispaniolan pine (*Pinus occidentalis*), about 1–3 m from the main trunk on a horizontal branch. In the Sierra de Bahoruco, 2 nests also have been found which were only 1.5 m and 2.0 m high in understory shrubs of the heath family (*Lyonia*). Breeds at least December to April, perhaps year-round.

Range: Endemic to Hispaniola.

Local names: DR: Pico Cruzado, Periquito; **Haiti:** Bèk-kwazé, Bec-Croise.

ANTILLEAN SISKIN—*Carduelis dominicensis* **Plate 55**

Breeding Resident—*Endemic*

Description: 11 cm long; 9 g. A small, chunky finch with a light yellow bill. Male is distinctive with a yellow body, olive green back, black head and throat, and two diffuse yellow patches on black tail. Female is olive green above and yellowish white below, with faint pale gray streaking, pale yellowish rump, and two yellowish wing bars.

Similar species: Female resembles female Village Weaver, which is larger and has a darker, more massive bill and yellow supercilium. Flat-billed Vireo, which is unstreaked below, has a darker, more slender bill. Female Yellow-faced Grassquit has a dark bill, lacks wing bars, and has no streaks on underparts.

Voice: Soft *chut-chut* when flushed and a higher-pitched *swee-ee*. Also a low, bubbling trill.

Hispaniola: Irregularly distributed in the interior hills and mountains, reported from an elevation range of 500–3,000 m. Typically found in pine forest and weedy clearings, seldom below 1,000 m or above 2,500 m, though there is some suggestion of postbreeding dispersal to lower elevations, and occasional movement to lower elevations at times of severe cold or frost. Also occurs regularly in montane moist broadleaf forest in Sierra de Bahoruco where it feeds on seasonal fruits. Reported as locally common in Cordillera Central and Sierra de Bahoruco, DR, and in Massif de la Selle, Haiti, but few reports from the foothills of Massif de la Hotte, west of the Jacmel-Fauché depression. The species seems to be irregular in distribution in the northwest of Haiti and in the Sierra de Neiba.

Status: Locally common resident. It surely has declined somewhat since 1930 because of the loss of pine forest.

Comments: Small flocks actively forage from tree to tree and in bushes or grassy patches.

Nesting: Lays 2–3 light greenish white eggs with brown spots in a small cup nest built in a bush or pine tree. Although it typically nests in pine forest, nests have been reported near the summit of Morne Tranchant, Haiti, some distance from any pine forest. Breeds in May and June.

Range: Endemic to Hispaniola.

Local names: DR: Canario, Ciguíta Amarilla; **Haiti:** Petit Serin, Ti Serin.

OLD WORLD SPARROWS: FAMILY PASSERIDAE

This Old World family, also known as weaver finches, is represented by only one introduced species in the Western Hemisphere. Drab birds with a conical finchlike bill and relatively large head, most species are gregarious and often roost in large flocks. Several build large, elaborate colonial nests.

HOUSE SPARROW—*Passer domesticus* **Plate 57**
Introduced Breeding Resident

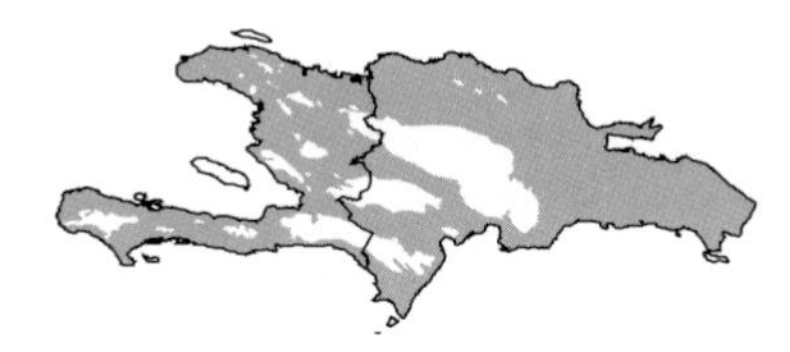

Description: 15 cm long; 28 g. A stocky, sparrowlike bird, with shorter legs and a thicker bill than emberizid sparrows. Male is brown above with blackish streaks and a single white wing bar; rump is grayish. Note the chestnut nape and gray crown, pale gray auriculars, and black patch on face and throat forming a bib. Underparts are grayish. Female and immature are similar to male but much duller; crown is gray-brown with a pale buffy supercilium, and there is a thin dusky eye-line. Underparts are dingy gray-brown, and black bib is lacking.
Similar species: Grasshopper Sparrow has a pale central crown stripe and buffy underparts. Adult Rufous-collared Sparrow has gray and brown stripes on head and a narrow black neck band.
Voice: Call notes include a loud distinctive *chirp* or *cheep,* also a husky rattle or chatter. Song is a monotonous series of chirp calls.
Hispaniola: Permanent resident spreading rapidly islandwide in the lowlands and beginning to penetrate some interior valleys, especially in towns. First reported late in 1976 near Monte Cristi, DR, it reached the National Botanical Garden in Santo Domingo by 1989. Recorded at Manabao at 850 m in the Cordillera Central, DR, by 1995, to date the highest elevation recorded. Common and widespread in western Haiti, at least from Port-au-Prince to Gonaïves and in the lower Artibonite River valley, and at Saint-Michel de L'Atalaye. To be expected in all urban areas.
Status: Local but increasingly common resident. Introduced to the West Indies probably as an incidental passenger on grain and tour ships. Population is growing and spreading rapidly.
Comments: Opportunistic feeder, primarily foraging on the ground. Diet includes cereal grains, weed seeds, and insects. Gregarious, it often flocks at all times of year.
Nesting: The only Hispaniolan reports are of 2 nests with clutches of 3 and 5 eggs, found in Santo Domingo in December 2003. At other West Indies localities, lays a clutch of 3–5 white eggs with spots or small splotches of gray or brown in a bulky nest of dried vegetation lined with feathers, sometimes string or paper. Nest sites typically are in nooks inside or on buildings or other humanmade structures. Breeds year-round but primarily March to September.
Range: Resident on all continents except Antarctica. In West Indies, common in Bahamas, Cuba, Hispaniola, and Puerto Rico; vagrant in Jamaica and St. Lucia; rare in Virgin Islands.
Local names: DR: Gorrión Cacero, Gorrión Doméstico; **Haiti:** Mwano Kay, Moineau Domestique.

WEAVERS: FAMILY PLOCEIDAE

This is a large Eastern Hemisphere family that consists primarily of heavy-billed, seed-eating birds similar to the finches. Many species are very colorful and make popular pets. As a result of extensive importations into the Western Hemisphere, individuals of several species have escaped and become established. Ploceids generally build domed nests.

VILLAGE WEAVER—*Ploceus cucullatus* Plate 57
Introduced Breeding Resident

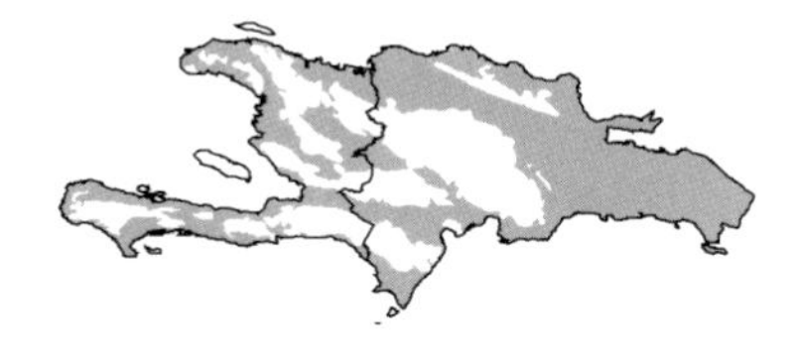

Description: 17 cm long; male 45 g, female 37 g. A chunky and heavy-billed finch. Male is a distinctive orange-yellow overall with a black hood, chestnut brown nape, and red iris. Female is generally yellowish green, brightest on face and breast, with darker wings and mantle, and yellow supercilium and wing bars. Eye is red. Immature is similar to female but iris is brownish.
Similar species: Adult male Baltimore Oriole is also black and orange-yellow but is slimmer and has a longer bill and tail. Female Antillean Siskin is smaller, has a paler and less massive bill, and lacks supercilium.
Voice: Steady high-pitched chatter with musical whistling calls.
Hispaniola: Known to occur in the entire coastal lowlands of the DR and Haiti, and up to 600 m elevation at Jumunucú and Jarabacoa in Cordillera Central, DR, as well as on Islas Catalina and Saona. Typically found in open dry forest, desert scrub, agricultural areas, second growth, vegetation near water, and rural villages. Also reported from arid areas of central Monte Cristi Province where abundant cactus fruit may substitute for other sources of water.
Status: Common introduced resident. Exact date of introduction is not known, but it is thought to have arrived with ships of the slave trade from West Africa. The earliest reports are primarily from western Haiti along the coast. Attempts to control its numbers in some areas have had little effect, but natural forces appear to be having an impact. Found to have steadily decreased in Pedernales, DR, and in towns around the border of Lago Enriquillo since the mid-1980s; not only has the number of colonies dropped, but the size of those remaining is smaller than previously.
Comments: Often flocks. During breeding, male performs a spread-wing, flapping display at the nest, sometimes hanging head down. This is a spectacular sight when several males display simultaneously while calling. Feeds on seeds and grain, including cultivated rice, where large flocks cause great damage. Nest is frequently parasitized by Shiny Cowbird.
Nesting: Eggs vary in color from light to dark blue-green, with light to moderate spotting. Clutch size is 2–4 eggs. Nests in noisy colonies with an average of 75 nests, but with up to 200 nests in some colonies. Each pair uses a woven spherical nest of plant fibers with a spout-like entrance, but individual males will build multiple nests in each colony. May be found breeding almost year-round, but in Haiti most nests are active primarily in May and June, extending to September, and in DR nesting activity is recorded mid-April to September, with the peak of activity from June to August.
Range: Native to much of sub-Saharan Africa. In West Indies, introduced to Hispaniola where resident islandwide, and more recently to Martinique.
Local names: DR: Madam Sagá, Cigua Haitiana, Chichiguao; **Haiti:** Madan Sara.

ESTRILDID FINCHES: FAMILY ESTRILDIDAE

This Old World family of small, finchlike birds is represented on Hispaniola only by introduced species. All are stocky with a conical bill, and are brightly plumaged. Most likely originated as escapees from captivity and have become established in the wild.

RED AVADAVAT—*Amandava amandava* **Plate 57**
Introduced Breeding Resident?

Description: 10 cm long; 10 g. Breeding male is primarily deep red overall with small white spots on wings, flanks, and sides. Adult female and non-breeding male are brownish above, paler below. Note red uppertail coverts and bill, white spots on wing, and dark eye-line. Immature is similar to adult female but lacks red and has buff-colored wing spots. Immatures typically occur in association with adults.
Similar species: See Nutmeg and Chestnut mannikins.
Voice: Call notes are a musical *sweet* and *sweet-eet*. The species has a variety of appealing songs, including melodious whistles and warbles.
Hispaniola: One DR report of a flock of 3 females and 4 males seen at Juan Dolio in January 1997 may possibly have spread from the population on Puerto Rico, or may have been released cage birds. Not known if they persist, but should be looked for primarily in grassy margins of freshwater swamps, borders of sugarcane fields, and along weedy drainage canals.
Status: Possibly introduced. Not recorded since 1997.
Comments: Typically flocks. Usually feeds on the ground under tall grass, making it more difficult to observe than other small seed-eaters. Its introduction in the West Indies likely resulted from pet birds escaping or being released.
Nesting: Unknown in the West Indies.
Range: Resident and native from Pakistan, India, and s. Nepal south through se. Asia; introduced to Hawaii, Fiji, Spain, Sumatra, and Singapore. In West Indies, introduced to Puerto Rico, reported in DR.
Local names: DR: Amandava Roja; **Haiti:** Bengali Wouj, Bengali Rouge.

NUTMEG MANNIKIN—*Lonchura punctulata* **Plate 57**
Introduced Breeding Resident

Description: 12 cm long; 14 g. Adult is deep brown above and whitish below, with a distinctive cinnamon-colored hood and scalloped black-and-white underparts. Immatures lack adult markings and are cinnamon-colored above, paler below.
Similar species: Immature can be confused with several other species, but heavy, blackish bill and light cinnamon coloration distinguish this species. Separated from immature Chestnut Mannikin by being less pale below.
Voice: Soft, plaintive, whistled *peet,* dropping in pitch and fading away at the end.
Hispaniola: Feral birds were first discovered near Guerra, DR, in January 1978 where a flock of 75–100 was found, indicating they were already firmly established, although local people reported that the birds had been present only 3–4 years. By 1996 the species had spread as far north as San Francisco de Macorís, and in 1997 it was found to be common to locally abundant in Pedernales Province, and was found on the northwestern coast of DR at Monte Cristi. The species is now found in Haiti, with reports since 1999 from Pignon, Deschapelle, and Port-au-Prince. Inhabits lowland open areas with seeding grass, such as the borders of

sugarcane plantations, agricultural areas, road edges, and parks in urban areas. Also reported from Navassa Island.

Status: Common introduced resident, spreading rapidly islandwide.

Comments: Typically flocks. Forages for seeds on the ground and from grass heads, and is a pest in rice-growing areas. The introduction of this species in the West Indies likely resulted originally from cage birds escaping or being released, but the population on Hispaniola was most likely a range expansion from Puerto Rico.

Nesting: Lays 6 white eggs in a bulky domed nest with a side entrance at a moderate height in a tree or shrub. Breeds primarily June to October. This species is a frequent host of the brood parasite Shiny Cowbird in Puerto Rico and can be expected to become so in Hispaniola.

Range: Resident and native from India east to East Indies and Philippines; introduced in Hawaii and Australia; vagrant to Florida. In West Indies, introduced and established though sometimes local in Cuba, Jamaica, Hispaniola, Puerto Rico, and Guadeloupe; vagrant to Virgin Islands and Martinique.

Local names: DR: Ciguíta Pechijabao; **Haiti:** Mannken Miskad, Capucin Damier.

CHESTNUT MANNIKIN or TRICOLORED MUNIA—*Lonchura malacca*

Introduced Breeding Resident **Plate 57**

Description: 12 cm long; 13 g. Adults are handsomely patterned with cinnamon-colored back and wings, full black hood, and white underparts with black belly patch. Bill is pale grayish. Immatures are cinnamon brown above and buffy below, with a dark bill.

Similar species: All immature mannikins are similar. This species is more cinnamon-colored and has a paler bill and underparts than Nutmeg Mannikin.

Voice: Call note is a thin, nasal honk, *neat,* less plaintive, clear, and melodious than that of Nutmeg Mannikin.

Hispaniola: The earliest record is a flock at Guerra, DR, in July 1982. The species is now primarily localized in the DR around Guerra, Bayaguana, and Yamasa. First recorded for Haiti at Bois Neuf, about 65 km northwest of Port-au-Prince, in 1999, and near Deschapelle in 2000. The first introductions may have been overlooked because this species frequently flocks with Nutmeg Mannikin. Typically found in grassy areas bordering sugarcane, marshes, rice plantations, and agricultural fields gone to seed at low elevations.

Status: Locally common introduced resident, spreading in lowlands.

Comments: Typically flocks. It is a potential pest to various grain crops. Forages on seeds both on the ground and from grass heads. Appears dominant to the other species of its genus where they come in contact. The populations on Hispaniola may have resulted from range expansion from other islands.

Nesting: Lays 4–5 white eggs in a bulky domed nest with a side entrance which is frequently placed in sugarcane, 1–3 m above ground. Breeds primarily June to September.

Range: Resident and native in c. and s. India and in Sri Lanka; introduced to Hawaii, Venezuela, and Japan; possible vagrant to Florida. In West Indies, introduced to Cuba, Jamaica, Hispaniola, Puerto Rico, and Martinique.

Local names: DR: Monjita Tricolor; **Haiti:** Mannken Twa Koulè, Capucin à Dos Marron.

Recent Additions to the Checklist

Below we list four species of migratory birds from North America that are new to the Hispaniola checklist and were reported just before publication.

GADWALL—*Anas strepera* **Plate 2**
A female Gadwall was seen and photographed in late-April 2005 at Letan Kokoye (Étang Cocotier) between Les Cayes and Camp Perrin. Although a number of earlier reports exist, this is the first confirmed record for the species.

EASTERN WOOD-PEWEE—*Contopus virens* **Plate 37**
A single bird was seen and photographed at Salinas de Baní in mid-October 2005. This was one of several vagrant species recorded on the same day following an extended period with weather conditions unfavorable for continued migration.

SWAINSON'S THRUSH—*Catharus ustulatus* **Plate 42**
A single bird was seen and photographed at Salinas de Baní in mid-October 2005. This was another of several vagrant species recorded on the same day following an extended period of weather that was unfavorable for continued migration.

SCARLET TANAGER—*Piranga olivacea* **Plate 53**
A first-year male Scarlet Tanager was photographed in November 2005 at Rabo de Gato on the north side of Sierra de Bahoruco National Park, constituting the first verified record for this species on Hispaniola.

APPENDIX A: Birdwatching on Hispaniola

Comprising a small island with 31 endemic species, and many more regional endemics, the Dominican Republic and Haiti are growing in popularity for the birdwatcher. The Sociedad Ornitológica de la Hispaniola, along with several private birding guides, offers a variety of tour ideas for birdwatchers. Here we outline a few of our favorite sites for sampling the island's diverse birdlife; these include sites suitable for day trips from Santo Domingo or Port-au-Prince and sites that may require more extended travel. We also provide guidelines on how to reach these locations. Please note, however, that access to remote areas of the island can be difficult for those unaccustomed to rural travel; we recommend that for the sake of convenience, safety, and the best birding, *please contact a local guide*. There are many well-trained and friendly birdwatching guides on the island who would love to help you discover Hispaniola's avifauna and tell you more about the island's habitats, plants, animals, history, culture, and conservation efforts. Contacts can be arranged through the Sociedad Ornitológica de la Hispaniola (http://www.soh.org.do) and the Société Audubon Haiti (http://www.societeaudubonhaiti.org).

Although the Dominican Republic's birding infrastructure is currently better developed than Haiti's, the latter country offers remarkable opportunities to encounter many Hispaniolan and Greater Antillean endemics. Haiti supports all populations of the Gray-crowned Palm-Tanager, and finding this rare species is virtually guaranteed in proper habitat. In addition, other Haitian sites can provide excellent views of rare endemic birds such as La Selle Thrush, Western Chat-Tanager, and Hispaniolan Highland-Tanager.

Recent conservation efforts by Société Audubon Haiti have shown this institution to be eminently capable of organizing trips to La Visite National Park and the Macaya Biosphere Reserve while making safety its top priority. However, because the number of birders visiting Haiti remains low, and the birding-tour infrastructure is still developing, the cost of undertaking such a trip would be unavoidably high for a single traveler. Many of Haiti's premier birding sites are in remote and difficult-to-access locations, thus tending to appeal to more adventurous and hardy birders. Regardless, the rewards of birding in Haiti are great, and the country's rich cultural heritage and friendliness of its people will be unquestionable highlights of your visit.

Birding Sites in the Dominican Republic

National Botanical Garden

The Jardín Botánico Nacional Dr. Rafael Moscoso, or National Botanical Garden, is in the northern part of Santo Domingo and is a fine place to encounter the common birds of the country. An early morning walk of about two hours, through wooded areas, a stream, and open palm savannas, will offer opportunities for seeing many species. The national bird, the Palmchat, will be seen readily, and the endemic Hispaniolan Parakeet is likely to be found even though it is often difficult to locate elsewhere on the island. The stream is home to Limpkins and Least Grebes, both of which are typically elusive elsewhere. Other more common birds to be found in the garden include the Black-crowned Palm-Tanager, Hispaniolan Woodpecker, Hispaniolan Lizard-Cuckoo, Mangrove Cuckoo, Antillean Palm-Swift, Gray Kingbird, Vervain Hummingbird, and Antillean Mango. Black-whiskered Vireos may be heard singing even if they are not seen easily. Many migratory warblers may also be seen in the Botanical Garden during the non-breeding season.

The garden is best approached by going north on Ave. Lincoln from central Santo Domingo. After crossing under the Ave. John F. Kennedy overpass, continue north until you reach a rotunda. Keep right on the rotunda and take the first right-hand exit. The garden entrance is a few hundred yards farther on your right. After entering the garden, go straight ahead to the base of the entry plaza where you may enter the wooded area often preferred by wintering warblers. To reach the stream area, follow signs for the Gran Cañada. The official hours of the Botanical Garden are 9:00 A.M. to 6:00 P.M., but birdwatchers may enter any time after 6:30 A.M. with identifying binoculars in place.

Salinas de Baní

The Salinas de Baní area is characterized by extensive sand dunes, interdunal swales, thickets, mangroves, salt-drying pans, lagoons, mudflats, and both sandy and rocky beaches. Many of these are not common habitats on Hispaniola, and the site therefore provides extremely important habitat for both migratory and nesting shorebird species. All egret and heron species present on the island may be seen here, as well as Clapper Rails and Whimbrels. Shorebirds are plentiful in the mudflats, coastal areas, and especially the salt pans. Nesting species at Salinas de Baní include Snowy Plover, Wilson's Plover, Least Tern, and Willet. Many warblers frequent the mangroves and thickets along the lagoons, and the bay often hosts boobies and other seabirds. Many Hispaniolan rarities have first appeared here, including Black-legged Kittiwake, Great Black-backed and Lesser Black-backed gulls, American Golden-Plover, Wilson's Phalarope, and Red-necked Phalarope.

Salinas de Baní is some 60 km west of Santo Domingo, or about a 1.5-hour drive, making it an ideal destination for a day trip from the city. To reach Salinas de Baní, take Ave. 27 de Febrero to the main highway leaving Santo Domingo to the west. After bypassing San Cristóbal, the first city you reach will be Baní. Shortly after crossing the Río Baní on a large bridge, look for a sign indicating a left turn for the village of Salinas. Take this road, which will pass through several small villages before reaching the town of Las Calderas and a Dominican naval base. At this point you begin to see extensive mudflats, mangroves, and sand dunes. Stop anywhere along here to scan for shorebirds, then continue to the main village of Salinas. Just past Salinas the road will fork, with the right fork crossing the salt pans on a paved road; take the left fork on a dirt road which very shortly dead-ends in an unofficial-looking parking area adjacent to the extensive salt pans. Park here and continue exploring on foot. A narrow dike will take you to the larger lagoons and the dunes and beach beyond. Arrive early to avoid the heat of the day; take plenty of sunscreen and drinking water.

Ebano Verde Reserve

Ebano Verde Reserve is an hour and a half from Santo Domingo on the easternmost slopes of the Cordillera Central. To get there you must drive north of Santo Domingo on the Autopista Duarte, the main south-to-north highway. Once past the town of Bonao, take the exit to Constanza, after which you will immediately start climbing up the mountain on what is known as the Carretera de Casabito. The road is narrow and winding, so exercise caution.

The reserve has two popular entrances. The first is at the very top of the Casabito road (there is a sign that marks the turn). This entrance should be used only with a four-wheel-drive vehicle (4WD), and if you have coordinated beforehand with the reserve personnel to open the gate. The second entrance is immediately after the second bridge on the right, El Arroyazo, and should be considered the main entrance, as the Fernando Aquino Visitor Center is located at the beginning of the trails here.

The best trail for birding is the Arroyazo Sendero de Nubes trail, which starts at the visitor center and ends at the top of Casabito. The trail extends 6 km through pine forest and second-growth areas. Halfway up the trail, you start walking beside El Arroyazo stream and the vegetation changes to riparian forest, characterized by the manacla palm (*Prestoea acuminata*) which typically grows in humid soil. The upper part of the trail, near the Casabito entrance, passes through beautiful, undisturbed cloud forest that is very much worth seeing. For most of its length, the trail is a relatively easy walk, although the final ascent to the top of Casabito Mountain is fairly steep.

Along the trail it is common to see Stolid Flycatcher, Greater Antillean Elaenia, Hispaniolan Pewee, Hispaniolan Trogon, Narrow-billed Tody, Hispaniolan Woodpecker, Caribbean Martin, Vervain and Emerald hummingbirds, Antillean Siskin, Rufous-collared Sparrow, Hispaniolan Spindalis, Black-crowned Palm-Tanager, Rufous-throated Solitaire, and Yellow-faced and Black-faced grassquits. In the upper cloud forest, you have a chance at briefly sighting the elusive Eastern Chat-Tanager as well.

Southeastern Dominican Republic: Del Este National Park, Isla Saona, Río Soco, Cumayasa, and Cochoprimo

Del Este National Park is on the extreme eastern end of Hispaniola, close to some of the island's most popular resort areas. The park contains an extensive area of dry forest where you are likely to find White-crowned Pigeon, Hispaniolan Parrot, Mangrove Cuckoo, Hispaniolan Lizard-Cuckoo, Broad-billed Tody, Antillean Piculet, Hispaniolan Woodpecker, Stolid Flycatcher, Flat-billed and Black-whiskered vireos, Greater Antillean Bullfinch, and other dry forest inhabitants. White-necked Crows may also be seen near Guaraguao. The main tourist attraction is the trip to Isla Saona, also part of the park system, although this is primarily a beach excursion. To visit the extensive Magnificent Frigatebird colony at Las Calderas on the way, it is advisable to contact the park office in the village of Bayahibe. Guides can also be arranged through the office of the local conservation organization, EcoParque.

The trip from Santo Domingo to Del Este takes about three hours. Head east from Santo Domingo to San Pedro de Macorís, then to La Romana. Follow signs for Higuey and watch for a right-hand turn to Bayahibe. The park office is in the village of Bayahibe where the tour buses park. Purchase tickets here for the walking trail, as well as for boat tours to Isla Saona. The trail is reached by returning a few kilometers toward the main road and taking a sharp right-hand fork toward the Dominicus Hotel Complex. After the last hotel on the right, Coral Canoa Hotel, follow the dirt road down to the beach and then turn left parallel to the sea until you reach the Guaraguao park entrance. The 3-km trail here is at first sandy but then passes over rough limestone. The rise in elevation, although slight, indicates a former marine shoreline as well as a small change in habitat type. Arrive early and carry plenty of water. When you reach El Puente Cave, look for the Taino petroglyphs as well as for an Ashy-faced Owl that sometimes roosts inside.

In addition to Del Este National Park, other interesting areas that offer good birding are found at Río Soco, Río Cumayasa, and inland at Ramón Santana and the Cochoprimo area, where scattered habitats include semihumid forest, dry scrub, wetlands, mangroves, salt marsh, and beaches. These sites are accessed from the main coastal highway east of San Pedro de Macorís. Where the highway crosses the Río Soco, explore south of the highway to the mouth of the river at Boca de Soco with its extensive wetlands and mangroves. Be sure to scan the nearby ocean too for seabirds such as Brown Booby, Magnificant Frigatebird, and Brown Pelican. Upstream along the Río Soco, in the vicinity of Ramón Santana, are the Cochoprimo wetlands, where egrets, herons, and bitterns may be found, as well as Limpkin and Greater Antillean Grackle.

Farther east along the coastal highway, the Río Cumayasa and the Cumayasa canyon are accessed by turning north on the gravel road just before the Cumayasa bridge. In the early morning, it is worthwhile to hike north along the river bed for at least 2 hours. This may turn up such species as Red-tailed Hawk, Antillean Piculet, Loggerhead Kingbird, Stolid Flycatcher, Red-legged Thrush, and Greater Antillean Bullfinch. South of the highway, the Cumayasa canyon can be explored further at the Rancho Cumayasa, where horses, kayaks, and cabins may be rented.

Finally, evening observations in this area are often successful in turning up Barn, Ashy-faced, and Short-eared owls and Double-striped Thick-knee. Scouting of potential pastures and sugarcane fields should be done in the daytime, but return at night, especially to fields where the cane has been recently cut. This is where nocturnal birds are likely to congregate, especially from January through July.

Los Haitises National Park

Los Haitises National Park is the last stronghold of the endemic and critically endangered Ridgway's Hawk. The park consists primarily of dense lowland broadleaf forest covering very hilly karst limestone formations called *mogotes*. Although much of the area has been previously deforested, the extremely steep mogotes were often left untouched, forming small islands of intact habitat. Today, with protection afforded by the park, the areas between the mogotes are regenerating with thick vegetation. Typical forest birds include White-crowned and Plain pigeons, White-necked Crow, Broad-billed Tody, Greater Antillean Pewee, Stolid

Flycatcher, Black-crowned Palm-Tanager, Hispaniolan Parrot, and Hispaniolan Oriole. Off the coast the mogotes form islands in the bay and are nesting sites for egrets, pelicans, frigatebirds, and Brown Boobies.

Access to Los Haitises National Park is difficult. The trip from Santo Domingo takes about three hours. From Santo Domingo head east, pass the Aeropuerto Internacional and Boca Chica, to San Pedro de Macorís. Then turn north to Hato Mayor and continue to Sabana de la Mar. In Sabana de la Mar, the park office is located at the north end of town near the pier where the boats leave for Samaná. The best way to visit the area is to contact the park office and hire a guide who is familiar with the park. One popular and pleasant route is to explore the margins of the park by boat from Caño Hondo, proceeding down a river through mangroves to the Bahía de Samaná and then stopping at various points to explore trails through lowland broadleaf forest. The other option is a difficult overland trail from Trepada Alta; this often produces Ridgway's Hawk sightings, but it requires the use of a local guide because of the difficulty of finding the trail.

Please note that the trails in Los Haitises National Park are not well maintained and can be lined with nettle and other irritant plants. Trails are often rocky and slippery because of the geological characteristics of the area. Afternoon rain is often the norm. Humidity levels can reach 95% in the forest, which will require you to carry sufficient drinking water.

Monte Cristi: Cayos Siete Hermanos, Laguna Saladilla, and Laguna Salinas

The town of Monte Cristi is in the northwestern corner of the country and is best known for the massive El Morro headland, seen and named by Christopher Columbus (known in Spanish as Cristóbal Colón). The main birding attractions here are the extensive mangroves which can be explored by boat, lagoons full of flamingos hidden in cactus forests (for which a local guide is essential), and the offshore islands, the Cayos Siete Hermanos. During the months of May through August, the islands are home to nesting seabirds, including large numbers of Brown Noddy and Sooty and Bridled terns. Besides flamingos, the mangroves host many herons, egrets, spoonbills, ibis, and other large waders, as well as shorebirds and waterfowl, including some hard-to-see species such as Ring-billed Gull and Gull-billed Tern. Although numbers of wintering ducks are reduced from historic highs, Monte Cristi is still one of the better places to find wintering and resident waterfowl.

Monte Cristi can be reached by taking Ave. John F. Kennedy in Santo Domingo to the Autopista Duarte and heading north. This is the main north-south highway in the country, and you will pass by Bonao, La Vega, and Santiago. After reaching Santiago, follow the highway signs to Monte Cristi. The drive should take about four and a half hours. Once in Monte Cristi, look for the park headquarters on the road out of town toward El Morro; here local boatmen offer trips to the Cayos Siete Hermanos.

Other interesting areas west of Monte Cristi are Laguna Saladilla and Laguna Salinas. To reach these sites, head west of Monte Cristi toward the town of Manzanillo. Laguna Salinas is reached by making a right-hand turn onto a dirt road close to a national park station situated on the opposite side of the road. This right-hand turn is just after the Nueva Judea sign on the left; if you reach the Los Conucos town sign on the right, you have missed the turn. After passing through a large tract of thorn scrub, you will find Laguna Salinas at the end of the dirt road, which opens onto a wide plain from where the lagoon can be seen in the close distance.

Laguna Saladilla can be reached after turning right toward Manzanillo at the Manzanillo-Dajabón intersection. Not far from there, make a left turn onto a dirt road that takes you to an aqueduct from which you can access the eastern bank of Laguna Saladilla. Unfortunately, dense reeds block most of the access to the lagoon, so any additional exploration of this freshwater habitat must be pursued by boat. One of the local fishermen will gladly take you out for a small fee.

Lago Enriquillo

This hypersaline lake, about the size of Manhattan, lies a remarkable 44 m *below* sea level. It is a remnant of the open marine channel that during much of the Pliocene and Pleistocene separated the southern paleo-island of the Sierra de Bahoruco, Massif de la Selle, and Massif de la Hotte from the rest of what is now Hispaniola. Lago Enriquillo is a surreal landscape, and it is home to hundreds of egrets, terns, herons, and flamingos, as well as crocodiles and iguanas. The sand- and mud-flats fringing the lake's shore can provide good shorebirding. White-necked Crows may be seen around the town of La Descubierta, and Palm Crows are found on Isla Cabritos, a large island in the middle of the lake.

Lago Enriquillo can be reached from Barahona by heading north until reaching the road to Duvergé and Jimaní at a major intersection. Turn west and continue until reaching the crossroad to Neiba, marked by a large statue of the Taino chief Enriquillo. Turn north again and continue to Neiba, then turn left and pass through a series of small towns until arriving at La Descubierta. Be aware of the many *policia acostado* ("sleeping policemen," or speed bumps) and small ditches intended to slow traffic speed in these towns. Access to the national park is a few kilometers east of La Descubierta on the north shore of Lago Enriquillo. From here boat trips can be arranged which proceed along the shore to see the birds and crocodiles, stopping on Isla Cabritos to view the iguanas. Access by land to the Los Borbollones area on the northern shore is via a dirt road in the village of Bartolomé. Ask for detailed directions at the park office. Access to Lago Enriquillo's south shore, where you can usually find Palm Crows, is best made from the Duvergé-Jimaní road, about 5 km west of the town of Baitoa. Look for a narrow dirt road that drops off to the right on a curve. Drive in a short way while listening for the crows. You may continue on this road about 2 km until you enter a large cactus grove and see the lake, but please *do not drive out on the sandflats* where nesting birds will be disturbed and you are likely to get your vehicle stuck in the soft sand.

Laguna de Oviedo

To experience a close look at Greater Flamingo, Roseate Spoonbill, and White Ibis, as well as other waders, shorebirds, gulls, terns, and pelicans, a boat trip on Laguna de Oviedo is recommended. This brackish lagoon is in southwestern Dominican Republic, halfway between Barahona and Pedernales. It is so shallow that after crossing the width of the lagoon to the mangrove edge, the boatman must get out and wade, pushing the boat by hand. By cutting the motor, it is usually possible to approach birds very closely. A longer trip down the length of the lagoon passes an island where White Ibis nest, and may also include a stop at another island to see the two local iguana species, one of which is endemic to Hispaniola.

To reach Laguna de Oviedo, take the main road south out of Barahona and continue for about 1.5 hours. Just before arriving at the lagoon, stop at the park house on your left to purchase a ticket and inquire about a tour. (In 2005, the cost was about $50 for a boatload of up to 8 people.) It is important to arrive at Laguna de Oviedo early in the morning because the wind typically picks up in late morning, making for a wet, though not dangerous, trip.

Sierra de Bahoruco, North Slope: Puerto Escondido, Rabo de Gato, La Placa, and Zapotén

The Sierra de Bahoruco is Hispaniola's premier birding area. Although some areas are remote and difficult to reach, requiring 4WD, other areas are more easily accessed. Of the 31 Hispaniolan endemics, only Ridgway's Hawk is not found in the area. It should also be noted that the Gray-crowned Palm-Tanager, endemic to Haiti, only rarely crosses into the Sierra de Bahoruco.

The outstanding feature of the Bahoruco's north slope is the montane broadleaf evergreen forest, or Dominican cloud forest, marked by large tree ferns above Zapotén at about 1,200 m elevation. This is the best site for high-elevation endemics such as Scaly-naped Pigeon, White-fronted Quail-Dove, Hispaniolan Emerald, Hispaniolan Trogon, Narrow-billed Tody, Rufous-throated Solitaire, La Selle Thrush, Green-tailed Ground-Tanager, Hispaniolan Highland-Tanager, Western Chat-Tanager, Hispaniolan Spindalis, Antillean Euphonia, and Antillean Siskin.

To reach the best birding areas on the north side requires a 4WD. From Barahona take the main road north toward Santo Domingo until reaching a major intersection and the road to Cabral, Duvergé, and Jimaní. Turn left here and continue to Duvergé (about one hour). In Duvergé look for signs for the left turn to Puerto Escondido. Outside Duvergé the road soon becomes dirt, and after about half an hour of steady climbing the small village of Puerto Escondido is reached. Continue straight through town until reaching a T-intersection. To reach Rabo de Gato, turn left at the T-intersection in Puerto Escondido and go past a chain-link fence surrounding the dam buildings. Then turn right, go down and then up a shallow dip in the road, cross a canal, and take the next right at the Rabo de Gato sign. Continue straight past a heavy fence on the left. Park here or continue up a small hill and park at its top. The fairly level trail can be walked in an hour or two. The area known as Rabo de Gato is an interesting narrow strip of riparian habitat with a surprising mix of birds. Both the Broad-billed and Narrow-billed todies occur here, as well as Hispaniolan Lizard-Cuckoo, Antillean Piculet, Hispaniolan Trogon, Flat-billed Vireo, White-necked Crow, Hispaniolan Oriole, and Antillean Euphonia, and on occasion Bay-breasted Cuckoo. Listen for the crows here, or return to the dam in the early morning or late afternoon when the birds may be more conspicuous.

To continue on to La Placa and Zapotén, turn right at the Puerto Escondido T-intersection (Rabo de Gato/Aguacate sign), pass by the guard post on the right, and then continue through the valley, which here is dominated by large avocado plantations on the left. About 10 km farther, you will enter an area of extensive dry broadleaf forest. Here look for a prominent yellow park sign on your right. This site, known as La Placa, is a reliable spot for Bay-breasted Cuckoo and Flat-billed Vireo. At dawn and dusk it is also one of the best places to see or hear Least Pauraque and Hispaniolan Nightjar. Continue up the road another 3–5 km until reaching the area known as Los Naranjos, which has traditionally provided the most dependable sightings of Bay-breasted Cuckoo. After about 3 km the road enters a (normally) dry riverbed full of loose stone; continue on carefully and the road begins to ascend as it enters the moist broadleaf zone. Stop at the military guard post at Aguacate, which lies just above the international border, and tell them you are a birdwatcher (*observador de aves*), at which point they will record your license plate number. Do not be intimidated by this procedure as this is part of regular protocol along the Dominican-Haitian border. In the daytime this is a spot to take a hard look at the badly deforested hills of Haiti.

The road enters a brief stretch of pine forest in another 5 km and then passes the potato market at Zapotén, which actually sits within the national park. Just short of 2 km farther on, stop at a wide curve and park well off the road. This is the beginning of the cloud-forest zone. The best birding strategy here is to walk up and down the road. Be advised, however, that in order to maximize the chance of seeing La Selle Thrush you should arrive before dawn, after which you might find birds foraging in the road. This means, however, leaving Barahona at 3:00 A.M. in order to reach this area at the break of dawn.

Beyond Zapotén the road continues to rise, crossing Loma de Toro and patches of excellent montane broadleaf forest, before descending through pine forest and agricultural areas of Los Arrollos to the border town of Pedernales where basic services are available. This road is frequently impassable, even with a 4WD, so always check local conditions before proceeding. From Pedernales, the south slope of the Sierra de Bahoruco is accessible via Aceitillar or the Alcoa road just east of town (see below).

Sierra de Bahoruco, South Slope: Cabo Rojo, Aceitillar, and Hoyo de Pelempito

Cabo Rojo and the Aceitillar sector of the Sierra de Bahoruco are about two hours southwest of Barahona, with a good paved road all the way. Portions of this road have many tight curves, and some sections have been washed away by floodwaters, so drive with caution. About half an hour west of Oviedo, just past the signs for Bahía de las Aguilas, you will cross a bridge overpass. The turn for the Alcoa road is a small dirt connector and the first right turn immediately after the bridge. This connector loops around to the paved Alcoa road; turn right (south toward the coast) to reach Cabo Rojo, or left (north toward the mountains) to reach Aceitillar and the pine zone.

The southern slope of the Sierra de Bahoruco is served by an excellent paved road that remains from the bauxite mining operations of Alcoa. At sea level, in the area known as Cabo Rojo, there is a small wetland across from the mine shipping port which attracts a fair number of waterfowl and shorebirds of all kinds, including the regional endemic White-cheeked Pintail, White Ibis, occasional Roseate Spoonbill, and wintering ducks and shorebirds. The mangroves here harbor good numbers of Yellow Warblers, an endemic subspecies, and many other non-breeding warblers. North from Cabo Rojo, the road climbs steadily through dry and broadleaf forests until reaching the Hoyo de Pelempito ticket station at Las Mercedes. Beyond this point the road reaches pine forest at about 1,100 m elevation. This portion of the park is known as Aceitillar, taking its name from a local grass. Just after entering the pine forests, look for a national park sign for La Charca, which is a small catch basin for water runoff. This is a great spot to look for Hispaniolan Emerald, Antillean Piculet, Caribbean Martin, and especially Golden Swallow and Hispaniolan Crossbill, as well as several species of warblers in season. The Hispaniolan Parrot, Hispaniolan and Olive-throated parakeets, and Plain Pigeon are also commonly seen here. Make frequent stops on the way down to find the piculet and other broadleaf specialists.

Another interesting stop is the new visitor center at Hoyo de Pelempito, at the end of the Aceitillar road. Tickets can be purchased at the Las Mercedes ticket station on your way up. The trail that leads from the parking lot may provide close views of some birds, and the visitor center offers breathtaking views of the karst cliffs and the eastern Bahoruco range. This is a favorite spot to view Hispaniolan Parrots in the late afternoon.

Birding Sites in Haiti

Trou Caïman

Trou Caïman (or Dlo Gaye) is a shallow, freshwater lake approximately 25 km northeast of Port-au-Prince. As part of the lowland Cul de Sac/Neiba corridor, Trou Caïman attracts a wide variety of birds, particularly migrants. In recent years, more than 100 species have been recorded at this site, including rare and vagrant birds such as Neotropic Cormorant, American Golden-Plover, Buff-breasted Sandpiper, Black Skimmer, American Pipit, and Bay-breasted Warbler. At least 30 species can be found at Trou Caïman on any given morning of the year.

Access to Trou Caïman is a relatively easy 45-minute trip by vehicle from Port-au-Prince. Begin by taking the road to Croix-des-Bouquet. This is the same road that leads to the Haitian-Dominican border at Malpasse and Jimaní. At the round-about intersection in Croix-des-Bouquet, take the road north that leads to Mirebalais (Plateau Central). Eventually the houses become scarce, more scrub land is evident, and the pavement abruptly ends just before a lone Texaco station. Continue on for less than 1 km and veer right onto a rough dirt road leading toward Thomazeau. After a few minutes of driving, you will see Trou Caïman in the distance below. Eventually the road passes very close to the lake itself at a bend in the road just before the tiny village of Trou Caïman. At this point, look for a small entry in the brush barely wide enough to accommodate a vehicle. Beware of thorns when entering, and also the soft ground under the grassy area next to the shoreline. From this point, you can view a portion of the cattail habitat and also a large portion of the lake.

The wide, flat, grass-fringed western edge of the lake is the best area for birdwatching since it is relatively clear of trees and scrub. This habitat is especially attractive to ducks, herons, egrets, plovers, sandpipers, gulls, and terns. Pied-billed Grebe, Common Moorhen, and American and Caribbean coots are often seen or heard. American Wigeon, Northern Shoveler, Lesser Scaup, and Ruddy Duck may be seen on the lake during the winter months. Walk south along the western edge for better viewing of the southwestern corner where bird activity is usually high. Make sure to check the scrubby areas for migrating and wintering warblers, as well as local birds such as Smooth-billed Ani, Broad-billed Tody, Hispaniolan Woodpecker, Hispaniolan Palm Crow, White-necked Crow, Gray Kingbird, and Village Weaver. From August to October, keep an eye overhead for migrating Osprey. During the winter months, Peregrine Falcon and Merlin can be observed hunting or resting. Large swaths of reeds dominate the northern part of the lake where one can find wintering waterfowl, Fulvous Whistling-Duck, and Least Bittern. Local fishermen will take birders there in a small skiff for a negotiable fee. There are no amenities in this area, so be prepared with water and food. Sun and mosquito protection, plus sturdy, waterproof footwear, are strongly advised. Best viewing is in the early morning, as an easterly wind usually picks up, making it difficult to hear and occasionally to hold binoculars and spotting scopes steady.

Near Port-au-Prince: Fermathe, Kenscoff, and Furcy

Approximately 20 km south of Port-au-Prince, this area offers an opportunity to see some of Haiti's high-elevation specialties such as Hispaniolan Emerald, Narrow-billed Tody, Greater Antillean Elaenia, Hispaniolan Pewee, Golden Swallow, Rufous-throated Solitaire, Red-legged Thrush, Hispaniolan Spindalis, and Antillean Siskin.

To reach the Fermathe, Kenscoff, and Furcy area, take the Route de Kenscoff out of the Port-au-Prince suburb of Petionville. This road is paved, but traffic can be heavy on the twisting road to Fermathe, so caution is advised. About 11 km from Petionville, stop at a small woodlot at the Baptist Haiti Mission in Fermathe, where you can find a variety of birds, especially during the winter months. Park your vehicle in the area next to the souvenir shop and restaurant. At the rear of the parking lot, pass through the gate on the right side of the church. Walk down the paved drive and look for a grassy playground lot on the left. A small zoo is on the right and is worthy of a visit itself. At the rear of the grassy area is an opening to the terraced woodlot which is perched on the edge of a steep slope. There is a decent path across the top of the lot, but the more adventurous will want to find the paths to the lower terrace levels which are vegetated with trees and dense shrubs. Green-tailed Ground-Tanager can be found in this area, as well as all three resident hummingbirds. Red-legged Thrush, Palmchat, Bananaquit, Black-crowned Palm-Tanager, Greater Antillean Bullfinch, and grassquits are commonly observed or heard. Listen for Black-whiskered Vireo during the spring and summer months. Regularly found non-breeding residents include Black-throated Blue, Cape May, and Black-and-white warblers, American Redstart, Ovenbird, and Louisiana Waterthrush. More unusual species that have been observed in the woodlot include Yellow-bellied Sapsucker, Rose-breasted Grosbeak, Baltimore Oriole, and even Gray Catbird.

Take a right turn out of the parking lot and follow the paved road a few kilometers to the village of Kenscoff. Pine trees line some sections of the road, so keep an ear open for singing Pine Warblers. At the police station in Kenscoff, stay on the paved road that veers to the left. At the next major fork in the road, stay right and continue on up the mountain toward Furcy. The road conditions are more rugged from this point on, the amount of pine cover increases, and houses become fewer. Finally, the road levels off briefly at the Ste. Helene (Sent Elèn) orphanage. The main unpaved road continues down into the tiny village of Furcy and enters more mixed pine woodland habitat. A narrow paved road veering off right at Ste. Helene leads to the top of the area's highest point where dozens of communication towers are situated. Golden Swallow can be seen anywhere in this habitat, Rufous-throated Solitaire can be heard singing early and late in the day, and Antillean Siskins congregate on the power line along the roadway.

La Visite National Park

La Visite National Park is southeast of Haiti's capital city of Port-au-Prince. Currently the best route is to drive west from Port-au-Prince and then south to Jacmel. Once in Jacmel, follow the road east of town, past Marigot, and then north up the mountain. You will need to ask farmers for directions to reach Fond Jean Noel and then Seguin. Past Seguin you will see Auberge La Viste on the left-hand side of the main road. Total driving time should be about five hours. Another popular route passes south of Port-au-Prince through Petionville and on to the town of Kenscoff. Beyond Kenscoff the road is in rough, typically impassable, condition, so visitors should plan to make a four- to five-hour hike in the direction of the town of Seguin. Before reaching Seguin, one must start asking for directions to Auberge La Visite, which is the only viable lodging facility in the area. The auberge can provide capable guides who know the trails in the area extremely well. Regardless of which route you select, bear in mind that many rural roads in Haiti are in very rough condition, so a 4WD is essential.

La Visite National Park has a system of unmarked trails, and birding this area requires quite a bit of hiking. Keep your eyes open when you pass any ravines with broadleaf vegetation, as these are the best places to find Western Chat-Tanager and La Selle Thrush. These birds can often be seen foraging on the ground in open areas, especially after dawn. Stands of pines near Pic La Visite regularly support small flocks of Antillean Siskins and Hispaniolan Crossbills, and Golden Swallows may be seen over open habitats. Other Hispaniolan endemics of montane broadleaf forests, such as Hispaniolan Trogon, Rufous-throated Solitaire, and Narrow-billed Tody, are restricted to the few remaining patches of this habitat type. Near some of the steep escarpments in La Visite are remnant breeding colonies of Black-capped Petrel, and these nocturnal birds may be heard during winter months.

Macaya Biosphere Reserve

Macaya Biosphere Reserve, consisting of 5,500 ha at the core of the Massif de la Hotte in extreme southwestern Haiti, supports a diversity of forest habitats, ranging from wet limestone forest at lower elevations to a complex mosaic of pine and cloud forest at upper elevations. The birdlife is correspondingly diverse, and many Hispaniolan endemics are found here. Reaching the reserve requires a serious commitment of time and effort, one that only the most hardy birders will likely be able to make. The distance from Port-au-Prince (six to seven hours driving) and remoteness (4WD and a willingness to hike are musts) present daunting logistic challenges. However, the landscape is magnificent, the local people friendly and helpful, and the birding rewards superlative.

Birding in Macaya is best accomplished by coordinating with the Macaya Guide Association and using the village of Kay Michel as a base. Several excellent and knowledgable local guides are available. It is recommended that the Société Audubon Haiti be the point of contact for arranging a trip. Kay Michel is reached via a steep climb over a rough road that begins just west of the town of Les Cayes. From Kay Michel there are several birding opportunities. One can spend a full day or more in the nearby karst limestone forest, where Hispaniolan Trogon, Golden Swallow, Hispaniolan Highland-Tanager, Gray-crowned Palm-Tanager, and Western Chat-Tanager are virtually guaranteed. The more adventurous and physically fit can also ascend with camping gear (most of which can be carried by local guides for a reasonable fee) into the higher mountains, where the pine and dense broadleaf forests also harbor Gray-crowned Palm-Tanager and Western Chat-Tanager, as well as Hispaniolan Crossbill. For those able to travel in winter, the highest reaches of Pic Formon and Pic Macaya may yield nocturnal encounters with Black-capped Petrels, which breed on steep cliffs and are noisily active at night.

APPENDIX B: Checklist of Birds of Hispaniola

This is a complete list, in phylogenetic order, of the 306 bird species known to have occurred on Hispaniola.

Anatidae

- ☐ White-faced Whistling-Duck ____
- ☐ Black-bellied Whistling-Duck ____
- ☐ West Indian Whistling-Duck ____
- ☐ Fulvous Whistling-Duck ____
- ☐ Canada Goose ____
- ☐ Wood Duck ____
- ☐ Gadwall ____
- ☐ Eurasian Wigeon ____
- ☐ American Wigeon ____
- ☐ Mallard ____
- ☐ Blue-winged Teal ____
- ☐ Northern Shoveler ____
- ☐ White-cheeked Pintail ____
- ☐ Northern Pintail ____
- ☐ Green-winged Teal ____
- ☐ Canvasback ____
- ☐ Redhead ____
- ☐ Ring-necked Duck ____
- ☐ Lesser Scaup ____
- ☐ Hooded Merganser ____
- ☐ Red-breasted Merganser ____
- ☐ Masked Duck ____
- ☐ Ruddy Duck ____

Phasianidae

- ☐ Red Junglefowl ____
- ☐ Ring-necked Pheasant ____
- ☐ Helmeted Guineafowl ____

Odontophoridae

- ☐ Northern Bobwhite ____

Podicipedidae

- ☐ Least Grebe ____
- ☐ Pied-billed Grebe ____

Procellariidae

- ☐ Black-capped Petrel ____
- ☐ Greater Shearwater ____
- ☐ Manx Shearwater ____
- ☐ Audubon's Shearwater ____

Hydrobatidae

- ☐ Wilson's Storm-Petrel ______
- ☐ Leach's Storm-Petrel ______

Phaethontidae

- ☐ White-tailed Tropicbird ______
- ☐ Red-billed Tropicbird ______

Sulidae

- ☐ Masked Booby ______
- ☐ Brown Booby ______
- ☐ Red-footed Booby ______

Pelecanidae

- ☐ Brown Pelican ______

Phalacrocoracidae

- ☐ Double-crested Cormorant ______
- ☐ Neotropic Cormorant ______

Anhingidae

- ☐ Anhinga ______

Fregatidae

- ☐ Magnificent Frigatebird ______

Ardeidae

- ☐ American Bittern ______
- ☐ Least Bittern ______
- ☐ Great Blue Heron ______
- ☐ Great Egret ______
- ☐ Little Egret ______
- ☐ Snowy Egret ______
- ☐ Little Blue Heron ______
- ☐ Tricolored Heron ______
- ☐ Reddish Egret ______
- ☐ Cattle Egret ______
- ☐ Green Heron ______
- ☐ Black-crowned Night-Heron ______
- ☐ Yellow-crowned Night-Heron ______

Threskiornithidae

- ☐ White Ibis ______
- ☐ Glossy Ibis ______
- ☐ Roseate Spoonbill ______

Ciconiidae

- ☐ Wood Stork ______

Cathartidae

- ☐ Turkey Vulture ______

Phoenicopteridae

- ☐ Greater Flamingo ______

Pandionidae

- ☐ Osprey ______

Accipitridae

- ☐ Swallow-tailed Kite ______
- ☐ Northern Harrier ______
- ☐ Sharp-shinned Hawk ______
- ☐ Ridgway's Hawk ______
- ☐ Broad-winged Hawk ______
- ☐ Swainson's Hawk ______
- ☐ Red-tailed Hawk ______

Falconidae

- ☐ American Kestrel ______
- ☐ Merlin ______
- ☐ Peregrine Falcon ______

Rallidae

- ☐ Black Rail ______
- ☐ Clapper Rail ______
- ☐ Sora ______
- ☐ Yellow-breasted Crake ______
- ☐ Spotted Rail ______
- ☐ Purple Gallinule ______
- ☐ Common Moorhen ______
- ☐ American Coot ______
- ☐ Caribbean Coot ______

Aramidae

- ☐ Limpkin ______

Burhinidae

- ☐ Double-striped Thick-knee ______

Charadriidae

- ☐ Black-bellied Plover ______
- ☐ American Golden-Plover ______
- ☐ Snowy Plover ______
- ☐ Wilson's Plover ______
- ☐ Semipalmated Plover ______
- ☐ Piping Plover ______
- ☐ Killdeer ______

Haematopodidae

- ☐ American Oystercatcher ______

Recurvirostridae

- ☐ Black-necked Stilt ______

Jacanidae

- ☐ Northern Jacana ______

Scolopacidae

- ☐ Greater Yellowlegs ______
- ☐ Lesser Yellowlegs ______
- ☐ Solitary Sandpiper ______
- ☐ Willet ______
- ☐ Spotted Sandpiper ______
- ☐ Whimbrel ______
- ☐ Hudsonian Godwit ______
- ☐ Marbled Godwit ______
- ☐ Ruddy Turnstone ______
- ☐ Red Knot ______
- ☐ Sanderling ______
- ☐ Semipalmated Sandpiper ______
- ☐ Western Sandpiper ______
- ☐ Least Sandpiper ______
- ☐ White-rumped Sandpiper ______
- ☐ Baird's Sandpiper ______
- ☐ Pectoral Sandpiper ______
- ☐ Dunlin ______
- ☐ Stilt Sandpiper ______
- ☐ Buff-breasted Sandpiper ______

- ☐ Short-billed Dowitcher ____
- ☐ Long-billed Dowitcher ____
- ☐ Wilson's Snipe ____
- ☐ Wilson's Phalarope ____
- ☐ Red-necked Phalarope ____

Laridae

- ☐ Pomarine Jaeger ____
- ☐ Parasitic Jaeger ____
- ☐ Long-tailed Jaeger ____
- ☐ Laughing Gull ____
- ☐ Franklin's Gull ____
- ☐ Bonaparte's Gull ____
- ☐ Ring-billed Gull ____
- ☐ Herring Gull ____
- ☐ Lesser Black-backed Gull ____
- ☐ Great Black-backed Gull ____
- ☐ Black-legged Kittiwake ____
- ☐ Gull-billed Tern ____
- ☐ Caspian Tern ____
- ☐ Royal Tern ____
- ☐ Sandwich Tern ____
- ☐ Roseate Tern ____
- ☐ Common Tern ____
- ☐ Forster's Tern ____
- ☐ Least Tern ____
- ☐ Bridled Tern ____
- ☐ Sooty Tern ____
- ☐ Black Tern ____
- ☐ Brown Noddy ____
- ☐ Black Skimmer ____

Columbidae

- ☐ Rock Pigeon ____
- ☐ Scaly-naped Pigeon ____
- ☐ White-crowned Pigeon ____
- ☐ Plain Pigeon ____
- ☐ White-winged Dove ____
- ☐ Zenaida Dove ____
- ☐ Mourning Dove ____
- ☐ Common Ground-Dove ____
- ☐ Key West Quail-Dove ____
- ☐ White-fronted Quail-Dove ____
- ☐ Ruddy Quail-Dove ____

Psittacidae

- ☐ Hispaniolan Parakeet ____
- ☐ Olive-throated Parakeet ____
- ☐ Hispaniolan Parrot ____

Cuculidae

- ☐ Black-billed Cuckoo ________
- ☐ Yellow-billed Cuckoo ________
- ☐ Mangrove Cuckoo ________
- ☐ Hispaniolan Lizard-Cuckoo ________
- ☐ Bay-breasted Cuckoo ________
- ☐ Smooth-billed Ani ________

Tytonidae

- ☐ Barn Owl ________
- ☐ Ashy-faced Owl ________

Strigidae

- ☐ Burrowing Owl ________
- ☐ Stygian Owl ________
- ☐ Short-eared Owl ________

Caprimulgidae

- ☐ Common Nighthawk ________
- ☐ Antillean Nighthawk ________
- ☐ Least Pauraque ________
- ☐ Chuck-will's-widow ________
- ☐ Hispaniolan Nightjar ________

Nyctibiidae

- ☐ Northern Potoo ________

Apodidae

- ☐ Black Swift ________
- ☐ White-collared Swift ________
- ☐ Chimney Swift ________
- ☐ Antillean Palm-Swift ________

Trochilidae

- ☐ Antillean Mango ________
- ☐ Hispaniolan Emerald ________
- ☐ Ruby-throated Hummingbird ________
- ☐ Vervain Hummingbird ________

Trogonidae

- ☐ Hispaniolan Trogon ________

Todidae

- ☐ Broad-billed Tody ______
- ☐ Narrow-billed Tody ______

Alcedinidae

- ☐ Belted Kingfisher ______

Picidae

- ☐ Antillean Piculet ______
- ☐ Hispaniolan Woodpecker ______
- ☐ Yellow-bellied Sapsucker ______

Tyrannidae

- ☐ Greater Antillean Elaenia ______
- ☐ Eastern Wood-Pewee ______
- ☐ Hispaniolan Pewee ______
- ☐ Great Crested Flycatcher ______
- ☐ Stolid Flycatcher ______
- ☐ Gray Kingbird ______
- ☐ Loggerhead Kingbird ______
- ☐ Scissor-tailed Flycatcher ______
- ☐ Fork-tailed Flycatcher ______

Vireonidae

- ☐ White-eyed Vireo ______
- ☐ Thick-billed Vireo ______
- ☐ Flat-billed Vireo ______
- ☐ Yellow-throated Vireo ______
- ☐ Warbling Vireo ______
- ☐ Red-eyed Vireo ______
- ☐ Black-whiskered Vireo ______

Corvidae

- ☐ Hispaniolan Palm Crow ______
- ☐ White-necked Crow ______

Hirundinidae

- ☐ Purple Martin ______
- ☐ Caribbean Martin ______
- ☐ Tree Swallow ______
- ☐ Golden Swallow ______
- ☐ Northern Rough-winged Swallow ______
- ☐ Bank Swallow ______
- ☐ Cliff Swallow ______
- ☐ Cave Swallow ______
- ☐ Barn Swallow ______

Regulidae

- ☐ Ruby-crowned Kinglet ____

Sylviidae

- ☐ Blue-gray Gnatcatcher ____

Turdidae

- ☐ Rufous-throated Solitaire ____
- ☐ Veery ____
- ☐ Bicknell's Thrush ____
- ☐ Swainson's Thrush ____
- ☐ Wood Thrush ____
- ☐ American Robin ____
- ☐ La Selle Thrush ____
- ☐ Red-legged Thrush ____

Mimidae

- ☐ Gray Catbird ____
- ☐ Northern Mockingbird ____
- ☐ Pearly-eyed Thrasher ____

Motacillidae

- ☐ American Pipit ____

Bombycillidae

- ☐ Cedar Waxwing ____

Dulidae

- ☐ Palmchat ____

Parulidae

- ☐ Blue-winged Warbler ____
- ☐ Golden-winged Warbler ____
- ☐ Tennessee Warbler ____
- ☐ Nashville Warbler ____
- ☐ Northern Parula ____
- ☐ Yellow Warbler ____
- ☐ Chestnut-sided Warbler ____
- ☐ Magnolia Warbler ____
- ☐ Cape May Warbler ____
- ☐ Black-throated Blue Warbler ____
- ☐ Yellow-rumped Warbler ____
- ☐ Black-throated Green Warbler ____
- ☐ Blackburnian Warbler ____
- ☐ Yellow-throated Warbler ____
- ☐ Pine Warbler ____

- ☐ Prairie Warbler ______
- ☐ Palm Warbler ______
- ☐ Bay-breasted Warbler ______
- ☐ Blackpoll Warbler ______
- ☐ Black-and-white Warbler ______
- ☐ American Redstart ______
- ☐ Prothonotary Warbler ______
- ☐ Worm-eating Warbler ______
- ☐ Swainson's Warbler ______
- ☐ Ovenbird ______
- ☐ Northern Waterthrush ______
- ☐ Louisiana Waterthrush ______
- ☐ Kentucky Warbler ______
- ☐ Connecticut Warbler ______
- ☐ Mourning Warbler ______
- ☐ Common Yellowthroat ______
- ☐ Hooded Warbler ______
- ☐ Wilson's Warbler ______
- ☐ Canada Warbler ______
- ☐ Yellow-breasted Chat ______

Coerebidae

- ☐ Bananaquit ______

Thraupidae

- ☐ Green-tailed Ground-Tanager ______
- ☐ Hispaniolan Highland-Tanager ______
- ☐ Black-crowned Palm-Tanager ______
- ☐ Gray-crowned Palm-Tanager ______
- ☐ Western Chat-Tanager ______
- ☐ Eastern Chat-Tanager ______
- ☐ Summer Tanager ______
- ☐ Scarlet Tanager ______
- ☐ Western Tanager ______
- ☐ Hispaniolan Spindalis ______

Emberizidae

- ☐ Yellow-faced Grassquit ______
- ☐ Black-faced Grassquit ______
- ☐ Greater Antillean Bullfinch ______
- ☐ Savannah Sparrow ______
- ☐ Grasshopper Sparrow ______
- ☐ Song Sparrow ______
- ☐ Lincoln's Sparrow ______
- ☐ Rufous-collared Sparrow ______

Cardinalidae

- ☐ Rose-breasted Grosbeak ______
- ☐ Blue Grosbeak ______
- ☐ Indigo Bunting ______

Icteridae

- ☐ Bobolink ____
- ☐ Tawny-shouldered Blackbird ____
- ☐ Greater Antillean Grackle ____
- ☐ Shiny Cowbird ____
- ☐ Hispaniolan Oriole ____
- ☐ Baltimore Oriole ____
- ☐ Orchard Oriole ____

Fringillidae

- ☐ Antillean Euphonia ____
- ☐ Hispaniolan Crossbill ____
- ☐ Antillean Siskin ____

Passeridae

- ☐ House Sparrow ____

Ploceidae

- ☐ Village Weaver ____

Estrildidae

- ☐ Red Avadavat ____
- ☐ Nutmeg Mannikin ____
- ☐ Chestnut Mannikin ____

SELECTED REFERENCES

American Ornithologists' Union. 1998. *Check-list of North American Birds*. 7th ed. American Ornithologists' Union, Washington, D.C.

American Ornithologists' Union. 2000. Forty-second supplement to the American Ornithologists' Union *Check-list of North American Birds*. *Auk* 117:847–858.

American Ornithologists' Union. 2002. Forty-third supplement to the American Ornithologists' Union *Check-list of North American Birds*. *Auk* 119:897–906.

American Ornithologists' Union. 2003. Forty-fourth supplement to the American Ornithologists' Union *Check-list of North American Birds*. *Auk* 120:923–931.

American Ornithologists' Union. 2004. Forty-fifth supplement to the American Ornithologists' Union *Check-list of North American Birds*. *Auk* 121:985–995.

Arendt, W. J. [with collaborators]. 1992. Status of North American migrant landbirds in the Caribbean region: a summary. Pp. 143–171 *in* J. M. Hagan III and D. W. Johnston (eds.), *Ecology and Conservation of Neotropical Migrant Landbirds*. Smithsonian Institution Press, Washington, D.C.

BirdLife International. 2000. *Threatened Birds of the World.* Lynx Edicions and BirdLife International, Barcelona, Spain, and Cambridge, U.K.

Bond, J. 1936. *Birds of the West Indies*. Academy of National Sciences, Philadelphia, PA.

Cory, C. B. 1885. *The Birds of Haiti and San Domingo.* Estes and Lauriant, Boston, MA.

Dávalos, L. M., and T. Brooks. 2001. Parc National La Visite, Haiti: a last refuge for the country's montane birds. *Cotinga* 16:36–39.

Dickinson, E. C. (editor). 2003. *The Howard and Moore Complete Checklist of the Birds of the World*. 3rd ed. Princeton University Press, Princeton, NJ.

Dod, A. S. 1978. *Aves de la República Dominicana.* Museo Nacional de Historia Natural, Santo Domingo.

Dod, A. S. 1981. *Guía de campo para las aves de la República Dominicana*. Editora Horizontes, Santo Domingo.

Dunning, J. B., Jr. (editor). 1993. *CRC Handbook of Avian Body Masses*. CRC Press, Boca Raton, FL.

Faaborg, J. 1985. Ecological constraints on West Indian bird distributions. Pp. 621–653 *in* P. A. Buckley, M. S. Foster, E. S. Morton, R. S. Ridgely, and F. G. Buckley (eds.), *Neotropical Ornithology*. Ornithological Monographs 36. American Ornithologists' Union, Washington, D.C.

Faaborg, J. R. 1980. The land birds of Saona and Beata Islands, Dominican Republic. *Caribbean Journal of Science* 15:13–19.

Fisher-Meerow, L. L., and W. S. Judd. 1989. A floristic study of five sites along an elevational transect in the Sierra de Baoruco, Prov. Pedernales, Dominican Republic. *Moscosoa* 5:159–185.

Food and Agricultural Organization. 1991. *Forest Resources Assessment 1990.* Food and Agricultural Organization of the United Nations, Rome, Italy.

Garrido, O. H., J. W. Wiley and A. Kirkconnell. 2005. The genus *Icterus* (Aves: Icteridae) in the West Indies. *Ornitología Neotropical* 16(4).

Garrido, O. H., G. M. Kirwan, and D. R. Capper. 2002. Species limits within the Grey-headed Quail-Dove *Geotrygon caniceps* and implications for the conservation of a globally threatened species. *Bird Conservation International* 12:169–187.

Garrido, O. H., and G. B. Reynard. 1993. The Greater Antillean Nightjar: is it one species? *El Pitirre* 7:5.

Garrido, O. H., G. B. Reynard, and A. Kirkconnell. 1997. Is the Palm Crow, *Corvus palmarum* (Aves: Corvidae), a monotypic species? *Ornitología Neotropical* 8:15–21.

Grupo Jaragua, Inc. 1994. *Estratégia para la conservación de la biodiversidad de la República Dominicana 1994–2003*. Centro Editorial, Santo Domingo.

Hardy, J. W., B. B. Coffey, Jr., and G. B. Reynard. 1988. *Voices of New World Nightbirds, Owls, Nightjars, and Their Allies*. 3rd ed. ARA Records, Gainesville, FL.

Hartshorn, G., G. Antonini, R. Dubois, D. Harcharik, S. Heckadon, H. Newton, C. Quesada, J. Shores, and G. Staples. 1981. *The Dominican Republic, Country Environmental Profile*. JRB Associates, McLean, VA.

Horn, S. P., K. H. Orvis, L. M. Kennedy, and G. M. Clark. 2000. Prehistoric fires in the highlands of the Dominican Republic: evidence from charcoal in soils and sediments. *Caribbean Journal of Science* 36:10–18.

Keith, A. R., J. W. Wiley, S. C. Latta, and J. A. Ottenwalder. 2003. *The Birds of Hispaniola: Haiti and the Dominican Republic*. British Ornithologists' Union, Tring, U.K.

Kepler, A. K. 1977. Comparative study of Todies (Todidae): with emphasis on the Puerto Rican Tody, *Todus mexicanus*. Nuttall Ornithological Club publication 16.

Klein, N. K., K. J. Burns, S. J. Hackett, and C. S. Griffiths. 2004. Molecular phylogenetic relationships among the wood warblers (Parulidae) and historical biogeography in the Caribbean Basin. *Journal of Caribbean Ornithology* 17:3–17.

Lahti, D. C. 2003. Cactus fruits may facilitate Village Weaver (*Ploceus cucullatus*) breeding in atypical habitat on Hispaniola. *Wilson Bulletin* 115:487–489.

Lanyon, W. E. 1967. Revision and probable evolution of the *Myiarchus* flycatchers of the West Indies. *Bulletin of the American Museum of Natural History* 136:329–370.

Latta, S. C. (editor). 1998. *Recent ornithological research in the Dominican Republic/Investigaciones ornitológicas recientes en la República Dominicana*. Ediciones Tinglar, Santo Domingo.

Latta, S. C. 2002. *Aves comunes de la República Dominicana/Common Birds of the Dominican Republic*. Sociedad Ornitológica de la Hispaniola, Santo Domingo.

Latta, S. C. 2003. Effects of scaley-leg mite infestations on body condition and site fidelity of migratory warblers. *Auk* 120:730–743.

Latta, S. C. 2005. Complementary areas for conserving avian diversity on Hispaniola. *Animal Conservation* 8:69–81.

Latta, S. C., and C. Brown. 1999. Autumn stopover ecology of the Blackpoll Warbler (*Dendroica striata*) in thorn scrub forest of the Dominican Republic. *Canadian Journal of Zoology* 77:1147–1156.

Latta, S. C., and J. Faaborg. 2001. Winter site fidelity of Prairie Warblers in the Dominican Republic. *Condor* 103:455–468.

Latta, S. C., and J. Faaborg. 2002. Demographic and population responses of Cape May Warblers wintering in multiple habitats. *Ecology* 83:2502–2515.

Latta, S. C., and R. Lorenzo (editors). 2000. *Results of the National Planning Workshop for Avian Conservation in the Dominican Republic*. Dirección Nacional de Parques, Santo Domingo.

Latta, S. C., C. C. Rimmer, and K. P. McFarland. 2003. Winter bird communities in four habitats along an elevational gradient on Hispaniola. *Condor* 105:179–197.

Latta, S. C., M. L. Sondreal, and C. R. Brown. 2000. A hierarchical analysis of nesting and foraging habitat for the conservation of the Hispaniolan White-winged Crossbill (*Loxia leucoptera megaplaga*). *Biological Conservation* 96:139–150.

Latta, S. C., M. L. Sondreal, and D. A. Mejía. 2002. Breeding behavior of the endangered Hispaniolan Crossbill (*Loxia megaplaga*). *Ornitología Neotropical* 13:225–234.

Latta, S. C., and J. M. Wunderle, Jr. 1996. The composition and foraging ecology of mixed-species flocks in pine forests of Hispaniola. *Condor* 98:595–607.

Latta, S. C., and J. M. Wunderle, Jr. 1996. Ecological relationships of two todies in Hispaniola: effects of habitat and flocking. *Condor* 98:769–779.

Latta, S. C., and J. M. Wunderle, Jr. 1998. The assemblage of birds foraging in native West Indian pine (*Pinus occidentalis*) forests of the Dominican Republic during the non-breeding season. *Biotropica* 30:645–656.

Lee, D. S. 2000. Status and conservation priorities for Black-capped Petrels in the West Indies. Pp. 11–18 *in* E. A. Schreiber and D. S. Lee (eds.), *Status and Conservation of West Indian Seabirds*. Special publication #1, Society of Caribbean Ornithology, Ruston, LA.

Lovette, I. J., and E. Bermingham. 2001. Mitochondrial perspective on the phylogenetic relationships of the *Parula* wood-warblers. *Auk* 118:211–215.

McDonald, M. A., and M. H. Smith. 1990. Speciation, heterochrony, and genetic variation in Hispaniolan Palm-Tanagers. *Auk* 107:707–717.

McDonald, M. A., and M. H. Smith. 1994. Behavioral and morphological correlates of heterochrony in Hispaniolan Palm-Tanagers. *Condor* 96:433–446.

National Audubon Society. 2002. WatchList 2002. www.audubon.org/bird/watchlist.
Núñez, F. 2003. Evaluación ecológica integrada: Parque Nacional Juan B. Pérez Rancier. Fundación Moscoso Puello and Secretaría de Estado de Medio Ambiente y Recursos Naturales, Santo Domingo.
Ottenwalder, J. A. 1988. La avifauna de los Parques Nacionales Armando Bermúdez y José del Carmen Ramírez, Cordillera Central. Reporte preparado para la Dirección Nacional de Parques y el Fondo PREI bajo contrato con Agridesa, S. A., Santo Domingo.
Ottenwalder, J. A., 2000. Medio ambiente y sostenibilidad del desarrollo. Pp. 65–98 *in Desarrollo Humano en la República Dominicana 2000*. Programa de las Naciones Unidas para el Desarrollo (PNUD), Santo Domingo.
Overton, L. C., and D. D. Rhoads. 2004. Molecular phylogenetic relationships based on mitochondrial and nuclear gene sequences for the Todies (*Todus*, Todidae) of the Caribbean. *Molecular Phylogenetics and Evolution* 32:524–538.
Paryski, P., C. A. Woods, and F. Sergile. 1989. Conservation strategies and the preservation of biological diversity in Haiti. Pp. 855–878 *in* C. A. Woods (ed.), *Biogeography of the West Indies: Past, Present, Future*. Sandhill Crane Press, Gainesville, FL.
Raffaele, H., J. Wiley, O. Garrido, A. Keith, and J. Raffaele. 1998. *A Guide to the Birds of the West Indies*. Princeton University Press, Princeton, NJ.
Ricklefs, R. E., and G. C. Cox. 1972. Taxon cycles in the West Indian avifauna. *American Naturalist* 106:195–219.
Ricklefs, R. E., and G. C. Cox. 1978. Stage of taxon cycle, habitat distribution, and population density in the avifauna of the West Indies. *American Naturalist* 112:875–895.
Rimmer, C. C., J. Almonte, E. Garrido, D. Mejía, M. Milagros, and P. R. Wieczoreck. 2003. Bird records in a montane forest fragment of western Sierra de Neiba, Dominican Republic. *Journal of Caribbean Ornithology* 16:55–60.
Rimmer, C. C., J. E. Goetz, and K. P. McFarland. 1998. Bird observations in threatened forest fragments of the Sierra de Neiba. *El Pitirre* 11:38–39.
Rimmer, C. C., K. P. McFarland, W. G. Ellison, and J. E. Goetz. 2001. Bicknell's Thrush (*Catharus bicknelli*). *In* A. Poole and F. Gill (eds.), *The Birds of North America*, no. 592. Birds of North America, Inc., Philadelphia, PA.
Rimmer, C. C., J. M. Townsend, A. K. Townsend, E. M. Fernandez, and J. Almonte. 2004. Ornithological investigations in Macaya Biosphere Reserve, Haiti, 7–14 February 2004. Unpublished report, Vermont Institute of Natural Science, Woodstock, VT.
Rimmer, C. C., J. M. Townsend, A. K. Townsend, E. M. Fernandez, and J. Almonte. 2005. Avian diversity, abundance, and conservation status in the Macaya Biosphere Reserve of Haiti. *Ornitología Neotropical* 16: 219–230.
Schreiber, E. A., and D. S. Lee. 2000. *Status and Conservation of West Indian Seabirds*. Special publication #1, Society of Caribbean Ornithology, Ruston, LA.
SEA/DVS. 1990. La diversidad biológica en la República Dominicana. Secretaría de Estado de Agricultura, Departamento de Vida Silvestre, Santo Domingo.
Selander, R. K. 1966. Sexual dimorphism and differential niche utilization in birds. *Condor* 68:113–151.
Stattersfield, A. J., M. J. Crosby, A. J. Long, and D. C. Wege. 1998. *Endemic Bird Areas of the World: Priorities for Bird Conservation*. BirdLife International, Cambridge, U.K.
Terborgh, J., and J. Faaborg. 1980. Factors affecting the distribution and abundance of North American migrants in the Eastern Caribbean region. Pp. 145–155 *in* A. Keast and E. S. Morton (eds.), *Migrant Birds in the Neotropics: Ecology, Behavior, Distribution, and Conservation*. Smithsonian Institution Press, Washington, D.C.
Tolentino, L., and M. Peña. 1998. Inventario de la vegetación y uso de la tierra en la República Dominicana. *Moscosoa* 10:179–203.
Wetmore, A., and B. H. Swales. 1931. *Birds of Haiti and the Dominican Republic*. U.S. National Museum Bulletin 155, Washington, D.C.
Wiley, J. W. 1979. The White-crowned Pigeon in Puerto Rico: status, distribution, and movements. *Journal of Wildlife Management* 43:402–413.

Wiley, J. W. 1998. Breeding-season food habits of Burrowing Owls (*Athene cunicularia*) in southwestern Dominican Republic. *Journal of Raptor Research* 32:241–245.

Wiley, J. W., and J. A. Ottenwalder. 1990. Birds of Islas Beata and Alto Velo, Dominican Republic. *Studies on Neotropical Fauna and Environment* 25:65–88.

Wiley, J. W., and B. N. Wiley. 1979. Status of the American Flamingo in the Dominican Republic and eastern Haiti. *Auk* 96:616–619.

Wiley, J. W., and B. N. Wiley. 1981. Breeding season ecology and behavior of Ridgway's Hawk (*Buteo ridgwayi*). *Condor* 83:132–151.

Wingate, D. B. 1964. Discovery of breeding Black-capped Petrels on Hispaniola. *Auk* 81:147–159.

Woods, C. A., and J. A. Ottenwalder. 1983. The montane avifauna of Haiti. Pp. 576–590 and 607–622 *in* A. C. Risser, Jr., and F. S. Todd (eds.), *Proceedings of the Jean Delacour IFCB Symposium on Breeding Birds in Captivity*. International Foundation for the Conservation of Birds, Los Angeles, CA.

Wunderle, J. M., Jr., and S. C. Latta. 1996. Avian abundance in sun and shade coffee plantations and remnant pine forest in the Cordillera Central, Dominican Republic. *Ornitología Neotropical* 7:19–34.

Wunderle, J. M., Jr., and S. C. Latta. 2000. Winter site fidelity of nearctic migrant birds in isolated shade coffee plantations of different sizes in the Dominican Republic. *Auk* 117:596–614.

Wunderle, J. M., Jr., and R. B. Waide. 1993. Distribution of overwintering nearctic migrants in the Bahamas and Greater Antilles. *Condor* 95:904–933.

Wunderle, J. M., Jr., and R. B. Waide. 1994. Future prospects for nearctic migrants wintering in Caribbean forests. *Bird Conservation International* 4:191–207.

INDEX OF LOCAL NAMES

INDEX OF ENGLISH AND SCIENTIFIC NAMES